AF862289

NONRESIDENT
TRAINING
COURSE

April 1999

Aerographer's Mate

Module 1—Surface Weather Observations

NAVEDTRA 14269

DISTRIBUTION STATEMENT A: Approved for public release; distribution is unlimited.

Although the words "he," "him," and "his" are used sparingly in this course to enhance communication, they are not intended to be gender driven or to affront or discriminate against anyone.

PREFACE

By enrolling in this self-study course, you have demonstrated a desire to improve yourself and the Navy. Remember, however, this self-study course is only one part of the total Navy training program. Practical experience, schools, selected reading, and your desire to succeed are also necessary to successfully round out a fully meaningful training program.

THE COURSE: This self-study course is organized into subject matter areas, each containing learning objectives to help you determine what you should learn along with text and illustrations to help you understand the information. The subject matter reflects day-to-day requirements and experiences of personnel in the rating or skill area. It also reflects guidance provided by Enlisted Community Managers (ECMs) and other senior personnel, technical references, instructions, etc., and either the occupational or naval standards, which are listed in the *Manual of Navy Enlisted Manpower Personnel Classifications and Occupational Standards*, NAVPERS 18068.

THE QUESTIONS: The questions that appear in this course are designed to help you understand the material in the text.

VALUE: In completing this course, you will improve your military and professional knowledge. Importantly, it can also help you study for the Navy-wide advancement in rate examination. If you are studying and discover a reference in the text to another publication for further information, look it up.

1999 Edition Prepared by
AGC(SW) Stephen M. Volpe

Published by
NAVAL EDUCATION AND TRAINING
PROFESSIONAL DEVELOPMENT
AND TECHNOLOGY CENTER

NAVSUP Logistics Tracking Number
0504-LP-026-8980

Sailor's Creed

"I am a United States Sailor.

I will support and defend the Constitution of the United States of America and I will obey the orders of those appointed over me.

I represent the fighting spirit of the Navy and those who have gone before me to defend freedom and democracy around the world.

I proudly serve my country's Navy combat team with honor, courage and commitment.

I am committed to excellence and the fair treatment of all."

CONTENTS

Nonresident Training Course Follows The Index

SUMMARY OF THE AEROGRAPHER'S MATE TRAINING SERIES

The following modules of the AG training series are available:

AG MODULE 1, NAVEDTRA 14269, *Surface Weather Observations*

This module covers the basic procedures that are involved with conducting surface weather observations. It begins with a discussion of surface observation elements, followed by a description of primary and backup observation equipment that is used aboard ships and at shore stations. Module 1 also includes a complete explanation of how to record and encode surface METAR observations using WMO and NAVMETOCCOM guidelines. The module concludes with a description of WMO plotting models and procedures.

AG MODULE 2, NAVEDTRA 14270, *Miscellaneous Observations and Codes*

This module concentrates on the observation procedures, equipment, and codes associated with upper-air observations and bathythermograph observations. Module 2 also discusses aviation weather codes, such as TAFs and PIREPs, and includes a chapter on surf observation procedures. Radiological fallout and chemical contamination plotting procedures are also explained.

AG MODULE 3, NAVEDTRA 14271, *Environmental Satellites and Weather Radar*

This module describes the various type of environmental satellites, satellite imagery, and associated terminology. It also discusses satellite receiving equipment. In addition, Module 3 contains information on the Weather Surveillance Radar-1988 Doppler (WSR-88D). It includes a discussion of electromagnetic energy and radar propagation theory, and explains the basic principles of Doppler radar. The module also describes the configuration and operation of the WSR-88D, as well as WSR-88D products.

AG MODULE 4, NAVEDTRA 14272, *Environmental Communications and Administration*

This module covers several of the most widely used environmental communications systems within the METOC community. It also describes the software programs and products associated with these systems. The module concludes with a discussion of basic administration procedures.

NOTE

Additional modules of the AG training series are in development. Check the NETPDTC website for details at http://www.cnet.navy.mil/netpdtc/nac/neas.htm. For ordering information, check NAVEDTRA 12061, Catalog of Nonresident Training Courses, which is also available on the NETPDTC website.

SAFETY PRECAUTIONS

Safety is a paramount concern for all personnel. Many of the Naval Ship's Technical Manuals, manufacturer's technical manuals, and every Planned Maintenance System (PMS) maintenance requirement card (MRC) include safety precautions. Additionally, OPNAVINST 5100.19 *(series), Naval Occupational Safety and Health (NAVOSH) Program Manual for Forces Afloat,* and OPNAVINST 5100.23 *(series), NAVOSH Program Manual,* provide safety and occupational health information. The safety precautions are for your protection and to protect equipment.

During equipment operation and preventive or corrective maintenance, the procedures may call for personal protective equipment (PPE), such as goggles, gloves, safety shoes, hard hats, hearing protection, and respirators. When specified, your use of PPE is mandatory. You must select PPE appropriate for the job since the equipment is manufactured and approved for different levels of protection. If the procedure does not specify the PPE, and you aren't sure, ask your safety officer.

Most machinery, spaces, and tools requiring you to wear hearing protection are posted with hazardous noise signs or labels. Eye hazardous areas requiring you to wear goggles or safety glasses are also posted. In areas where corrosive chemicals are mixed or used, an emergency eyewash station must be installed.

All lubricating agents, oil, cleaning material, and chemicals used in maintenance and repair are hazardous materials. Examples of hazardous materials are gasoline, coal distillates, and asphalt. Gasoline contains a small amount of lead and other toxic compounds. Ingestion of gasoline can cause lead poisoning. Coal distillates, such as benzene or naphthalene in benzol, are suspected carcinogens. Avoid all skin contact and do not inhale the vapors and gases from these distillates. Asphalt contains components suspected of causing cancer. Anyone handling asphalt must be trained to handle it in a safe manner.

Hazardous materials require careful handling, storage, and disposal. PMS documentation provides hazard warnings or refers the maintenance man to the Hazardous Materials User's Guide. Material Safety Data Sheets (MSDS) also provide safety precautions for hazardous materials. All commands are required to have an MSDS for each hazardous material they have in their inventory. You must be familiar with the dangers associated with the hazardous materials you use in your work. Additional information is available from you command's *Hazardous Material Coordinator.* OPNAVINST 4110.2 (series), *Hazardous Material Control and Management,* contains detailed information on the hazardous material program.

Recent legislation and updated Navy directives implemented tighter constraints on environmental pollution and hazardous waste disposal. OPNAVINST 5090.1 (series), *Environmental and Natural Resources Program Manual,* provides detailed information. Your command must comply with federal, state, and local environmental regulations during any type of construction and demolition. Your supervisor will provide training on environmental compliance.

Cautions and warnings of potentially hazardous situations or conditions are highlighted, where needed, in each chapter of this TRAMAN. Remember to be safety conscious at all times.

INSTRUCTIONS FOR TAKING THE COURSE

ASSIGNMENTS

The text pages that you are to study are listed at the beginning of each assignment. Study these pages carefully before attempting to answer the questions. Pay close attention to tables and illustrations and read the learning objectives. The learning objectives state what you should be able to do after studying the material. Answering the questions correctly helps you accomplish the objectives.

SELECTING YOUR ANSWERS

Read each question carefully, then select the BEST answer. You may refer freely to the text. The answers must be the result of your own work and decisions. You are prohibited from referring to or copying the answers of others and from giving answers to anyone else taking the course.

SUBMITTING YOUR ASSIGNMENTS

To have your assignments graded, you must be enrolled in the course with the Nonresident Training Course Administration Branch at the Naval Education and Training Professional Development and Technology Center (NETPDTC). Following enrollment, there are two ways of having your assignments graded: (1) use the Internet to submit your assignments as you complete them, or (2) send all the assignments at one time by mail to NETPDTC.

Grading on the Internet: Advantages to Internet grading are:

- you may submit your answers as soon as you complete an assignment, and
- you get your results faster; usually by the next working day (approximately 24 hours).

In addition to receiving grade results for each assignment, you will receive course completion confirmation once you have completed all the assignments. To submit your assignment answers via the Internet, go to:

http://courses.cnet.navy.mil

Grading by Mail: When you submit answer sheets by mail, send all of your assignments at one time. Do NOT submit individual answer sheets for grading. Mail all of your assignments in an envelope, which you either provide yourself or obtain from your nearest Educational Services Officer (ESO). Submit answer sheets to:

COMMANDING OFFICER
NETPDTC N331
6490 SAUFLEY FIELD ROAD
PENSACOLA FL 32559-5000

Answer Sheets: All courses include one "scannable" answer sheet for each assignment. These answer sheets are preprinted with your SSN, name, assignment number, and course number. Explanations for completing the answer sheets are on the answer sheet.

Do not use answer sheet reproductions: Use only the original answer sheets that we provide—reproductions will not work with our scanning equipment and cannot be processed.

Follow the instructions for marking your answers on the answer sheet. Be sure that blocks 1, 2, and 3 are filled in correctly. This information is necessary for your course to be properly processed and for you to receive credit for your work.

COMPLETION TIME

Courses must be completed within 12 months from the date of enrollment. This includes time required to resubmit failed assignments.

PASS/FAIL ASSIGNMENT PROCEDURES

If your overall course score is 3.2 or higher, you will pass the course and will not be required to resubmit assignments. Once your assignments have been graded you will receive course completion confirmation.

If you receive less than a 3.2 on any assignment and your overall course score is below 3.2, you will be given the opportunity to resubmit failed assignments. **You may resubmit failed assignments only once.** Internet students will receive notification when they have failed an assignment--they may then resubmit failed assignments on the web site. Internet students may view and print results for failed assignments from the web site. Students who submit by mail will receive a failing result letter and a new answer sheet for resubmission of each failed assignment.

COMPLETION CONFIRMATION

After successfully completing this course, you will receive a letter of completion.

ERRATA

Errata are used to correct minor errors or delete obsolete information in a course. Errata may also be used to provide instructions to the student. If a course has an errata, it will be included as the first page(s) after the front cover. Errata for all courses can be accessed and viewed/downloaded at:

http://www.cnet.navy.mil/netpdtc/nac/neas.htm

STUDENT FEEDBACK QUESTIONS

We value your suggestions, questions, and criticisms on our courses. If you would like to communicate with us regarding this course, we encourage you, if possible, to use e-mail. If you write or fax, please use a copy of the Student Comment form that follows this page.

For subject matter questions:

E-mail: n315.products@cnet.navy.mil
Phone: Comm: (850) 452-1001, Ext. 1713
DSN: 922-1001, Ext. 1713
FAX: (850) 452-1370
(Do not fax answer sheets.)
Address: COMMANDING OFFICER
NETPDTC (CODE N315)
6490 SAUFLEY FIELD ROAD
PENSACOLA FL 32509-5000

For enrollment, shipping, grading, or completion letter questions

E-mail: fleetservices@cnet.navy.mil
Phone: Toll Free: 877-264-8583
Comm: (850) 452-1511/1181/1859
DSN: 922-1511/1181/1859
FAX: (850) 452-1370
(Do not fax answer sheets.)
Address: COMMANDING OFFICER
NETPDTC (CODE N331)
6490 SAUFLEY FIELD ROAD
PENSACOLA FL 32559-5000

NAVAL RESERVE RETIREMENT CREDIT

If you are a member of the Naval Reserve, you will receive retirement points if you are authorized to receive them under current directives governing retirement of Naval Reserve personnel. For Naval Reserve retirement, this course is evaluated at 6 points. (Refer to *Administrative Procedures for Naval Reservists on Inactive Duty,* BUPERSINST 1001.39, for more information about retirement points.)

COURSE OBJECTIVES

In completing this nonresident training course, you will demonstrate a knowledge of the subject matter by correctly answering questions on the following subjects: surface observation elements, equipment, codes, and plotting standards.

Student Comments

Course Title: *Aerographer's Mate, Module 1—Surface Weather Observations*

NAVEDTRA: 14269 **Date:** ____________

We need some information about you:

Rate/Rank and Name: ____________ SSN: ____________ Command/Unit ____________

Street Address: ____________ City: ____________ State/FPO: ________ Zip ________

Your comments, suggestions, etc.:

Privacy Act Statement: Under authority of Title 5, USC 301, information regarding your military status is requested in processing your comments and in preparing a reply. This information will not be divulged without written authorization to anyone other than those within DOD for official use in determining performance.

NETPDTC 1550/41 (Rev 4-00)

CHAPTER 1

SURFACE OBSERVATION ELEMENTS

INTRODUCTION

In this chapter, we give you an overview of Surface Weather Observations and provide references for observation procedures. We also discuss some of the important values that weather observers calculate from observed data. These values include both physiological indicators and aircraft performance indicators. Physiological indicators are values that help estimate the effects of weather on the human body, and aircraft performance indicators are values that allow aviators to assess the effects of weather on aircraft.

CONDUCTING OBSERVATIONS

LEARNING OBJECTIVES: Identify measurement systems and time standards used while conducting surface weather observations. Recognize the general order in which elements are routinely observed.

Throughout the Navy and Marine Corps, Aerographer's Mates, Quartermasters, and Marine Corps weather observers use similar techniques and procedures to determine the current weather conditions. Accurate and timely submission of environmental observations are basic to the development of oceanographic and meteorological forecasts and tactical indices used in support of fleet operations. The methods used aboard ship differ slightly from those used at shore stations. In this section, we discuss procedures used both aboard ship and ashore.

Some weather elements are observed by using different criteria, depending on the recording format and reporting code used. As of July 1996, all U.S. Navy weather activities have adopted the Aviation Routine Weather Report (METAR) and the Aviation Selected Special Weather Report (SPECI) codes for weather observation procedures. The criteria for these codes are covered in detail in chapter 3 of this module.

The criteria for U.S. Navy surface aviation weather observations ashore are contained in Surface *METAR Observations User's Manual,* NAVMETOC-COMINST 3141.2. There are minor differences in observation criteria between U.S. shore stations and activities located outside of the continental United States (OCONUS). These differences are highlighted in the manual. *United States Navy Manual For Ship's Surface Weather Observations,* NAVMETOC-COMINST 3144.1, is used for shipboard weather observations. Except when necessary, we will not repeat information covered in those manuals, but will refer you to the manual.

Before discussing the procedures or methods used to observe weather elements, let's review some basics about observing and measuring the elements.

MEASUREMENT SYSTEMS

In the mid-1970's, the United States began switching to the metric system for weights and measures. In the field of military meteorology and oceanography, it is common to measure an element by using units from the old system and then converting the measurement to the metric system. Because of this, weather observers should be well versed in both systems and be able to convert units of length, volume, temperature, pressure, and mass. Appendix II of this module contains tables and conversion factors to convert from one system to another. Weather observers make temperature conversions most frequently.

The three temperature scales used are the *Fahrenheit, Celsius,* and *Kelvin* scales. The United States and several other countries still use the Fahrenheit scale, which fixes the freezing point of water at 32°F and the boiling point at 212°F. Most of the world uses the Celsius scale, which fixes the freezing point of water at 0°C and the boiling point at 100°C. In meteorology and oceanography, both temperature scales are used, with frequent conversions between the two. Conversions may be made by using a conversion table or by using the following formulas:

$$F = \frac{9}{5}C + 32$$

or

$$C = \frac{5}{9}(F - 32)$$

where F is degrees Fahrenheit, and C is degrees Celsius.

Many calculations in meteorology and oceanography use the Kelvin, or Absolute, temperature scale. A *kelvin* is exactly equal to a Celsius degree in scale, but the starting point of measurement on the Kelvin scale (0 kelvin) is *absolute zero,* or -273.16°C. That is the temperature at which, theoretically, all molecular motion would stop. Water freezes on the Kelvin scale at 273.16 K and boils at 373.16 K. Conversions may be easily made between the Kelvin and the Celsius scales by addition or subtraction using the following formulas:

$$C = K - 273.16$$

or

$$K = C + 273.16,$$

where K is kelvin and C is Celsius degrees.

TIME STANDARDS

Though the hour, minute, and second convention is universally used in keeping time, various time zones are also used. In North America, eastern standard time (EST), central standard time (CST), mountain standard time (MST), and Pacific standard time (PST) are used. Standard time zones generally cover strips of the globe, extending north and south parallel to the longitude lines. Each time zone covers about 15° longitude centered on 0° longitude, with all longitudes evenly devisable by 15. Time zone boundaries that cross land masses have been adjusted by local agreement, and often zig-zag.

Standard time zones for the world are provided in Appendix III. Throughout most of the world, standard time is 1 hour earlier for each time zone to the west and 1 hour later for each time zone to the east. A list of countries, provinces, and states, with their local standard time departure from the 0° longitude standard zone is provided in the *Nautical Almanac,* published each year by the U.S. Naval Observatory. When standard time is used, it is referred to as *local standard time* (LST) or by a standard zone designation, such as Eastern Standard Time (EST) or Ppacific Sstandard Time (PST).

Daylight savings time or *summer time* is the convention adopted by most regions in North America. On the first Sunday in April at 0200, the clocks are set ahead 1 hour. On the last Sunday in October at 0200, they are set back 1 hour. During the summer, time in these regions is called *daylight time;* for example, Eastern Daylight Time (EDT) or Pacific Daylight Time (PDT). Other regions of the world have also adopted this practice.

Another measurement of time sometimes used is local mean time (LMT). *Local mean time* is time measured in 24-hour days relative to the movement of the sun. When the sun is highest in the sky, local mean time is 1200 noon. Within a time zone, local mean time may be off standard time by up to several hours. Local mean time changes by 4 minutes for every degree of longitude.

To prevent confusion between the different zones and types of time, meteorological and oceanographic records, charts, and reports use *Coordinated Universal Time* (UTC). UTC time is kept by using the 24-hour clock. UTC is the local mean time at the Royal Greenwich Observatory in East Sussex, England, at 0° longitude. This time is the same all over the world, regardless of local time conventions. All times in UTC are suffixed with a Z for identification. Because of this, UTC time is sometimes referred to as "Zulu Time." The term *Coordinated Universal Time* and the abbreviation *UTC,* by international agreement, have replaced the older term *Greenwich mean time* and the older abbreviation *GMT.*

ORDER OF OBSERVATION

Surface weather observations are completed and transmitted every hour. The various weather elements are actually observed from 5 to 15 minutes before the hour, for routine observations. The time to begin monitoring the elements should be adjusted as the observer gains speed and experience.

As a general rule, first observe the elements from outside the weather office, and then from the equipment inside the office. Pressure elements and those elements changing quickly should be observed last. Even when automatic observation equipment is used, these general rules apply. Most of the necessary observation data will come from equipment located within the office spaces, but outside measurements, such as cloud type and visibility, should be done first.

Although many observation records and reporting codes are in use, the formats have many elements in common. For example, all surface weather observations include data for state-of-the-sky, visibility, and temperature. Some observation formats require additional data, such as sea condition, seawater temperature, and sea ice conditions. The following sections cover the various data types and the methods used to obtain the data. These data types are not arranged in any particular code format, but are generally arranged in observation order.

REVIEW QUESTIONS

Q1. Which two manuals contain detailed instructions regarding the Navy's surface weather observation program?

Q2. What are the three temperature scales in use today? Fahrenheit, Celsius, Kelvin

Q3. Each time zone covers approximately how many degrees of longitude? 15

Q4. What does the abbreviation UTC mean? Cordinated universal time

Q5. What observation element should be observed last? Pressure elements and quickly changing elements

SKY CONDITION

LEARNING OBJECTIVES: Define sky condition and state-of-the-sky. Describe cloud form, cloud genera, cloud species, and cloud variety. Identify the 10 cloud forms and their characteristics. Identify significant supplemental cloud features. Define orographic clouds. Explain cloud layer coverage and total sky coverage. Define cloud ceiling and explain how cloud layer height and ceiling height are measured.

As an observer, your interpretation of the sky condition may determine whether a pilot should fly under instrument flight rules (IFR) or visual flight rules (VFR). In some circumstances, your judgment of sky conditions might even prevent a pilot from getting off the ground. Sky condition reports also reflect developing weather conditions; these reports help predict weather over the next several days.

The term *sky condition* includes all cloud parameters estimated or measured by weather observers. *State-of-the-sky* is a specific term that equates to one of the 27 internationally recognized sky states. These 27 states-of-the-sky are represented by code numbers that identify the type of cloud or combinations of clouds present in the sky, and the changes in the clouds over the past few hours. Refer to Appendix IV, WMO codes 0513, 0515, and 0509.

The following sections discuss the identifying features of cloud types. Both NAVMETOCCOMINST 3141.2 and NAVMETOCCOMINST 3144.1 have detailed descriptions of the 27 states-of-the-sky. Two other manuals useful in cloud identification are the *International Cloud Atlas,* WMO Publication 407, and the full-color edition of *Cloud Types For Observers,* Her Majesty's Meteorological Office Publication 716. Posters and charts are also available from the National Weather Service.

CLOUD IDENTIFICATION SCHEME

Clouds may be identified by using very general terms or very specific names. We can classify cloud identification terms from the most general to the most specific as cloud form, cloud genus, cloud species, and cloud variety. Some clouds also prominent supplementary features that are considered either odd or significant enough to be identified with a specific name.

How specific must the observer be in identifying clouds? In the METAR/SPECI code, the observer is not usually required to identify the clouds by name; only the amount of sky covered by clouds in the various cloud layers is recorded and encoded. However, certain cloud genera and supplementary features are considered significant enough to report as Remarks. In the Land and Ship Synoptic codes, the observer must identify the proper cloud genera, species, and variety in order to select the proper state-of-the-sky code.

The following information on cloud identification is presented to help you understand the various reportable cloud "types."

Cloud Forms

There are three different general cloud forms: cumuliform, stratiform, and cirriform. Cumuliform clouds are puffy, with distinct elements or cells. The puffy appearance of these clouds is caused by moist air rising within the cloud cell. Stratiform clouds develop in uniform layers and present a smoother appearance. Cirriform clouds are the thin, wispy, hairlike clouds.

The primary factor that determines cloud form is the stability of the air. *Unstable* air tends to rise on its *own. Stable* air tends to remain at the same level in the atmosphere. And *conditionally unstable* air will retain its level until some force provides initial lift, and then it will continue to rise on its own.

Cumuliform clouds form because moist, conditionally unstable air is initially forced upward by some lifting mechanism, and becomes unstable. The moist, unstable air cools gradually as it rises, reaches saturation, and condenses to form a visible cloud at a certain level in the atmosphere. The continued addition of moist air maintains the cloud base at that certain

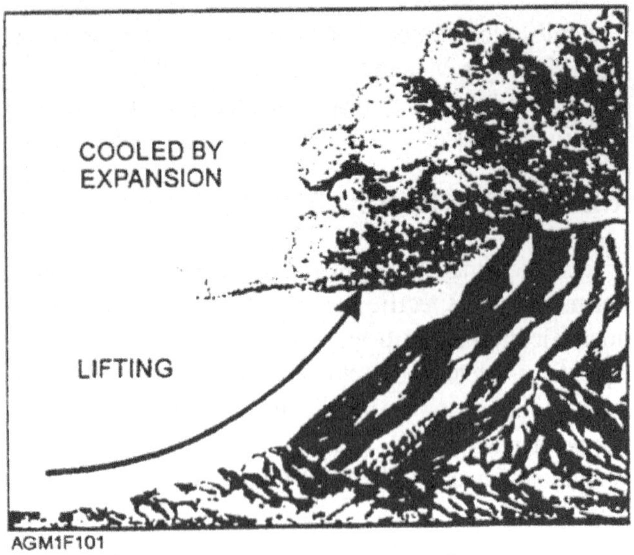

Figure 1-1.—Orographic lift.

height, but the saturated air within the cloud continues to rise, forming the puffy, cumuliform buildups.

Stratiform clouds may form where stable air is brought to saturation by either the addition of moisture or by cooling the air. Most stratiform clouds, however, form when a layer of stable air is forced upward by a lifting mechanism. The entire layer cools as it is lifted, reaches saturation, and forms a cloud layer.

There are four processes that cool the air by lifting the air mass: mechanical lift, convective lift, convergence, and vorticity.

Mechanical lift is a process by which a physical barrier forces air aloft. The barrier may be a sloping plain, a rising coastline, or a mountain. Those land barriers cause a type of mechanical lift called *orographic* lift (fig. 1-1). The barrier may also be air masses of different density; for instance, when fast-moving, warm air overrides the slower moving, cooler air in a warm front, or when fast-moving, cold air forces slower moving warm air aloft in a cold front. Frontal barriers cause a type of mechanical lift known as *frontal lift* (figs. 1-2 and 1-3). *Turbulent lift* is mechanical lift caused by friction between the earth's surface and the air moving above it or between adjacent layers of air in which wind speed (rig. 1-4) or direction is different. Turbulent lift appears to be the key factor in the development of cloud layers with both stratiform and cumuliform characteristics at all levels in the atmosphere.

Convective lift is a process that occurs when cool air is heated from the surface and rises (fig. 1-5). Convective lift is the key factor in cumuliform cloud development within an air mass.

Convergence occurs when windflow at a particular level forces air to "pile up" in a general area, which creates a lifting action. For instance, where straight-line winds of higher speed decrease, more air is transported into an area than is carried away, and a mass of air builds up vertically. This is known as *speed convergence.* Alternatively, *directional convergence* occurs when winds of different directions come together and merge at a certain location. Convergence plays a key role in cumuliform cloud development in the tropical regions.

The last lifting mechanism is *vorticity.* Vorticity is the rotational motion of molecules in the atmosphere,

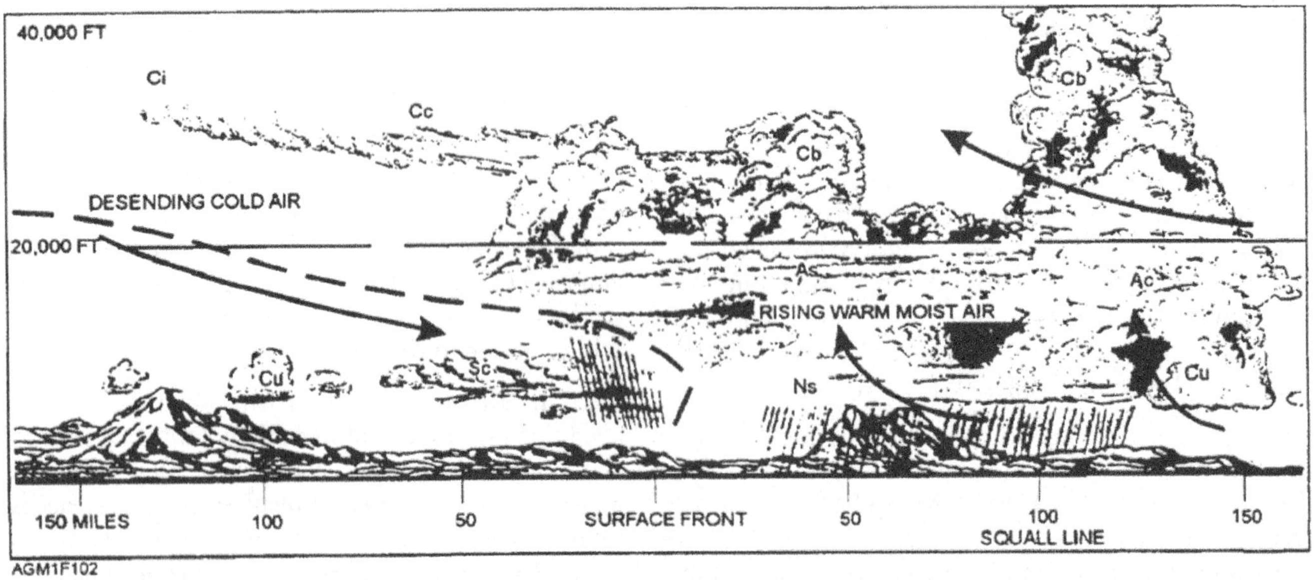

Figure 1-2.—Frontal lift—conditionally unstable air causing cumuliform cloud development along a cold front.

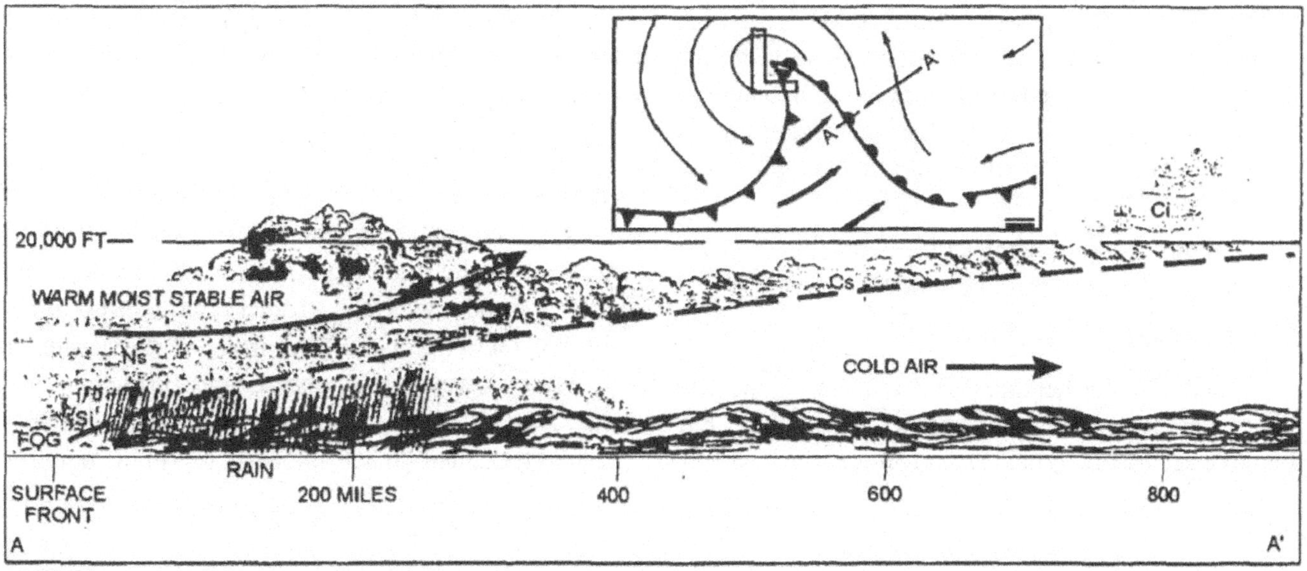

Figure 1-3.—Frontal lift-stable air causing stratiform cloud development along a warm front.

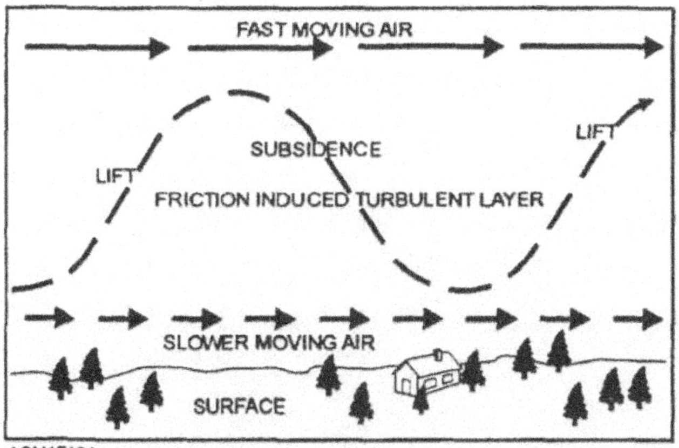

Figure 1-4.—Turbulent lift.

caused by the spinning of the earth and the path of each molecule through the atmosphere. Although the theory is complex, the net result is that air following a cyclonically curved path through the atmosphere tends to rise, while air following an anticyclonically curved path tends to subside. Vorticity is the leading cause of most middle- and high-level, nonfrontal cloudiness.

Air may become saturated without being lifted. At night, air may cool by radiation of heat in a process known as *radiational cooling*. Warm, moist air may move over a cooler surface, such as a cool body of land, and cool to saturation by conduction. *Conduction* is the transfer of heat energy through contact. Finally, air may also be brought to saturation by the addition of moisture

Figure 1-5.—Convective lift.

to the air mass through evaporation. Moisture may evaporate into an air mass from precipitation falling through the air mass and by evaporation from bodies of water. These processes generally are only significant in cloud and fog formation within several hundred feet of the earth's surface.

You will study these cloud formation mechanisms in detail in the later training modules.

Cloud Genera

Clouds are further identified with more specific names that are commonly referred to as cloud type, but are more accurately termed *cloud genera.* The basis for the cloud genera identification is the level at which the cloud forms, as well as the way the cloud formed.

Cloud Etage

With respect to clouds, the atmosphere is broken down into three layers or *etages.* In the middle latitudes or temperate region, the low-etage is from the surface to 6,500 feet; the mid-etage, from 6,500 feet to 23,000 feet; and the high-etage, from 16,500 feet up to near 45,000 feet (fig. 1-6). The limits of the etages are generally lower in the polar regions (mid-etage from 6,500 to 13,000 feet and high-etage from 10,000 to 25,000 feet), and higher in the tropics (mid-etage from

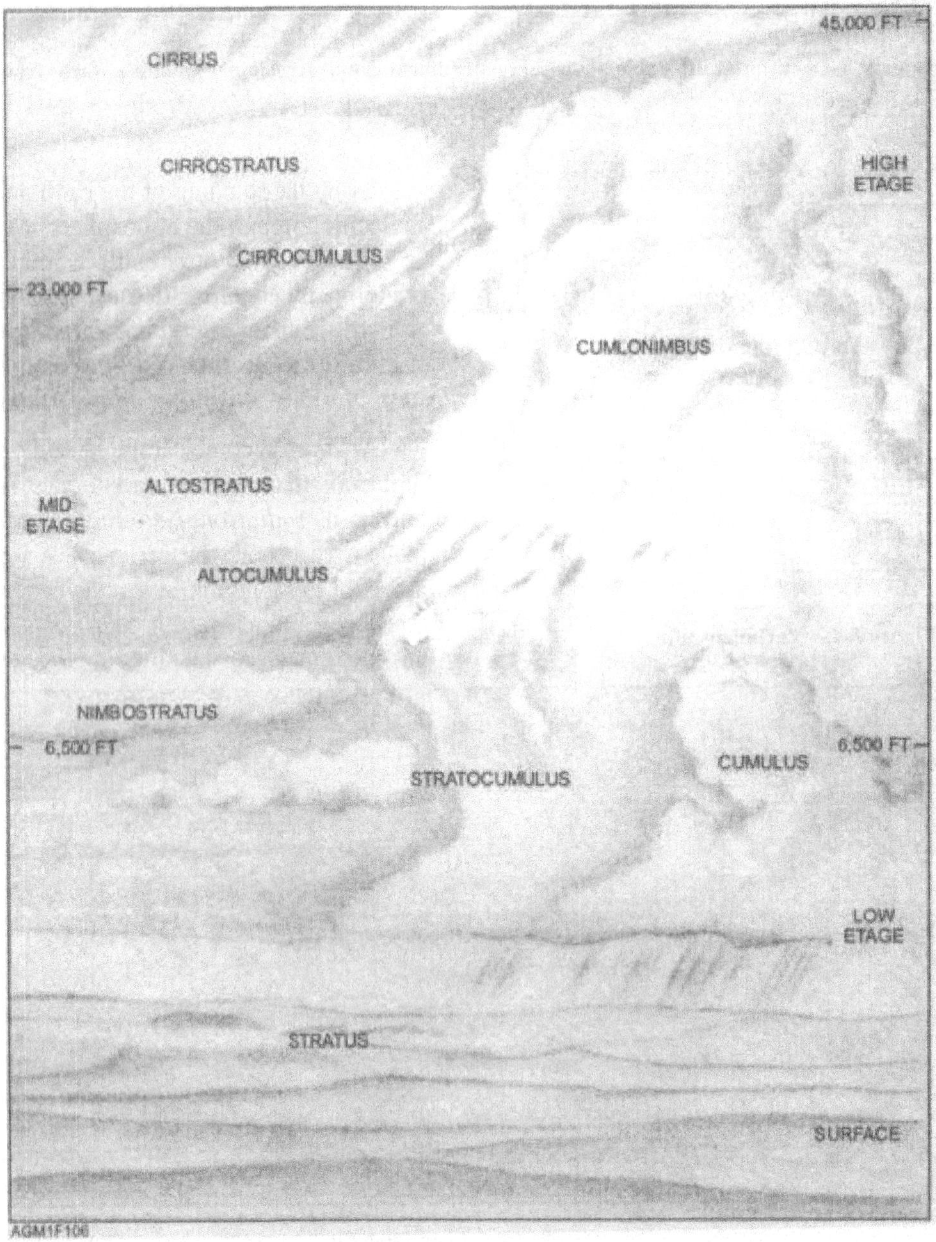

Figure 1-6.—Temperate Region cloud genera—cloud forms in the three etages.

Table 1-1.—WMO Cloud Classification by Cloud Etnge, Form, Genus, Species, and Variety

ETAGE	FORM	GENUS	SPECIES	VARIETY
LOW-ETAGE	Cumulifrom	cumulus	humilis mediocris congestus fractus	
		cumulonim bus	capillatus calvus	
	Stratiform	stratocumulus	floccus castellanus stratiformis lenticularis	opacus translucidus undulatus radiatus perlucidus
		stratus	nebulosus fractus	opacus translucidus undulatus
MID-ETAGE		nimbostratus		
		altostratus		opacus translucidus undulatus radiatus duplicatus
	Cumuliform	altocumulus	castellanus floccus stratiformis lenticularis	opacus translucidus undulatus radiatus perlucidus duplicatus lacunosus
HIGH-ETAGE	Cirriform	cirrus	uncinus spissatus floccus castellanus	radiatus duplicatus intortus vertebratus
	Stratiform	cirrostratus	nebulosus fibratus	duplicatus
	Cumuliform	cirrocumulus	stratiformis floccus castellanus lenticularis	undulatus duplicatus lacunosus

6,500 to 25,000 feet and high-etage from 20,000 to 60,000 feet).

The low-etage cloud genera may be cumuliform, such as the cumulus or cumulonimbus (identified by their size and extent of development); stratiform, such as the stratus; or have mixed characteristics, such as the stratocumulus. The mid-etage cloud genera are mostly identified with the prefix *alto.* The mid-etage contains the cumuliform clouds, such as altocumulus, and the stratiform clouds, such as altostratus and nimbostratus. The high-etage cloud genera contain the prefix *cirro.* Cumuliform clouds in this etage are called cirrocumulus, while stratiform clouds are called cirrostratus. Another form of cloud found only in the high-etage is the cirriform clouds that are the normally thin, wispy, or hairlike ice-crystal clouds that can be defined as neither cumuliform nor stratiform, but are simply called cirrus clouds.

Cloud Species

Besides the identification of clouds by genera, most cloud forms may be further identified by *cloud species.* The species identifies the size, shape, or form of the elements within a cloud layer. Table 1-1 lists the cloud etages, cloud genera, cloud species, and cloud varieties used to identify clouds. To improve your understanding of the many cloud types, be sure to locate the specific classification of each cloud in table 1-1 during the following discussion.

Cloud Variety

In the 27 *states-of-the-sky,* many of the differences between different "cloud-states" are not based solely on genera, but on a combination of the genera, species, and variety. Cloud *variety* identifies the specific appearance of the arrangement of elements within a cloud layer, the thickness of the layer, or the presence of multiple layers. The nine different cloud varieties are used to further identify cloud species by specific appearance. The variety name is appended after the species name to further identify a cloud. An example is "stratus nebulosus opaqus," which is a low-etage stratiform cloud (genus, stratus) without distinct features (species, nebulosus) but dense enough to obscure the sun (variety, opaqus). State-of-the-sky codes usually do not name the cloud variety, but give a description of the dominant cloud genus, species, and variety. The nine varieties in table 1-1 are defined as follows:

- **Opaqus:** A sheet, layer, or patch of clouds the greater part of which is sufficiently dense to obscure the sun or moon. Opaqus is used to modify low- and mid-etage stratiform cloud layers, particularly those of the species stratiformis. It is not used with the species cirrus spissatus, since spissatus is inherently opaque.

- **Perlucidus:** Clouds of the genus alto- or stratocumulus, usually of the species stratiformis, in which the distinct spaces between the cloud elements allow blue sky, the sun, moon, stars, or higher clouds to be clearly seen.

- **Translucidus:** A sheet, layer, or patch of clouds the greater part of which is sufficiently translucent to reveal the position of the sun or moon. The term is used to modify low- and mid-etage stratiform cloud layers, particularly those of the species stratiformis. It is not used with any of the high-etage cloud names, since these clouds are inherently translucent.

- **Duplicatus:** Two or more sheets, layers, or patches of cloud of similar type at different levels in the atmosphere, commonly overlapping. This situation is usually associated with the species fibratus, uncinus, stratiformis, and lenticularis.

- **Undulatus:** Elements or cells in a sheet, layer, or patch of clouds arranged in parallel rows and forming a wavelike pattern similar to swell waves in the ocean. The popular names for these cloud patterns are "billow clouds," "wind row clouds," and "wave clouds." This wavelike pattern is principally found in the genera cirrocumulus, altocumulus, altostratus, and stratocumulus, but is rarely associated with stratus. When distinct rows and columns are apparent in the pattern of cloud elements in a single layer, the term *bi-undulatus* may be used.

- **Radiatus:** A cloud pattern, similar to undulatus, in which cloud elements in the rows are merged together so that parallel bands of clouds are formed. Due to the effect of perspective, these straight parallel bands seem to merge together near the horizon. The popular name for this cloud pattern is Abraham's Tree. This pattern is frequently associated with the genera cirrus, altocumulus, altostratus, and stratocumulus, and usually associated with the species stratiformis.

- **Lacunosus:** A cloud pattern in which the rounded holes between the clouds form a honeycomb or netlike pattern is the dominant feature. The clouds may be equated to the wax of a honeycomb or the cord in a net. This pattern is associated with the genera cirrocumulus and altocumulus, and is usually used to further define the species stratiformis, castellanus, or floccus.

- **Intortus:** Associated only with the genus cirrus, this term is used when the cirrus fibers or filaments are entangled, curved, bent and irregular, or form a zig-zag pattern.

- **Vertebratus:** Associated mainly with the genus cirrus, this term is used when the cloud fibers extend outward from an elongated central core and are suggestive of vertebrae, ribs, or a fish skeleton.

Supplementary Features

Clouds may also be identified by the presence of *supplementary cloud features.* Supplementary cloud features are specific portions of a larger cloud. Most supplementary features are associated with cumulonimbus clouds. Virga, tuba (funnel clouds, tornadoes, waterspouts), incus (anvil tops), arcus (roll clouds), wall clouds, mamma (formerly called mammatus), pileus, velum, and pannus are all supplementary features of cumulonimbus clouds. Virga may also be associated with many cumuliform clouds, and pannus with nimbostratus clouds. *Cloud*

Types For Observers has excellent photographs and descriptions of these supplementary features.

CLOUD IDENTIFICATION

Since the only method at present to identify cloud type is by visual identification, you must be familiar with the characteristics of the various clouds. Although NAVMETOCCOMINST 3141.2 and NAVMETOC-COMINST 3144.1 present a good description of the clouds present in the 27 states-of-the-sky, you should be thoroughly familiar with the identification features of each cloud type. Let's take a closer look at some of the important identification features of cloud genera.

Cumulus (CU)

Cumulus, translated from Latin, means "heap." Heap aptly describes a cumulus cloud in most of its stages. Since cumulus clouds form by convective action, the height of their base above the surface is directly related to the amount of moisture near the surface. The higher the moisture content, the lower the cloud base. Although the water droplets in cumulus are very numerous, they are very small in the cloud's early stages. As a cumulus cloud continues to grow, the number of large drops within the cloud increases. These large drops may be precipitated from the cloud or may continue to be suspended by the vertical air currents within the cloud. Precipitation in the form of showers occurs with cumulus clouds of moderate development. Although this precipitation may be of moderate intensity, its duration is usually short lived. These clouds do not produce the heavy rain and high winds that are associated with their bigger brothers, the cumulonimbus.

Occasionally, the showers from cumulus clouds evaporate before they reach the ground. This situation is known as *virga* and is characterized by a dark "fuzzy" area immediately below the nearly uniform base of the cumulus cloud. This darker fuzzy area, caused by the precipitation, decreases in intensity beneath the cloud until it disappears (complete evaporation). When virga consists of snow or ice crystals, the virga is not dark and appears more "wispy." This is due to the greater influence of the wind on the snow and ice crystals; the precipitation trails appear to be bent by the wind.

Cumulus clouds produced by convective heating develop in a distinct sequence. This is the primary means by which convective clouds form within an air mass. The cumulus clouds first appear as cumulus humilis, and then develop into cumulus mediocris and cumulus congestus. Although cumulus congestus may continue to develop into cumulonimbus, the cumulonimbus clouds are identified as a separate cloud genus.

When cumulus clouds are formed by mechanical lift, the sequence is the same. However, early stages of development may not be apparent, especially if stratiform clouds are already present.

CUMULUS HUMILIS.—In the earliest stage of development, cumulus usually forms in, and indicates, good weather. A typical cumulus cloud is shown in figure 1-7 in its formative stage. Point A shows the clearly defined outline, the distinct white color, and the characteristic "bulging" appearance. At point B, notice the characteristic flatter and darker base. In this stage, the cumulus is often called a "fair-weather cumulus" or *cumulus humilis.*

CUMULUS MEDIOCRIS.—When cumulus clouds continue to develop vertically, and reach a moderate vertical extent, they are called *cumulus mediocris* (fig. 1-8). Cumulus mediocris have small cauliflower-like buildups, but rarely produce precipitation other than virga.

CUMULUS CONGESTUS.—Cumulus clouds that continue to develop and to reach moderate vertical extent are called *cumulus congestus. Congestus* generally means the sky is filled with clouds vertically, rather than horizontally. These clouds are identified by several layers of puffy, cauliflower-like buildups

Figure 1-7.—Cumulus humilis cloud.

Figure 1-8.—Cumulus mediocris cloud.

extending upwards from the base (fig. 1-9, view A), and also by their overall size. It is not uncommon to see one or two cumulus congestus cells in the sky surrounded by cumulus mediocris cells.

Rapidly building cumulus congestus cloud cells may produce clouds of great vertical extent with relatively small base areas so that they appear to be in the form of large towers. The cloud is still classified a cumulus congestus cloud, but this appearance is commonly called *towering cumulus* (TCU). Towering cumulus clouds normally do not develop as fast horizontally as they do vertically. A rule of thumb for identifying a cumulus congestus cloud as a towering cumulus is that the height appears to be twice the width of the base (fig. 1-9, view B).

Cumulus congestus cells, and especially towering cumulus (congestus) clouds may produce light to moderate showers. Over warm ocean waters, towering cumulus may produce waterspouts. When a large cumulus congestus cloud begins to produce either a wispy cirrus blow-off or a well-defined anvil-shaped top (the upper portion of the cloud column begins bulging horizontally) or if lightning is seen or thunder is heard, the cloud is automatically classified a different type of cloud: the cumulonimbus.

Cumulonimbus (CB)

Cumulonimbus clouds are generated from large cumulus congestus clouds. These clouds cells are distinguished from cumulus congestus by their massive appearance and extensive vertical development. The presence of thunder, lightning, or an anvil top automatically classifies the cloud a cumulonimbus.

Although cumulonimbus may develop cirrus blow-off in the polar regions or during the winter in the mid-latitudes at 20,000 feet, most commonly the cirrus blow-off or top of the anvil will be somewhere between 25,000 to 45,000 feet in the mid latitudes. Tops of the larger cumulonimbus cells have been measured in the tropics in excess of 60,000 feet.

Figure 1-9.—Cumulus congestus. (A) Typical, (B) towering cumulus appearance.

CUMULONIMBUS CAPILLATUS.—The presence of the anvil top (incus) with or without cirrus blow-off identifies a *cumulonimbus capillatus* (fig. 1-10), which is the most recognizable form of cumulonimbus. These clouds, sometimes called "anvil" clouds or "thunderheads," frequently produce thunder, lightning, surface hail, strong and gusty surface winds, and heavy rain.

CUMULONIMBUS CALVUS.—A second species of cumulonimbus, the *cumulonimbus calvus,* lacks the familiar anvil top. Typically, it looks like an extremely large cumulus congestus cell, with less developed cumulus clouds surrounding it and appearing to form a horizontal extension from the base of the larger buildup (fig. 1-11). Although an anvil top, thunder, or lightning need not be observed, the cloud is classified by its size, development, and ominous appearance. Typically, cumulonimbus calvus cells have very dark gray bases. These clouds may later develop an anvil top to become cumulonimbus capillatus. If conditions are not favorable for continued vertical development, cumulonimbus calvus clouds may produce moderate to heavy shower activity as the upward air currents in the cloud loose intensity.

REVIEW QUESTIONS

Q6. How many states-of-the-sky are internationally recognized?

Q7. List the three general cloud forms.

Q8. Stable air is normally associated with what general cloud form?

Q9. Describe the four processes that cool air by lifting.

Q10. Mid-etage clouds in the temperate regions of the earth are found between what levels?

Q11. Define cloud variety.

Figure 1-10.—Cumulonimbus capillatus (anvil cloud—note that cirrus blow-off has not occurred.

Figure 1-11.—Cumulonimbus calvus cloud.

Q12. Most supplementary cloud features are associated with what type of cloud?

Q13. The height of the base of cumulus clouds is directly related to what factor?

Q14. What is a simple rule of thumb for classifying cumulus congestus clouds as towering cumulus?

Q15. Cumulonimbus capillatus clouds are identified by what distinguishing feature?

SUPPLEMENTAL FEATURES.—Many supplemental cloud features of cumulonimbus clouds indicate the stage of development or the potential severity of the "storm."

Light precipitation can begin to fall while the cloud is still increasing in size. Heavy precipitation indicates the cell has slowed or stopped increasing in height. With the beginning of heavy precipitation, the cloud base becomes rougher and less clearly defined. Smaller, ragged, or fragmented clouds are frequent under the base of the dissipating CB cell. These ragged cloud elements are the species *cumulus fractus* (fig. 1-12) if they are more or less in individual elements, or *stratus fractus* if they form a mostly continuous layer. Collectively, the layer of fragmented elements is called *pannus*—a supplemental cloud feature. Cumulus fractus or stratus fractus clouds, often called "scud," usually are associated with falling precipitation or found in the vicinity of numerous showers. Pannus may exhibit very rapid movement under the CB cell base, with individual elements moving in a radically different direction than that of the CB cell. For example, it would not be uncommon to see a CB cell moving toward the northeast, with a low layer of pannus moving toward the south.

Figure 1-12.—Cumulus fractus cloud in a pannus layer.

The upper portion of the CB cell becomes broader and the top may become very indistinct as cirrus and dense cirrus clouds develop from the ice crystals blowing downwind from the top portion of the cell. The cirrus blow-off shield may extend well downstream of the CB cell. Then, the shield may thicken downward, becoming dense enough to blot out the shape of the sun (fig. 1-13). Sometimes the dense cirrus blow-off *(cirrus*

Figure 1-13.—Dense cirrus blow-off (cirrus spissatus) from a thunderstorm.

spissatus) will continue to thicken downward enough to be classified altostratus. Occasionally, rounded bulges may develop, protruding from the base of the dense cirrus blow-off. Although found on the dense cirrus or altostratus blow-off from a cumulonimbus cell, these rounded bulges are called *cumulonimbus mamma,* or simply *mamma* (fig. 1-14). Mamma is a strong indicator that conditions are favorable for severe weather to occur under and near the cumulonimbus base. Mamma should NOT be interpreted as developing funnel clouds.

During the dissipation stage, there are more pronounced downdrafts within the cloud than updrafts. The remaining updrafts may become more concentrated and originate from the rear portion of the cloud cell. Severe weather, such as strong or gusty surface winds, heavy rain, hail, and tornadoes, usually is confined to the dissipating stage.

The beginning of the dissipation stage may be marked by the sudden onset of strong, cold downdrafts, *known* as a *downrush,* exiting the base of the cloud. These winds are deflected by the ground to create very gusty, sometimes dangerously strong winds blowing outward from under the cloud cell. The leading edge and upward boundary of the bubble of gusty winds form a distinct boundary known as the *gust front* or the *outflow boundary. The* outflow boundary—a sharply defined separation between the wind flowing toward the base of the cloud cell and the strong, cold, outward flowing wind-typically is associated with *low-level wind shear* (LLWS), a dangerous phenomenon that has caused many fatal aircraft crashes. A particularly strong, concentrated downrush from the base of the cloud is *known* as a *microburst.* A microburst produces concentrated, extremely strong straight-line winds blowing outward from the base of the CB cell. Typically, these winds produce great damage and are often initially and incorrectly reported by the public as a tornado.

A *roll cloud (arcus)* may form along the leading edge of the cumulonimbus cloud base during a downrush. The ragged bottom edge of the roll cloud is usually much lower than the more uniform base of the

Figure 1-14.—Cumulonimbus mamma on mid-etage altostratus cloud layer.

CB cell (fig. 1-15). It may extend along the entire outer edge of the base of the CB cell, or may angle out slightly ahead of the CB base with the top of the outflow boundary. Although with weaker gust fronts, the roll cloud may appear rough, ragged, or bumpy, under certain conditions the roll cloud may appear very smooth. Roll clouds indicate that thunderstorm downrush has occurred and that LLWS may be present.

The turbulent action along the cold air outflow boundary may produce small-scale vortices on the ragged base of the roll cloud. These vortices sometimes take on the appearance of small, ragged funnel clouds. The public commonly mistakes these vortices for funnel clouds and occasionally reports them as funnel clouds. These vortices are known as *cold-air funnels.* Rarely, a cold-air funnel will develop sufficiently to reach the surface. It does not have the strength of a true funnel cloud or tornado, and is about as powerful as a strong dust devil. By itself, it may be able to pick up objects, such as trash cans or to shake a small camp trailer. Usually, the damage associated with sightings of cold-air funnels is caused by the much more powerful straight-line winds in and behind the outflow boundary (gust front) that produced the cold-air funnel.

Another phenomenon associated with the outflow boundary is a *dust cloud* on or near the earth's surface. This phenomenon is frequent in desert areas and is fairly common in other areas after a dry spell. The dust often appears to be rolling outward and upward from the ground as it moves over an observer. It is associated with the first gust of a thunderstorm's gust front.

A *wall cloud,* a sometimes hollow, generally circular patch of cloud with a ragged bottom edge, may lower from the base of a CB cell. A wall cloud is usually much smaller than the base of the CB cell, and will usually form in the right rear quadrant of the CB cell with respect to the CB's movement. When viewed from the side of the CB, the wall cloud usually is under the rear portion of the cell, where the billowy cloud tops appear to be tapering from the anvil top downward toward the rear portion of the cell. A slowly rotating or spinning wall cloud is an indication of a very strong thunderstorm and impending funnel-cloud development.

Funnel clouds (tuba) usually form on the edge of the wall cloud or near the wall cloud at the rear of the storm. In the early stages of development, the funnel cloud may be only a rotating rounded bulge extending

Figure 1-15.—Roll cloud formation on cumulonimbus.

from the base of the CB cell. As the funnel develops, it will gradually assume the more common cone-shaped appearance. When the funnel cloud (fig. 1-16) extends downward from the CB base to a point where its circular rotation reaches the ground, it is then called a *tornado* (fig. 1-17). Funnel clouds and tornados contain destructive, concentrated, cyclonic winds in excess of 150 knots. The force of the wind is amplified by the rapid change in the direction of the winds as the tornado passes over an area. Estimates based on damage equate the force of tornados to straight-line winds of near 500 miles per hour. Funnel clouds can be seen due to the visible moisture from the parent cloud. Under a funnel cloud, the rapidly circulating winds may be invisible until the circulation picks up dust and debris from earth's surface.

Waterspouts (fig. 1-18) develop over warm ocean or bay waters more frequently than overland. They have been observed from the bases of rapidly building towering cumulus cells, often without any precipitation occurring. They are generally weaker than tornadoes, but still contain dangerous, destructive winds.

When conditions are favorable for tornado development, waterspouts may be assumed to be as strong as a tornado. But when conditions are NOT favorable for tornadic development, then any waterspouts that form are usually less powerful. A Great Lakes freighter, S.S. *Edmond Fitzgerald,* was lost

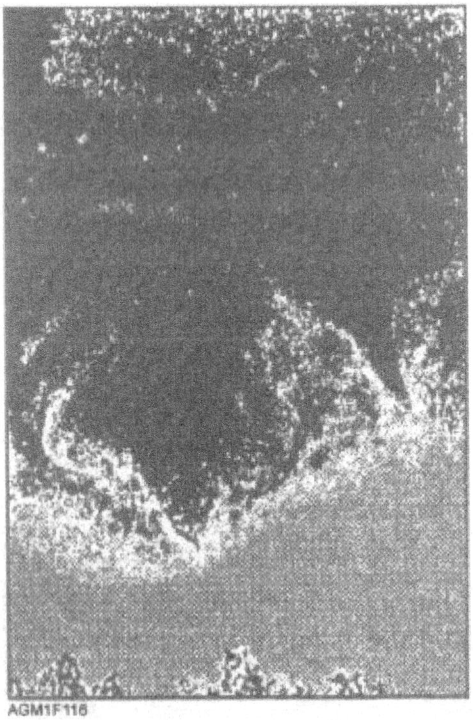

Figure 1-16.—Funnel clouds.

Figure 1-17.—Tornado.

Figure 1-18.—Waterspout.

with all hands after being struck by a waterspout. This strong waterspout was associated with a tornado outbreak over the Great Lakes region. Off the west coast of Mexico in late 1989, a cruise liner was hit by a weaker waterspout that broke many windows and caused some minor structural damage. This weaker waterspout formed in conditions that were NOT favorable for tornado development. Nontornadic waterspouts are thought to be well-developed, cold-air funnels.

Hail may begin forming in the building stage as water drops are carried above the freezing level. It may fall from the cloud base as the updrafts weaken, or become more concentrated during the dissipating stage. Hail frequently is thrown out the top and sides of building and mature CB cells, and may travel up to 25 miles from the cell. This hail rarely reaches the surface before melting. During the dissipation stage, however, the stronger, colder downdrafts tend to bring the hail to the surface UNDER the cloud base before it has a chance to melt. Most frequently, hail reaches the ground in the general vicinity of the wall cloud. The size of the hail is dependent on many factors within the cloud, which you will study later in preparation for AG2.

Another accessory cloud associated with large cumulonimbus buildups are the layers of altocumulus or altostratus clouds, often seen developing from the mid-portion of the cumulonimbus column. These clouds are formed by the spreading out of moisture into a thin, somewhat more stable layer of atmosphere as the unstable column of air in the building cumulus punches through the layer. These cloud layers are considered part of the cumulonimbus cell and are called *velum.*

Occasionally, a rapidly building cumulus will approach a layer of mid-level or upper-level stratiform clouds and appear to push the cloud layer upwards, forming a cap-shaped feature above the cauliflower top of the cumulus or cumulonimbus cell. Until the cumulus cell makes contact with the bulging cloud layer, the cap-shaped feature is called *pileus.*

Stratocumulus (SC)

Clouds in the genus stratocumulus appear to be a combination of the smooth, even, stratiform cloud and the puffy cells of the cumuliform clouds. Stratocumulus clouds are distinguished from cumulus clouds by their flatter appearance. Stratocumulus typically appear to be tightly-packed, flattened cumulus cells but with less distinct edges (fig. 1-19). As stratocumulus cloud elements merge into a continuous layer, they appear gray with dark areas. These dark areas are the thicker portions of the SC clouds.

Stratocumulus is sometimes mistaken for altocumulus, which is the same type of cloud form in the

Figure 1-19.—Stratocumulus cloud.

mid-etage (fig. 1-20). Because of their greater height, the elements of altocumulus clouds appear smaller than stratocumulus cloud elements. The best way to judge whether a cloud is altocumulus or stratocumulus is by obtaining the height from the ceilometer or other cloud-height sensors. Alternatively, an estimate of the size of the individual elements is a reliable indicator. Individual elements of stratocumulus clouds are wider than 5°, while the individual elements of altocumulus are narrower than 5°. The width of your three middle fingers held at arm's length is about 5°.

Precipitation rarely occurs with stratocumulus clouds. When it does occur, it tends to be very light and intermittent. Light snow showers (snow flurries) are the most common form ofprecipitation from stratocumulus clouds. Due to the light intensity of stratocumulus-produced precipitation, more often than not any precipitation will be in the form of virga.

The major species of stratocumulus clouds are stratocumulus floccus, stratocumulus castellanus, and stratocumulus stratiformis.

STRATOCUMULUS FLOCCUS.—*Sirato-cumulus floccus* clouds appear as tightly-packed, flattened cloud cells in a layer, with higher clouds or blue sky visible in the area between the cells. Individual cells have very indistinct, feathery edges.

STRATOCUMULUS CASTELLANUS.—In the species *stratocumulus castellanus,* the individual cloud cells develop in a stratiform layer. These cells usually have rather indistinct edges and scattered cumuliform buildups protruding upward from the generally flatter tops of the stratiform cloud layer. These buildups are commonly in the form of small towers or tufts.

STRATOCUMULUS STRATIFORMIS.—When a layer of cumulus humilis or cumulus mediocris cloud cells begins to lose the clear definition on the edges, such as is seen beginning at point C in figure 1-7, the layer is classified *stratocumulus stratiformis.* As development continues, these cells may merge, completely tilling in the clear gaps between the cloud cells.

Stratus (ST)

Occasionally, a layer of stratocumulus is mistaken for a stratus cloud. The base of a stratus cloud has a very smooth and uniform appearance, often appearing fuzzy or indistinct. A stratus cloud base may have slightly darker areas, but the darker patches are not arranged in any regular pattern.

Stratus usually forms very close to earth's surface, and is called fog when it is in contact with or within 50 feet of the surface. Stratus clouds may form in conjunction with other higher clouds.

Figure 1-20.—Lower mid-etage altocumulus cloud.

Stratus is capable of producing only very light precipitation, such as drizzle or snow grains; or during extremely cold temperatures, ice crystals. Heavier precipitation, such as showers, accompanied by a very dark portion of the stratus layer, indicates the presence of embedded or higher level cumuliform clouds. Rain or snow would be a strong indication that the cloud layer is NOT stratus, but actually the mid-level cloud, nimbostratus. Another factor that indicates that the cloud layer is nimbostratus rather than stratus is the presence of stratus fractus clouds.

The appearance of the sun through the cloud layer is another indication of the cloud type. If a sharp, well-defined outline of the sun can be seen through a cloud layer, the cloud layer may be stratus. If, however, the outline of the sun is blurred, fuzzy, or appears to be viewed through ground glass, the cloud layer may be altostratus.

One of the key indicators of stratus clouds lies in the previous observation record. Stratiform cloud layers normally do not suddenly develop or move over an area. They are associated with stable layers in the atmosphere and evolve slowly. A hazy layer aloft or at the surface will gradually thicken with time to form stratus or fog. When moving over an area, a very thin or hazy layer will gradually become denser as the stratiform layer movement progresses. The exception to this is the fog banks found over large bays and coastal waters. These fog banks may have very distinct boundaries. As they move overland, they may stay on the ground as fog or may lift slightly, forming a low stratus layer. The transition from clear skies and unrestricted visibility to low overcast stratus and poor visibility in fog may be very sudden. The observation record, however, would note the presence of the moving fog bank.

STRATUS NEBULOSUS.—When stratus forms in a layer with no distinct features or denser portions, it is termed *stratus nebulosus.* Stratus nebulosus is the most common form of stratus (fig 1-21).

STRATUS FRACTUS.—*Strutus fractus* clouds form in more or less continuous layers. They present a ragged appearance, as if shredded by the wind. Stratus fractus clouds are generally indicators of bad weather and are usually found below layers of nimbostratus clouds. As with cumulus fractus clouds, a layer of clouds of the species stratus fractus is called "pannus."

Altostratus (AS)

This mid-etage cloud has features similar to stratus, as we have previously discussed. Although the height of the cloud base is the primary difference between stratus and altostratus, the cloud composition is another important clue. Altostratus clouds frequently form above the freezing level. In North America, the freezing level during the winter may be at the surface; but during summer the freezing level may range from 4,000 feet to 16,000 feet, depending on location and weather patterns. The range between 8,000 to 10,000 feet is a fair mean freezing level for the continental United States. When altostratus clouds form above the freezing level, they consist of ice crystals and super-cooled water droplets (water at or below freezing that has not crystallized). Ice crystals give the altostratus clouds their grayish or bluish color and the customary fibrous appearance. The ice crystals also diffuse light more, such that the sun will appear as though viewed through ground glass.

Figure 1-21.—Stratus cloud.

Another indicator used to differentiate between altostratus, stratus, and the higher cirrostratus clouds is the presence and type of optical phenomena. We must consider two types of optical phenomena at this point: the corona and the halo.

A *corona* is a reddish or brownish ring of small diameter seen around the sun or the moon when viewed through clouds. It is often easier to see in a reflection off calm water than by direct observation. A corona may occasionally display very pale rainbow colors, but red will normally predominate and show in the outermost ring. The corona is produced by refraction of light in liquid water droplets, such as the super-cooled droplets found in altocumulus clouds. It is rare for even super-cooled water droplets to exist at too high an altitude, so a corona usually indicates a low or mid-etage cloud form. Large droplets produce a small corona, while smaller droplets produce a larger corona.

A *halo* is a 22° diameter ring seen encircling the sun or moon when viewed through clouds. The ring may show pale colors of the spectrum. Occasionally a secondary ring of 46° diameter may be visible encircling the 22° ring. Bright spots, which are called *mock suns,* may appear on the halo in a horizontal plane with the sun. A bright horizontal line may appear to connect the mock suns and the actual sun, which is called the *parhelic circle.* Occasionally, vertical columns of light, or *pillars,* may appear above and below the sun or moon at low elevation angles, which intersect and form bright spots on the 22° halo.

In the mid latitudes, the corona usually indicates that the cloud is a mid-etage cloud. It is only rarely observed in higher low-etage stratus clouds and the lowest high-etage stratiform clouds. The presence of a halo, on the other hand, will indicate that the cloud is a high-etage cloud form, most often cirrostratus, and it may occasionally occur with cirrus. It does not occur with altostratus clouds.

During the day, an indicator of altostratus is the absence of shadows on the ground. If the sun is seen through a stratiform cloud and shadows are present on the ground, the cloud could be either altostratus or cirrostratus. However, if the cloud is dense enough to prevent shadows from forming, it should be classified as altostratus. Cirrostratus is never dense enough to prevent shadows during the daylight hours.

The height of the base of the altostratus clouds may range from 6,500 feet to 23,000 feet. The density of the stratiform cloud is the primary determining factor of stratiform cloud typing in the 18,500 to 23,000 foot range, while the presence of the corona and halo may be used as reliable secondary indicators.

There are no species associated with altostratus clouds, although there are several different varieties.

Nimbostratus (NS)

Usually formed from altostratus clouds thickening downward, nimbostratus, commonly called "the rain cloud," ranges in color from medium to very dark gray, with a diffuse, indefinite base. It is always thick enough to obscure the sun and is almost exclusively found near frontal zones. Stratus fractus clouds are commonly found under nimbostratus cloud layers, especially just prior to the start of precipitation and during light precipitation. The stratus fractus tend to dissipate during heavier precipitation. Nimbostratus clouds may also form from dissipating cumulonimbus clouds.

Although nimbostratus is classified as a mid-etage cloud, its base often lowers well into the low-etage. With approaching occluded and warm frontal systems, nimbostratus may lower to within several hundred feet of the ground. Nimbostratus bases with stationary fronts tend to be slightly higher.

Normally, altostratus is reclassified as nimbostratus when the cloud base becomes very dark or stratus fractus clouds are observed under the base of the layer. Altostratus clouds <u>must</u> be reclassified as nimbostratus when precipitation begins or when bases drop to less than 6,500 feet. Nimbostratus clouds are usually distinguished from opaque altostratus clouds by the more diffuse, but denser and darker appearance of the base, which is often described as appearing "wetter" than altostratus.

The genus nimbostratus has no distinct species or varieties.

Altocumulus (AC)

Altocumulus clouds are composed of super-cooled water droplets and ice crystals when located above the freezing level. Altocumulus clouds look very similar to stratocumulus clouds; the primary difference in their classification is by height, which may be inferred by the size of the elements in the cloud. We have already discussed how to differentiate between stratocumulus and altocumulus clouds based on the size of the elements. Unfortunately, the altocumulus clouds in the middle to upper portion of the mid-etage (fig. 1-22) and the still higher cirrocumulus clouds of the high etage (fig. 1-23) also look very similar. If the cloud elements are larger than the width of one finger held at arm's length, the cloud should be classified as altocumulus. If the individual cloud elements are smaller than the width of a finger held at arm's length, the cloud should be classified as cirrocumulus. Do not use this method unless the cloud in question is more than 30° above the horizon.

Altocumulus clouds appear white to light gray, or mottled with shadings between white and light gray. When altocumulus clouds do not present a uniform appearance, you should consider other identifying features. Virga may occur from altocumulus clouds, but the trails appear shorter than those associated with stratocumulus. Based on the known height of the freezing level, the bent virga trails associated with frozen precipitation may indicate whether the cloud is high enough to be altocumulus.

The presence of a corona is most frequently associated with altocumulus clouds, even more so than

Figure 1-22.—Typical altocumulus cloud.

with altostratus clouds. Mock suns or pillars, without the surrounding halo, indicate high level altocumulus clouds composed mainly of ice crystals. *Irisation, the* pastel shading of cloud element edges with colors of the spectrum, occurs in ice crystal clouds, which, based on the known freezing level, may help rule out stratocumulus clouds. Irisation may occur with cirrocumulus clouds.

Altocumulus clouds are classified in four species, and have more varieties than any other cloud form. The four species of altocumulus clouds are altocumulus castellanus, altocumulus floccus, altocumulus stratiformis, and altocumulus lenticularis. The

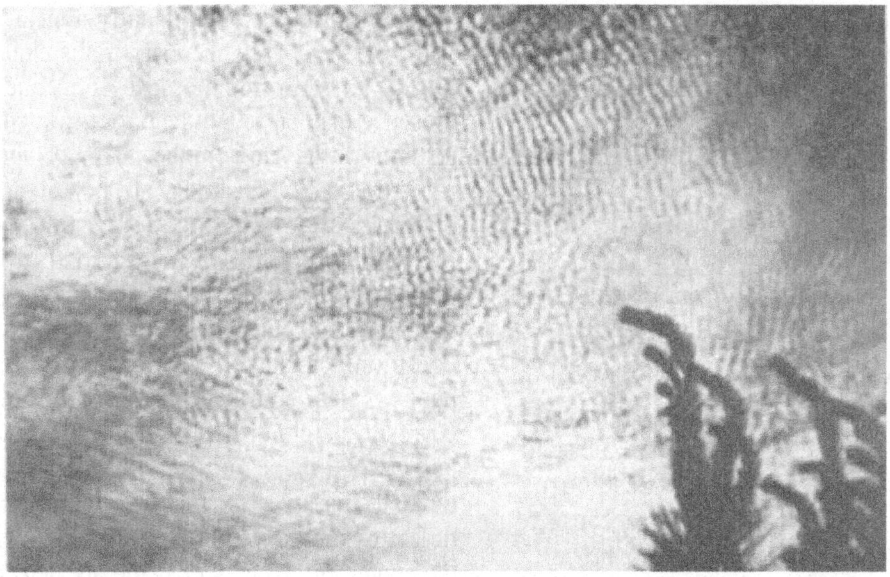

Figure 1-23.—Typical cirrocumulus cloud.

lenticularis species is discussed in a following section on "Orographic Clouds."

ALTOCUMULUS CASTELLANUS.—*Altocumulus castellanus* cloud elements build upward from the base to form towers, tufts, or "turrets." The tops and edges of the buildup may appear ragged, and not have the smoother rounded appearance or cauliflower-like tops. The ragged tops are sometimes described as looking similar to the turrets on a medieval castle, which gives some reference for the name "castellanus." Usually originating in the lower portion of the mid-etage, these clouds may build upwards to moderate or great extent, and are similar to towering cumulus, except with high bases and smaller appearing elements. Continued development may, in rare situations, produce cumulonimbus clouds. Virga is common, and even light precipitation reaching the ground is not uncommon with altocumulus castellanus.

ALTOCUMULUS FLOCCUS.—Related to the altocumulus castellanus is the less developed altocumulus floccus cloud form. *Altocumulus floccus* resemble small, ragged cumulus humilis and typically appear as small tufts of white cloud with rounded or bulging tops. They often have small fibrous trails of virga extending from ragged bases. Both altocumulus castellanus and altocumulus floccus indicate approaching frontal systems with conditions favorable for thunderstorm activity.

ALTOCUMULUS STRATIFORMIS.—*Altocumulus stratiformis* is by far the most common form of altocumulus. In this form we typically see an extensive layer of cloud with smooth, evenly spaced rounded cells or just a cell-like pattern in a generally stratiform layer. Figure 1-22, a typical altocumulus cloud, is a good example of the species altocumulus stratiformis.

Cirrus (CI)

Cirrus clouds, a high-etage cloud type, form generally between 16,500 feet and 45,000 feet in the mid-latitudes. Cirrus clouds are composed of ice crystals.

CIRRUS UNCINUS.—The most common form of cirrus is the thin strand-like wisps of cloud filaments, often curved at on end and described as hook-shaped and called *cirrus uncinus* (fig. 1-24). These cirrus clouds are popularly called "mare's tails" because of their resemblance to the tail of a galloping horse.

CIRRUS SPISSATUS.—The dense blow-off from the top of a cumulonimbus, which looks similar to stratus or altostratus clouds, is called *cirrus spissatus,* but is often referred to as dense cirrus (refer to figure 1-13). Although this cloud typically presents a stratiform appearance, it is not called cirrostratus, because, by definition, cirrostratus is never dense enough to hide the sun. Cirrus spissatus usually forms in a single large patch with a distinct edge. This cloud is reclassified as altostratus when the base lowers to less than 23,000 feet.

A different variety of cirrus spissatus also forms from phenomena that have nothing to do with cumulonimbus blow-off or dissipating cumulonimbus cells. When dense cirrus is formed by other means than by cumulonimbus blow-off or dissipating cumulonimbus clouds, it will frequently be seen as many dense patches at different levels (cirrus spissatus duplicatus), often mixed with thin cirrus filaments. Another variety, cirrus spissatus intortus, is sometimes described as looking like "entangled sheaves" of cirrus clouds. When viewed toward the sun, the denser patches often have gray bases.

CIRRUS FLOCCUS AND CIRRUS CASTELLANUS.—Patches of dense cirrus may take on the form of *cirrus floccus,* with the upper portion of the patch forming rounded tufts, and the base portion becoming ragged. Dense cirrus patches may also grow turrets or battlements and become *cirrus castellanus.* Both cirrus floccus and cirrus castellanus may have ice crystal virga trails showing from the base of the cloud patch, and may be slightly larger than the standard 1°. (Also see cirrocumulus floccus and cirrocumulus castellanus.)

Cirrostratus (CS)

Cirrostratus clouds usually appear as a thin white veil over the sky. If the cirrostratus clouds are very thin and of uniform thickness, the only indication of their presence may be a faint halo, or a whitish tint to the sky. As long as the sun is higher than 30° above the horizon, cirrostratus clouds should not be able to block the sun; shadows should be apparent from sunlight shining through this cloud. When cirrostratus is low on the horizon, it tends to block the blue color of the sky more thoroughly because it is viewed on an angle, and it is commonly mistaken for denser altostratus. This also happens near sunrise and sunset with low sun angles. Cirrostratus cloud layers appear to move very slowly, and change shape very slowly. Typically, the edge of a cirrostratus layer is so indistinct that it is difficult to detect where the blue sky ends and the cloud begins. If movement or changes in shape are detectable during the observation period, the cloud near the horizon may well be altostratus.

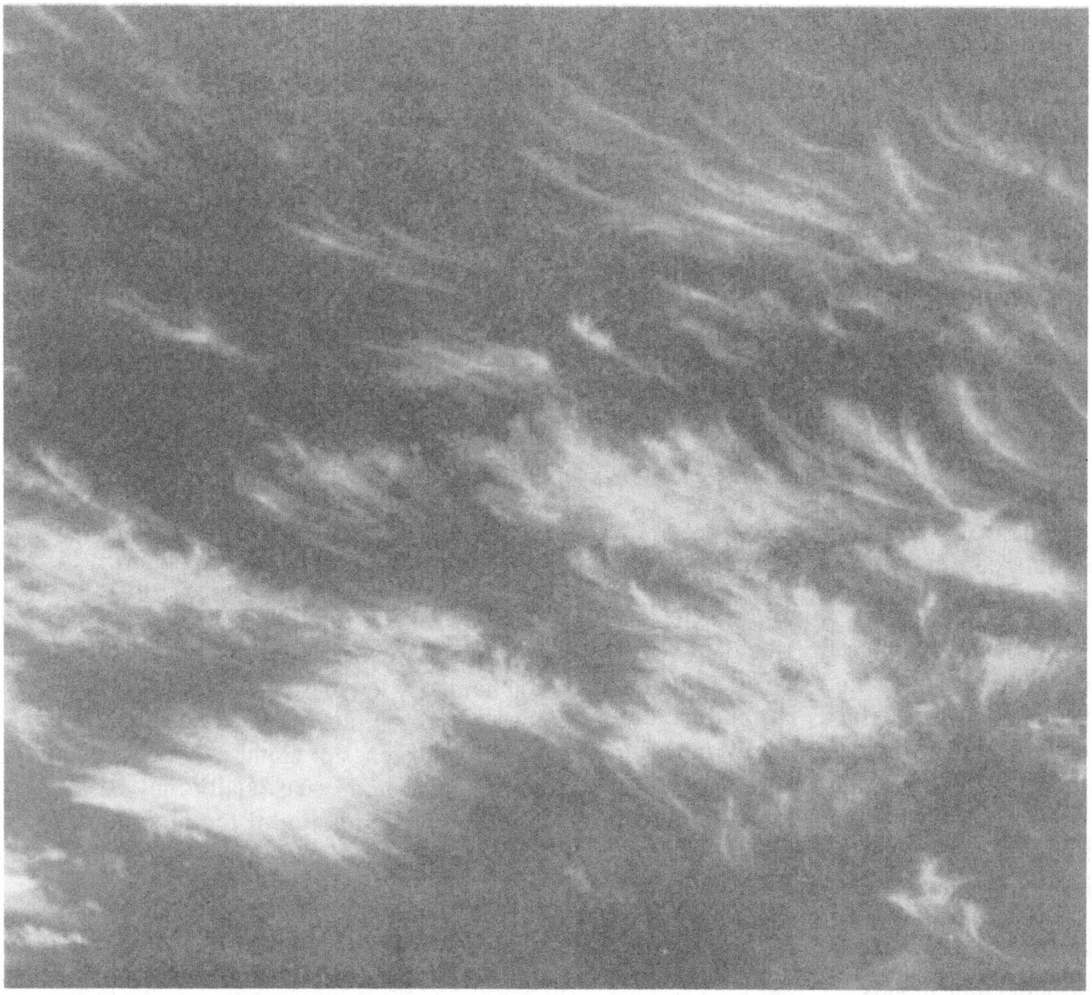

Figure 1-24.—Cirrus uncinus (mare's tails).

CIRROSTRATUS NEBULOSUS.—*Cirrostratus nebulosus* often appears as a thin veil over the sky, without any distinguishable features. This cloud is sometimes mistaken for haze. Haze, however, will have a yellowish or brownish color as opposed to the milky appearance of cirrostratus. A halo in an otherwise clear sky indicates cirrostratus nebulosus.

CIRROSTRATUS FIBRATUS.—Occasionally a cirrostratus layer contains fibrous filaments. When this occurs, the entire cloud layer is classified *cirrostratus fibratus.*

Cirrocumulus (CC)

Cirrocumulus clouds are very similar in appearance to the high altocumulus clouds. The small white cells of cirrocumulus clouds, however, cover less than 1° of the sky, which means that they may be hidden by one finger held at arm's length. The small cloud elements are usually arranged in tightly packed rows. An area of cirrocumulus clouds (refer to figure 1-23) is frequently described as looking like a honeycomb, a fish net, or like the scales of a fish. The last description gave rise to the popular name for these clouds, the mackerel sky.

CIRROCUMULUS STRATIFORMIS.—Cixrocumulus clouds in an extensive sheet or layer are identified by the species *cirrocumulus stratiformis.* Typically, however, the area covered by cirrocumulus does not cover the entire sky, but only small patches of the sky.

CIRROCUMULUS FLOCCUS.—When the cirrocumulus elements show rounded tops and ragged bases, often with short virga trails, the *cirrocumulus* *floccus* species name is used. When the virga trails make the entire element larger than 1°, the cloud must be classified cirrus floccus.

CIRROCUMULUS CASTELLANUS.—*Cirrocumulus castellanus* identifies the cirrocumulus cloud layers where each element shows a cumuliform buildup such that the height of the element is greater than the width of its base, and each element is smaller than 1°. If the vertical development of the cloud elements progresses such that the element becomes larger than 1°, the cloud is reclassified *cirrus castellanus. Now* let's consider some cloud types we haven't covered in table 1-1.

Orographic Clouds

Several species of low-, mid-, and high-etage clouds are associated only with moist airflow over mountainous areas.

These clouds usually form during Mountain Wave conditions, when strong winds blow across mountain ranges. The presence of these clouds is significant in that they may be associated with dangerous turbulence. All of the orographic cloud forms are unique in that they are stationary over a particular area and do not move with the wind flow. Slow changes in the arrangement of elements or the cloud pattern may be noted as the upper wind direction or intensity changes. The significant orographic cloud forms are the rotor cloud, the cap cloud, and the lenticular clouds.

The *rotor cloud* (fig. 1-25) is formed downwind from the mountain range. The rotor cloud is formed as the strong winds moving across the mountains set up a wavelike action in the winds downstream from the mountain. The waves or eddies in the air may extend in a series downstream from the mountain for hundreds of miles. The upward moving air in the waves or eddies, if moist enough, is brought to saturation as it rises, forming the rotor clouds.

Rotor clouds are cap-shaped, with smooth rounded tops, flattened or concave bases, ragged up-wind edges, and very ragged downstream edges. The up-wind edge continuously forms, while the downstream edge continuously dissipates. Rotor clouds may form as a few isolated elongated elements, or in succeeding bands elongated parallel to the mountain range. When you are observing the sky condition, rotor clouds are usually classified as cumulus clouds, and a note about the presence of rotor clouds may be added in the remarks section.

A special type of stratocumulus cloud, called the *Foehnwall* or *cap cloud,* may form on the top of mountain ranges, resembling a "cap" on the mountain. It is formed as moist air is forced upward by the mountain top, and dissipates on the leeward side of the mountain as the moving air descends. Cloud particles and denser patches of cloud may be seen moving over the mountain and occasionally may be seen flowing down the leeward side of the mountain, giving the appearance of a waterfall. The cloud itself is stationary on the mountain top.

Altocumulus lenticularis (abbreviated ACSL) clouds are typically described as lens-shaped, almond-shaped, or cat-eye shaped, and usually have a windswept appearance (fig. 1-26). Although the cloud elements may grow or shrink in size, they are usually stationary. The size of the individual cloud elements is usually quite large. The leading or windward edge

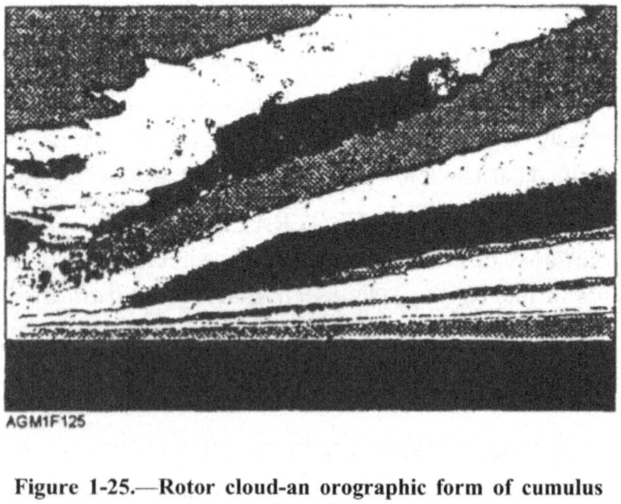
AGM1F125

Figure 1-25.—Rotor cloud-an orographic form of cumulus cloud.

AGM1F126

Figure 1-26.—Standing lenticular cloud.

continuously forms, while the trailing or leeward edge continuously dissipates.

These clouds usually form downwind from a mountain range or over a mountain in a small portion of the sky at one level, but may form in different levels and appear stacked on top of the lower level cloud elements. They may also form in a layer downwind from a mountain range with individual cloud elements well separated from each other. Lenticularis clouds need not form in conjunction with other orographic clouds, the rotor clouds, and cap clouds.

The process that forms altocumulus lenticularis clouds also on occasion forms the same clouds closer to the ground in the low-etage, and frequently forms similar clouds in the high-etage. The difference in the classification of stratocumulus lenticularis, altocumulus lenticularis, and cirrocumulus lenticularis stems solely from the height at which they form, and should not be based on the apparent size of the cloud, which, in this case, may be very misleading. Studies have shown that although lenticularis clouds usually form with bases in the mid-etage range, they can form downstream of larger mountain ranges, with the base of the lowest lenticularis in the 20,000- to 30,000-foot range. Bases of the higher elements of stacked lenticularis may be as high as 35,000 feet, with cloud tops near 40,000 feet.

REVIEW QUESTIONS

Q16. Describe two potentially dangerous wind phenomena associated with an outflow boundary.

Q17. A wall cloud will usually form in what location of a CB cell?

Q18. Stratus fractus clouds generally form in conjunction with what other type of cloud?

Q19. When must altostratus clouds be reclassified as nimbostratus?

Q20. What might the formation of altocumulus castellanus or altocumulus floccus indicate?

Q21. What are cirrus clouds composed of?

Q22. Explain the formation of a rotor cloud.

Q23. Explain the formation of a cap cloud.

CLOUD AMOUNTS

The amount of cloud cover is the second determination you must make in observing the overall condition of the sky. Clouds of the types we have just discussed may form at various levels in the atmosphere. It is not uncommon to have different layers of low-etage clouds, along with several layers of mid- and high-etage clouds.

Estimation by the observer is the primary method used to determine cloud amounts. However, automatic weather systems can measure the amount of clouds at each level, and these inputs may be used as a supplemental tool by observers for cloud layer coverage and total sky cover.

Generally, there are two different types of cloud amount measurements necessary for an observation. The more difficult measurement is *cloud layer coverage.* The easier measurement is *the total sky* cover. In cloud layer coverage, the amount of cloud in each layer must be estimated. Both are estimated in eighths *(oktas)* of *the celestial dome* (the total area of sky or the dome of the sky). Layer coverage is used to determine the *cloud ceiling,* which is the lowest layer or layers that block 5/8 or more of the celestial dome from being seen.

Layer Coverage

A cloud *layer* is defined as "clouds and/or obscuring phenomena aloft, either continuous or composed of detached elements, that have bases at approximately the same level." Both continuous and detached elements may combine to form a layer, and all layers and obscuring phenomena are to be considered opaque. The essential requirement is that bases be at approximately the same level. The upper portions of cumulonimbus clouds are often spread horizontally by the wind, and form a layer of cirrus spissatus, dense altostratus, or dense altocumulus clouds, Velum may also be present. These horizontal extensions are regarded as separate layers if their bases appear horizontal, and they cover 1/8 of the sky or more. A layer may be a combination of cloud types or a combination of obscuring phenomena as long as the bases are all at approximately the same level. For example, cumulus mediocris and cumulus congestus may be considered as the same layer if their bases are the same height.

When observing layer coverage, you must not only estimate the amount of clouds in each layer, but also consider phenomena that hide 1/8 or more of the sky as a layer. (A partial obscuration hiding less than 1/8 of the sky is ignored.) Obscurations may be surface-based or aloft and include phenomena such as rain, snow, fog, smoke, or haze. However, liquid or frozen water particles falling through the atmosphere are never classified as obscurations aloft. In the past, you may have looked at the sky and seen thin fog or haze on the horizon that blocked your view of the clouds. Sometimes the portion of the sky that is hidden from view extends only a few degrees above the horizon, while at other times, the phenomenon may extend well above the horizon. When the phenomenon is thin enough to allow the sun, clouds, the moon, or stars to be seen overhead but not seen near the horizon, the phenomenon is termed a *partial obscuration* (fig. 1-27). If the phenomenon is dense enough to hide even the portion of the sky directly overhead, it is called a *total obscuration.* An obscuring phenomenon frequently extends around the entire horizon circle, 360° of azimuth, to completely surround the observation site.

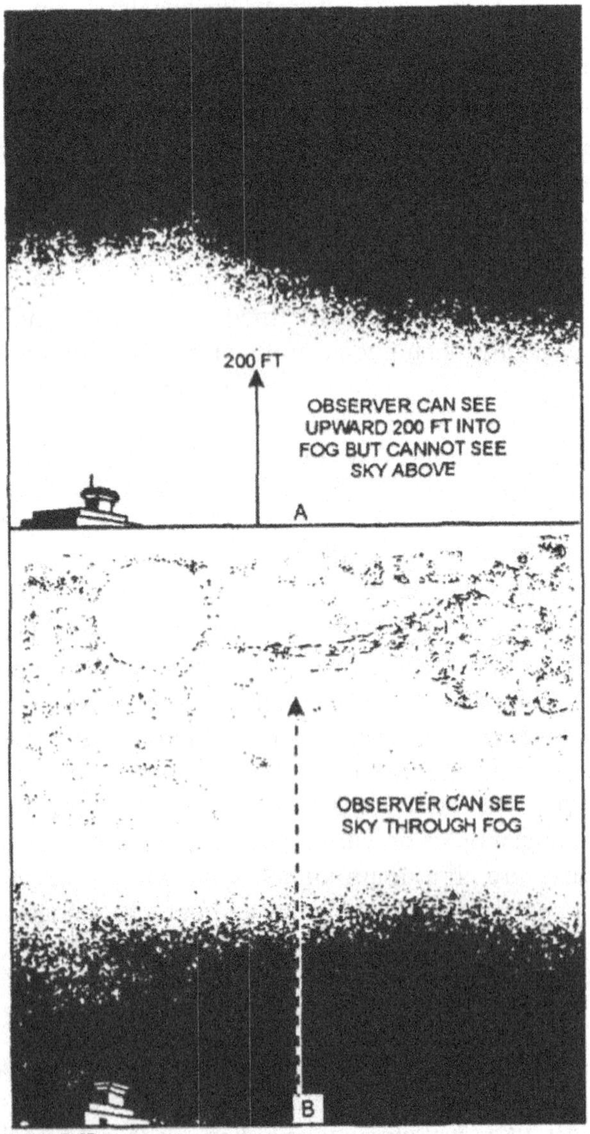

Figure 1-27.—Obscuration. (A) Total obscuration-sky completely hidden, (B) partial obscuration-higher cloud, sun, moon, or sky may be seen.

The best method to determine how much of the sky is hidden by a partial obscuration is to measure the elevation angle by using a clinometer (see chapter 2). The top of the partial obscuration is considered to be the point where the outline of higher clouds, the sun or the moon, or the light from stars is visible. Tables in NAVMETOCCOMINST 3141.2 and NAVMETOC-COMINST 3144.1 are used to convert elevation angle to eighths of sky coverage. Let's review a few of the important concepts involved when observing layer amounts.

When you are observing cumuliform clouds in a layer and blue sky is visible between the elements, the blue sky is not included as part of the layer coverage. For example, if many small cumulus clouds covered the sky from horizon to horizon, but the blue sky visible between each of the cloud cells is about the same size as the cloud cells, the layer coverage would only be 4/8. If however, the cumuliform cloud elements are joined by very thin cloud, even if the thin cloud is transparent and higher clouds can be seen through the thin cloud, the cloud layer is considered to cover the area. This is commonly the situation with altocumulus and cirrocumulus clouds. For example, if half of the sky is covered by a layer of altocumulus cloud with the denser cells being opaque, and the area between the cells is filled with a transparent cloud through which a very pale blue sky or higher clouds may be seen, the layer coverage is the same as the entire portion of the sky covered, or 4/8. The portion of a higher cloud layer visible through lower transparent clouds is treated as if it were not visible, that is, as if the lower cloud layer were opaque.

In a METAR/SPECI observation, when observing layers of clouds that are stacked on top of other layers, only count those clouds visible to you in the layer amount. If, for example, a layer of stratocumulus clouds covers half of the sky, and directly above the stratocumulus layer the observer can see only the edges of an altocumulus layer that seems to cover the same area as the stratocumulus layer, the most the observer could report is 4/8 stratocumulus and 1/8 altocumulus. Similarly, if an altocumulus layer covers 6/8 of the sky in a dense sheet, and 1/8 of cumulus is located below the altocumulus layer, the observer "knows" that the altocumulus layer covers the entire layer extending across 6/8 of the sky, even though 1/8 of the layer is hidden from view by the cumulus cloud. The cloud layers must still be reported as 1/8 cumulus and 5/8 altocumulus, since the observer cannot actually see the remaining 1/8 of altocumulus cloud. The maximum number of reportable layers is limited to six. OCONUS stations are limited to three reportable layers unless CB or TCU are present; in which case, a fourth layer may be reported.

Total Sky Coverage

To determine *total sky coverage,* simply add together the amount of clouds and/or obscurations in each layer. Total sky cover cannot exceed 8/8.

Summation Sky Coverage

Summation sky coverage is a concept used to determine cloud ceiling. Summation sky cover is determined for each layer of clouds by adding the coverage in the cloud layer to the coverage of all layers below it. In the example that follows, this has been done.

LAYER NUMBER	LAYER AMOUNT	SUM OF SKY COVER	SUMMATION LAYER COVER
1	1/8	1/8	FEW
2	3/8	4/8	SCATTERED
3	3/8	7/8	BROKEN
4	1/8	8/8	OVERCAST

Summation coverage for each layer is converted into plain language terms. They are *clear,* for no clouds present; few, for greater than 1/8 to 2/8; *scattered,* for 3/8 to 4/8 clouds; *broken,* for 5/8 to 7/8 clouds; and *overcast,* for 8/8 clouds, as shown in table 1-2. The term thin is not used in METAR/SPECI observations.

Table 1-2.—Cloud Ceiling Summation Coverage and Terms

Contraction	Meaning	T e x t T e x t
SKC	Sky clear	0/8
FEW	trace	>0/8-2/8
SCT	Scattered	3/8-4/8
BKN	Broken	5/8-7/8
OVC	Overcast	8/8
VV	Vertical Visibility	8/8

CEILING DETERMINATION

The terms *ceiling* and *cloud ceiling* are defined as the height-above-ground level of the lowest broken or overcast layer. If the sky is totally obscured, the height of the vertical visibility (VV) is used as the ceiling height. Table 1-2 relates the measurement of sky cover as observed in eighths, to the terms used to discuss sky cover and ceilings. In the example we just covered, the ceiling would be at layer No. 3, the broken layer.

In addition to determining layer amounts, total sky coverage, summation coverage, and the ceiling based on the summation coverage, you must also identify cloud layer heights and ceiling height.

CLOUD LAYER HEIGHTS/CEILING HEIGHT

The height must be determined for the base of all layers of clouds. The height for the lowest broken or overcast layer is used as the ceiling height. A surface-based obscuration also constitutes a ceiling; vertical visibility into the obscuration is used as ceiling height.

Cloud layer height and ceiling height may be determined by several methods. Estimation is the most frequently used method for cloud layer height. It is acceptable for determining heights of low scattered cloud layers or higher cloud layers. When the cloud layer forms a ceiling, especially if the layer height is below 3,000 feet, one or more of the more accurate methods should be used. The following list of cloud layer height determination methods is generally ordered from the most accurate to the least accurate:

- Measurement by comparison to known heights of structures or landmarks within 1 1/2 miles of the runway
- Measurement by ASOS or SMOOS cloud-height sensor
- Measurement by rotating beam ceilometer for heights less than 10 times the baseline
- Measurement by ceiling light and clinometer
- Estimation by rotating beam ceilometer for heights more than 10 times the baseline
- Estimation by pilot during ascent or descent
- Estimation by ceiling balloon
- Estimation by comparison to terrain or structures more than 1 1/2 miles from the runway
- Estimation by convective-cloud-height diagram
- Estimation using the Skew-T, Log P diagram
- Estimations from conventional weather/Doppler radar
- Estimation using other station reports in the vicinity
- Estimation by observational experience

Obviously, not all methods may be used at all times, and some methods work better than others in different situations. When a cloud layer height value falls halfway between two reportable valves, round down to the nearest reportable increment given in table 1-3.

Table 1-3.—Reportable Values for Cloud Layer Height and Ceiling Height

HEIGHT IN FEET	REPORTABLE VALUE
≤50	0
5,000 or less	Nearest 100 feet
5,001 to 10,000	Nearest 500 feet
Above 10,000	Nearest 1,000 feet

NAVMETOCCOMINST 3141.2 and NAVMETOCCOMINST 3144.1 discuss in detail the various methods and procedures used to determine cloud height and ceiling height.

REVIEW QUESTIONS

Q24. How is the amount of *cloud layer coverage and total sky coverage measured?*

Q25. Define cloud layer.

Q26. Define cloud ceiling.

Q27. Define total obscuration.

Q28. Define summation sky coverage.

Q29. Given: Layer No.1 is 1/8 fog (SFC)

Layer No.2 is 2/8 cumulus 3,000 ft

Layer No.3 is 2/8 altocumulus 12,000 ft

Layer No.4 is 5/8 cirrus 20,000 ft

What is the ceiling height?

Q30. A cloud height of *7,550 feet would be reported as what height?*

VISIBILITY

LEARNING OBJECTIVES: Describe prevailing visibility, sector visibility, and differing level visibility. Define runway visual range.

Visibility, as well as ceiling height, aids in decisions involving air traffic control. For this reason, the observation of visibility must be timely, accurate, and representative. There are four types of visibility that you must observe: (1) prevailing visibility, (2) sector visibility, (3) differing level visibility, and (4) runway visual range. Both NAVMETOCCOMINST 3141.2 and NAVMETOCCOMINST 3144.1 provide thorough and detailed guidance on visibility observations. Ashore, visibility is observed in statute miles. Aboard ship, visibility is observed in nautical miles. Observing stations located outside the United States report visibility in meters.

When observing visibility, you should note the distance as follows:

- To the nearest 1/16 mile when visibility is less than 3/8 mile
- To the nearest 1/8 mile when it is between 3/8 and 2 miles
- To the nearest 1/4 mile, between 2 and 3 miles
- To the nearest mile, from 3 to 15 miles
- In 5-mile increments, above 15 miles

When the visibility falls between two values, the lower value is always used. For example, a measured visibility of 3 3/4 miles is called "3 miles." See Appendix II of this module for a visibility conversion table.

PREVAILING VISIBILITY

Prevailing visibility is the greatest distance that known objects can be seen and identified throughout half or more of the horizon circle. The most reliable method for determining prevailing visibility is by the eye of a trained observer. The sensors provided with automatic observing systems provide only an approximation of prevailing visibility based on the sampling of obstructions-to-vision present in only a small area around the sensor. To aid in the determination of prevailing visibility, observation stations are required to maintain a visibility chart. The visibility chart identifies each daytime and nighttime visibility marker with direction and distance to the marker. Daytime markers are generally dark, prominent objects that stand out when viewed against the lighter sky. Nighttime markers are usually unfocused lights of moderate intensity, such as radio tower lights or channel marker lights.

At sea, since the ship is usually moving, fixed visibility markers are not available. The Combat Information Center (CIC), however, maintains tracks on other ships in the area, as well as coastal formations and prominent objects ashore. Coastal formations and prominent objects ashore may be used as visibility markers. Direct coordination between the observer and CIC is necessary to obtain timely and accurate distances to observable objects. This may be done through one of the Lookouts equipped with a sound-powered phone, on the "JL" circuit. All ships and most large coastal objects may also be used as distance markers at night; their navigation lights should be clearly seen. CIC will also be able to inform you of the probable light patterns and colors that various objects may be showing.

Radar returns from landmasses or isolated rain showers may also provide a valuable indication of visibility range. The distance to the horizon also plays an important part in visibility observations at sea. (The distance to the horizon in nautical miles is 1.15 times the square root of the height, in feet, of your eyes above the water.) From the deck of a small boat, the horizon is only about 3 nautical miles away. From ships with weather decks about 30 feet above the water, an observer sees the horizon at about 7 nautical miles. And from the flight deck of an aircraft carrier (average 65 feet above the water), a standing observer's eyes are about 70 to 71 feet above the water, and the horizon is seen at just under 10 nautical miles. These distances usually limit what an observer may be able to see. A table of the distance an object may be seen based on the height of the observer's eyes above sea versus the height of the object is provided in NAVMETOCCOMINST 3144.1.

In certain situations, prevailing visibility may fluctuate up and down during the observation period. In those cases, average visibility is used and is called *variable visibility*. The observer must note the lowest and highest visibility for entry on the observation record.

SECTOR VISIBILITY

A sector is any portion of the area surrounding the station out as far as the horizon. When the visibility surrounding the station is not uniformly equal in all directions and the difference is operationally significant, then each area with a different visibility is designated a sector. The size of the sector, extending in a pie-slice out from the observation point, is as large or as small as is required to describe the area affected by the different visibility, but must be limited to 1/8 (45°) of the horizon circle. Sector visibility is commonly used at air stations that have lakes, rivers, or swamps nearby, which favor fog development. Visibility in fog over a swamp area may, for example, be 3 miles, while the remainder of the area has 7 miles visibility in haze. Smoke and localized rain showers are other phenomena that commonly cause poorer visibility in a sector.

Once an area of lower or higher visibility is identified, the directions of the S-point compass (N, NE, E, SE, S, SW, W, and NW) are usually used to identify the sector. Each sector, using the 8-point compass, covers 45° of azimuth centered on the compass point identified.

Sector visibility is reported in the observation only when it differs from the prevailing visibility, and either the prevailing or sector visibility is less than 3 miles.

DIFFERING LEVEL VISIBILITY

Differing level visibility is any prevailing visibility observed from an elevation or location other than the official observation site. Differing level visibility is commonly evaluated from the aircraft control tower by certified tower visibility observers. In this case the prevailing visibility is usually called *tower visibility*. Tower visibility may differ from the airfield-level prevailing visibility based on the type of obstruction-to-vision present. Differing level visibility is only reported when the prevailing visibility is 4 miles or less.

RUNWAY VISUAL RANGE

The runway visual range, abbreviated RVR, is an instrument measurement of the distance the pilot can see down the runway as an aircraft touches down during landing. RVR is observed at shore stations using the AN/GMQ-32 transmissometer when the prevailing visibility or sector visibility falls below 2 miles, but is only reportable when the prevailing visibility is 1 mile or less or the RVR value for the runway is 6,000 feet or less. NAVMETOCCOMINST 3 141.2 further outlines procedures for reporting RVR.

When the prevailing visibility falls to less than 7 miles, the reason that the visibility is restricted must be noted in the observation. Any phenomenon that reduces visibility is called an "obstruction to vision." The occurrence of "weather," such as precipitation, also may reduce visibility. In the next section, we will cover weather and obstructions to vision.

REVIEW QUESTIONS

Q31. What are the four types of *visibility that may be observed?*

Q32. Define prevailing visibility.

Q33. Where can shipboard observers obtain visibility range information at sea?

Q34. When is sector visibility reported?

Q35. When is differing level visibility reported?

WEATHER AND OBSTRUCTIONS TO VISION

LEARNING OBJECTIVES: Identify six types of lithometeors. Compare the condensation/sublimation and precipitation forms of hydrometeors. Explain wind-blown forms of hydrometeors. Describe two types of electrometeors. Define and list four types of photometeors.

The occurrence of weather and the presence of obstructions to visibility are directly related to sky condition and the visibility. Observing the type of weather occurring and the presence of any obstructions to visibility is usually the third task undertaken in an observation. A pilot may use this information to determine the impact of the conditions at a station on the aircraft being flown. In this discussion, we use the term *weather* to refer to any particles suspended in or precipitating from the atmosphere, or to the process that causes these particles to precipitate from the atmosphere. Observable weather elements may be broken down into four groups: lithometeors, hydrometeors, electrometeors, and photometeors.

LITHOMETEORS

A *lithometeor* is any dry particle suspended in or falling from the atmosphere. The particles are usually formed on earth's surface and then are carried aloft by either wind or thermal currents. Haze, smoke, dust, dust-devils, ash, and sand are all lithometeors.

Haze

Haze is composed of suspended dust, plant pollen, or salt particles that are so small that they cannot be seen by the unaided eye. It is opalescent, reducing visibility. Haze typically produces a bluish tinge when viewed against a dark background. It produces a dirty yellow or orange tinge when viewed against a brighter background because of the scattering of light. When haze is present and the sun is well above the horizon, its light may have a silvery tinge. Haze particles are *hygroscopic*—they attract moisture. Because they attract moisture, they are good *condensation nuclei.* When conditions are favorable, haze may attract sufficient moisture to thicken into fog as the sun sets and the temperature drops.

Smoke

Smoke is composed of fine ash particles and other by-products of combustion. When concentrated at its source, smoke may appear white to bluish-black, or yellow to brown, depending on its composition and the amount of water vapor present. After it is dispersed in the atmosphere, smoke is distinguished from haze by its characteristic reddish tinge, especially near the horizon at sunrise and sunset.

Dust

Dust is composed of fine solid matter uniformly distributed in the air. It typically imparts a tan or gray hue to distant objects. The sun's disk may appear pale and colorless, or may have a yellow tinge when viewed through dust. Although dust and haze appear similar, when the visibility is less than 7 miles, dust may be differentiated from haze or fog by the low relative humidity associated with dusty conditions. In certain areas of the world, suspended dust may reduce visibility to less than a mile. Normally, the lower visibility associated with dust is limited to *blowing dust*—dust picked up and carried by the wind. The term *dust storm* usually refers to blowing dust reducing visibility to 5/16 to 5/8 of a mile, while the term *heavy* or *severe dust storm* is reserved for use with blowing dust that restricts visibility to less than 5/16 mile.

Dust/Sand Whirl

Dust/sand whirls or *dust devils,* are rotating columns of dust or sand-laden air, caused by intense solar radiation. They are best developed on calm, hot, clear afternoons and in desert regions. Warm, ascending air in a dust devil may carry leaves and other small debris to a height of a few feet or a few hundred feet.

Ash

The phenomenon called *ash* in a surface meteorological observation usually refers to the heavier volcanic ash particles falling from a volcanic cloud. It may also be used to identify heavier solid particles precipitating and falling from an industrial smoke plume, the smoke from a forest fire, or the debris falling from a nuclear mushroom cloud.

Sand

Sand particles may be picked up from dry surfaces by the wind at wind speeds as low as 21 knots and carried to moderate heights. Stronger winds may carry sand to extreme heights. The term *sand storm* refers to blowing sand that reduces visibility from 5/16 to 5/8 mile, while the term *heavy or severe sand storm* means that visibility is less than 5/16 mile.

The only hazard to aviation caused by haze and smoke is reduced visibility. But dust, ash, and sand can also clog engine intakes and be very abrasive to moving components. Aircraft flying through these conditions may experience fatal engine failure.

HYDROMETEORS

Hydrometeors are liquid or solid water particles falling through, suspended in, or condensing/subliming from the atmosphere, as well as solid or liquid water blown from the surface by wind. The term refers to all forms of condensation, such as clouds, fog, dew, and frost; all forms of precipitation, such as rain, drizzle, snow and hail; and all forms of moisture blown about by the wind.

Condensation/Sublimation Forms

Many of the weather elements identified as hydrometeors are formed by the condensation or sublimation of water vapor in the air or on surfaces. Clouds and fog are hydrometeors of suspended liquid or solid moisture suspended in the air. Dew and frost are hydrometeors of moisture that condense or sublime directly on surfaces or on the ground.

CLOUDS.—Clouds are the visible form of water vapor, and consist of minute suspended droplets of liquid water or ice particles. Fog is a cloud on the earth's surface. Liquid water droplets develop from gaseous water vapor by the process of condensation. Solid water particles or ice crystals develop by the process of sublimation. During sublimation, gaseous water vapor bypasses the liquid state and goes directly from a gas to a solid, thereby releasing heat into the atmosphere. Three factors are necessary for cloud formation: sufficient moisture, hygroscopic nuclei, and a cooling process.

Moisture is supplied by evaporation and is distributed vertically by convection currents and horizontally by winds.

Hygroscopic nuclei are small particles on which water vapor can condense or sublime. Hygroscopic nuclei actually attract water vapor. The most effective hygroscopic nuclei are the by-products of combustion, sulfuric acid and nitric acid particles, and salts (such as sodium chloride raised from the sea surface). Dust particles may contain sufficient salts or acids to become hygroscopic nuclei, but dust particles in general are not effective hygroscopic nuclei. The presence of hygroscopic nuclei is a must for water vapor to condense. Air has been super-saturated in laboratories to over 400% before condensation began in the absence of hygroscopic nuclei. In actual conditions, in the presence of abundant hygroscopic nuclei, condensation may begin at relative humidities near 70%. Saturation of the air is reached when the relative humidity reaches 100%. At this point, the evaporation rate from liquid water droplets to water vapor equals the condensation rate from water vapor to liquid water, or theoretically, the sublimation rates from gas-to-solid and from solid-to-gas are exactly equal.

Hygroscopic nuclei are also called *condensation nuclei* and *sublimation nuclei* when referring to the specific process of condensation or sublimation.

A *cooling process* aids in condensation, since it increases the humidity of the air without increasing the amount of water vapor present. The higher the humidity, the easier condensation proceeds. The cooling process most frequently associated with condensation is *adiabatic expansion.* When a parcel of air is lifted higher in the atmosphere (where the pressure is lower), it expands and its temperature decreases. Another important cooling process is radiational cooling. Simply put, as the sun goes down, the air cools because the heat source, the sun, is no longer available to maintain the heating.

FOG.—Fog is a suspension of small visible water droplets (or ice crystals) in the air that reduces horizontal and/or vertical visibility at the earth's surface. Fog is a stratus cloud on the surface of the earth. It is distinguished from smoke, haze, or dust by its dampness and gray appearance. Fog usually does not form or exist when the difference between the air temperature and the dew-point temperature is greater than 4 Fahrenheit degrees (2 Celsius degrees). However, at temperatures below -2O°F (-29°C), *freezing fog,* or ice fog, may form when the dew-point temperature is as much as 8°F (4 Celsius degrees) lower than the air temperature. Freezing fog is composed entirely of ice crystals that sparkle brilliantly in light. When the air temperature is between 32°F and -20°F, fog may exist as super-cooled liquid droplets. This situation may produce delicate needle or platelike ice crystals on exposed surfaces, *known* as *hoarfrost. Rime ice,* a smooth, milky, white ice coating, or *glaze ice,* a smooth coating of clear ice, may also be produced by fog when the temperature is below freezing.

Fog is sometimes identified by the physical process by which it forms. Examples are *radiation fog,* formed by radiational cooling; *advection fog,* formed by moist air moving over a cooler surface; *steam fog,* formed when cold air moves over a warm body of water; *upslope fog,* caused by air cooling as it rises up a hill or mountain; and *frontal fog,* formed by the evaporation of rain in a colder air mass. You will study these later in preparation for AG2. To observe the presence of fog, you need not know how fog forms-only if it is present. The terms used to record fog that you, as the observer, must be familiar with are as follows:

- *Fog*—The vertical depth of fog is greater than 20 feet and the prevailing visibility is reduced to less than 5/8 mile (1,000 meters OCONUS).

- *Mist*—A fog condition that reduces prevailing visibility to between 5/8 mile (1,000 meters) and 6 miles (9,000 meters). The vertical depth of mist is greater than 20 feet.

- *Ground fog*—This term applies to fog that has little vertical extent, i.e., normally greater than 6 feet but less than 20 feet. This is a local phenomenon, usually formed by radiational cooling of the air. Ground fog can further be described as shallow, partial, or patchy.

- *Shallow fog*—This descriptor of fog applies to ground fog that covers the station and visibility at eye level is 7 miles or more, but the apparent visibility in the fog layer is still less than 5/8 mile. Shallow fog does not extend above 6 feet.

- *Partial fog*—This descriptor of fog applies to ground fog that covers a substantial part of the station and visibility in the fog is less than 5/8 mile, and visibility over the uncovered parts of the station is 5/8 mile or more. The vertical extent ofpartial fog is greater than 6 feet but less than 20 feet. This type of ground fog may be coded even when the prevailing visibility is 7 miles or more.

- *Patchy fog*—This descriptor of fog applies to ground fog that covers portions of the station, the apparent visibilty in the fog patch or bank is less than 5/8 mile, and visibility over the uncovered portions of the station is 5/8 mile or greater. The vertical extent of patchy fog is greater than 6 feet but less than 20 feet. This type of ground fog may be coded even when the prevailing visibility is 7 miles or greater.

DEW.—*Dew* is moisture that condenses directly on surfaces. Dew will form during the evening or late at night, usually when the winds are light. After the sun sets, the ground and objects near the ground cool by radiational cooling; they radiate heat energy as infrared radiation. When the ground or objects cool to the *dew-point temperature,* water vapor condenses out of the air onto the object's surface. *White dew* is dew that has frozen after the water condenses. It is recognizable as small beads or a beaded layer of clear ice, or sometimes milky-colored ice on surfaces.

FROST.—*Frost* is a layer of milky white ice crystals that sublime directly on the ground or on other surfaces. The crystals are commonly in the shape of needles, scales, feathers, or fans. Frost forms when radiational cooling lowers the temperature of objects below the freezing level. Since many objects cool faster than the air surrounding the objects, frost may form with ambient air temperatures above freezing, as high as 37°F. For frost to form, the object must be cooled to the *frost-point temperature,* which is also referred to the "dew-point temperature with respect to ice." This is the dew-point temperature calculated on the "low temperature" side of the Psychrometric Computer. Thicker deposits of needle or platelike frost, up to several inches thick, form in fog with ambient air temperatures below freezing. This form of frost is known as *hoarfrost.*

REVIEW QUESTIONS

Q36. What type of lithometeor produces a yellow or orange tinge when viewed against a brighter background?

Q37. Where are dust/sand whirls most likely to develop?

Q38. What does the term "heavy sandstorm" mean?

Q39. Define sublimation.

Q40. What three factors are necessary for cloud formation?

Q41. Fog may form when the temperature-dewpoint spread is how many Celsius degrees?

Q42. Fog formed by moist air moving over a cooler surface is known by what term?

Q43. Define the term "mist."

Q44. Explain the formation of frost.

Precipitation Forms

Precipitation includes all forms of moisture that fall to the earth's surface, such as rain, drizzle, snow, and hail. Precipitation is observed and classified by form, type, intensity, and character.

PRECIPITATION FORM.—*Precipitation form* is the state that the moisture is in: liquid, freezing, or *frozen. Liquid precipitation* is any precipitation that falls as a liquid and remains liquid after striking an object, such as the earth's surface or the skin of an aircraft. Rain and drizzle are the only two types of liquid precipitation.

Freezing precipitation is any precipitation that falls as a liquid and freezes upon contact with an object, such as freezing rain or freezing drizzle. In this form of precipitation, the liquid water may be a *super-cooled* liquid and freeze upon contact with an object, or the water droplet may have an above freezing temperature and freeze upon contact with an object that has a temperature below freezing. (Super-cooled liquids have a temperature below their normal freezing temperature, but still exist in the liquid state.) Small freezing drizzle particles form a milky white ice coating, typically referred to as *rime ice,* especially on aircraft in flight. Larger freezing drizzle and freezing rain drops form a transparent ice coating known as *clear ice* on aircraft in flight or as *glaze ice* on the ground, power lines, or trees.

Frozen precipitation is any precipitation of water that falls in its solid state, such as snow, hail, or ice pellets. Different forms of precipitation may occur together, such as mixed rain and snow; but such an occurrence is simply a mixture of forms, not a separate form of precipitation.

PRECIPITATION TYPE.—*Precipitation type* is the term used to identify various precipitation. Discussion of the types of precipitation follows:

- *Rain*—Liquid precipitation that has a water droplet diameter of 0.02 inch (0.5 mm) or larger. If the water droplets freeze upon contact with a surface, the phenomenon is called *freezing rain.*

- *Drizzle*—Liquid precipitation that consists of very small and uniformly dispersed droplets of liquid water that appear to "float" while following air currents. Drizzle usually falls from low stratus clouds and is frequently accompanied by fog. A slow rate of fall and the small size of the droplets (less than 0.02 inch) distinguish drizzle from rain. When drizzle freezes on contact with the ground or other objects, it is referred to as *freezing drizzle.* Drizzle usually restricts visibility.

- *Snow*—Precipitation that consists of white or translucent ice crystals. In their pure form, the ice crystals are highly complex, hexagonal, branched structures. Snow falls as a combination of individual crystals, fragments of crystals, or clusters of crystals. Warmer conditions tend to favor larger crystal sizes and clusters of crystals. Snow must form in cloud temperatures below freezing, though it may fall through air at above freezing temperatures for a short period of time before melting.

- *Snow Pellets/Small Hail*—White, opaque, round (or occasionally conical) kernels of snowlike consistency, 0.08 to 0.2 inch in diameter. They are crisp, easily compressible, and may rebound or burst upon striking a hard surface. Snow pellets occur almost exclusively in snow showers.

- *Snow Grains*—Very small, white, opaque grains of ice similar in crystal structure to snow. Whereas the crystal structure of snow has very fine, needlelike branches, the crystal structure of snow grains has thicker, denser elements, with the space between hexagonal branched commonly completely filled. Snow grains do not bounce or shatter on hard surfaces. They usually fall in small quantities, mostly from stratus clouds and never as showers.

- *Ice Pellets*—Transparent or translucent particles of ice that are either round or irregular (rarely conical) and have a diameter of 0.2 inch or less. They usually rebound upon striking hard surfaces and make a sound upon impact. The term *ice pellets* describes two different types of similar looking solid precipitation. One type is composed of hard grains of ice formed from freezing rain or the refreezing of melted snowflakes. It falls as continuous precipitation and is sometimes referred to as *sleet.* Another type is composed of pellets of snow encased in a thin layer of ice. It is formed from the freezing of water droplets intercepted by snow pellets or by the refreezing of a partially melted snow pellet. This type falls as showery precipitation and is usually associated with thunderstorms.

- *Hail*—A clear to opaque ball of hard ice, ranging in diameter from 1/8 inch or so to 5 inches or larger. Hailstone size is measured and reported in inches, but hailstones are usually compared to common objects when reported to the public by television or radio, such as pea size, walnut size, golf-ball size, baseball size, or softball size. Hail frequently displays a layered appearance of alternate opaque and clear ice. It is produced only in thunderstorms, but may be ejected from the top or sides of a thunderstorm to fall and strike the ground without a cumulonimbus cloud directly overhead.

- *Ice crystals*—Tiny unbranched crystals of ice in the form of needles, hexagonal columns, or plates. They are often so small that they may be suspended in air and are sometimes referred to as diamond dust. Ice crystals are visible mainly when they glitter in the sunlight or in spotlights at night. Although common in polar regions, this phenomenon occurs only during very cold temperatures in stable air masses. Ice crystals may fall from any type of cloud or from clear air. As moist air cools below -40°F, the water vapor may sublime directly to form ice crystals, and precipitate, without ever forming a cloud.

PRECIPITATION INTENSITY.—*Precipitation intensity* is an approximation of the rate of fall or the rate of accumulation of precipitation. During an observation, intensity for each type of precipitation (other than hail and ice crystals) must be determined. NAVMETOCCOMINST 3141.2 and NAVMETOCCOMINST 3144.1 provide valuable information on determining intensity by visibility, accumulation rate, size of the rain drops, sound on the roof, height of splashes, and the rate at which puddles form. The primary indicator for snow and drizzle is visibility. Table 1-4 summarizes the indicators to aid in your understanding of the term *precipitation intensity.*

Direct observation is the best method of determining the type of precipitation occurring. However, in chapter 2, you will study observation equipment that provides valuable indicators of precipitation intensity.

PRECIPITATION CHARACTER.—*Precipitation character* is a term used to describe how precipitation falls. Three terms are used to describe character: *continuous, intermittent,* and *showery. The* term *continuous precipitation* means that the precipitation falls for a long period of time over a specific area. When the system producing the precipitation is moving, use of the term implies that the area covered by the rain is extensive. Continuous precipitation falls from stratiform clouds, especially nimbostratus. Continuous precipitation changes intensity only slowly, and may be of light or moderate intensity, rarely heavy. When used alone, the terms *rain, drizzle,* and snow refer to either continuous or intermittent precipitation.

The term *intermittent precipitation* is used to describe precipitation that occurs for brief periods of time (lasting less than 1 hour). Intermittent precipitation changes intensity slowly, and is usually light. Although the overall area affected by intermittent precipitation is usually very large, at any given time only a portion of the area is actually receiving precipitation. Like continuous precipitation, intermittent precipitation usually falls from stratiform clouds, especially nimbostratus.

Sudden starting or stopping ofprecipitation or rapid changes in the intensity of precipitation indicate *showery precipitation.* Showery precipitation, or showers, fall from cumuliform clouds, especially cumulonimbus. Showers cover only a relatively small area at a given time, and, unless the cumuliform cloud is stationary, showers last only a brief time before moving on. Rain falling from cumuliform clouds is called a "rain shower," and a cumuliform shower of snow is called a "snow shower." The public popularly calls a very light snow shower "snow flurries."

PRECIPITATION THEORY.—Several valid theories have been formulated in regard to the growth of raindrops. The theories most widely accepted today are treated here in a combined form.

It is believed that most precipitation in the mid- and high-latitudes starts as ice crystals. The crystals melt and fall as liquid precipitation only when it passes through an above-freezing stratum of air. Due to the low freezing level in these regions, the abundance of water vapor in the atmosphere is found at, near, or below freezing temperatures. In clouds below freezing temperatures, water coexists in all three states: solid, liquid, and gas. Both solid and liquid particles are present within most clouds. The higher vapor pressure for the liquid droplets compared to the low vapor pressure for the solid ice crystals tends to cause a net evaporation of gaseous water from the liquid droplets. In turn, there is a corresponding net sublimation of the gaseous vapor on to the solid ice crystals. This tends to retard the growth of the liquid drops while aiding the growth of the crystals. When the ice crystals become too large (too heavy) to remain suspended in the atmosphere, they fall as precipitation.

In the low-latitudes (tropics) and much of the mid-latitudes during the warmer months, the freezing level in the atmosphere is generally much higher. The abundance of moisture in the above freezing portions of the lower atmosphere allows the majority of the precipitation to form initially as liquid water droplets.

Table 1-4.—Precipitation Intensity Indicators

TYPE	LIGHT	MODERATE	HEAVY	
Rain, Snow Freezing rain	trace to 0.10" trace to 2.5 mm	0.11" to 0.30" 2.8 mm to 7.6 mm	>.30" >7.6 mm	**ACCUMULATION PER HOUR**
Drizzle	trace to 0.01" trace to 0.3 mm	>0.01" to 0.02" >0.3 mm to 0.5 mm	>0.02" >0.5 mm	
Rain, Snow Freezing rain	trace to 0.01" trace to 0.3 mm	>0.01" to 0.03" >0.3 mm to 0.8 mm	>0.03" >0.8 mm	**ACCUMULATION PER 6 MINUTES**
Ice pellets	Little/none	Slow	Rapid	**ACCUMULATION**
Drizzle, Snow grains, Snow pellets and snow	2 5/8 mi 2 0.55 nm 2 1,000 m	5/16 to ≤1/2 mi 0.25 to 0.50 nm 500 to 1000 m	≤ 1/4 mi ≤ 0.20 nm ≤ 400 m	**VISIBILITY**
Rain	Easily seen	Not easily seen	Unidentifiable Rain in sheets	**DROPLET IDENTIFICATION**
Rain	Hardly noticeable	Noticeable	Heavy, several inches high	**SPRAY OVER HARD SURFACES**
Rain	Forms slowly	Forms rapidly	Forms very rapidly	**PUDDLES**

After rain droplets and/or ice crystals form, growth in size is aided by the process of *accretion*—the fusing together of small droplets that collide. Droplets may also collide with ice crystals and freeze upon contact to make a larger crystal. Turbulence within a cloud may increase the rate of accretion, while strong updrafts within cumulus clouds may keep the crystals or droplets, which are continuously increasing in size, suspended for longer periods of time and allow the growth of very large drops or crystals. Chapter 9 of the text *Meteorology Today* contains more detailed information on precipitation processes.

Wind-blown Forms

A few reportable hydrometeors are simply moisture picked up from the ground or ocean surface and carried by the wind. Blowing and drifting snow, dust, sand, and blowing spray are hydrometeors of this group.

BLOWING SNOW/DUST/SAND.—The hydrometeor *blowing snow, dust, or sand* exists only when strong winds lift snow, dust or sand from the surface to a height of 6 feet or greater, and the snow, dust, or sand reduces the visibility to less than 7 miles.

LOW DRIFTING SNOW/DUST/SAND.—This phenomena exists only when strong winds lift snow, dust, or sand from the surface to *less* than 6 feet, and the snow, dust, or sand *does not* reduce the visibility below 7 miles.

BLOWING SPRAY.—A hydrometeor that occurs only in very high winds, where water is lifted from the ocean by the wind and reduces visibility at eye level to 6 nautical miles or less.

ELECTROMETEORS

Our discussion includes lightning and auroras, which are the only significant electrometeors.

Lightning

Lightning is the most frequently observed electrometeor. This massive electrical discharge from rapidly growing cumuliform clouds is a very dangerous phenomenon that kills an average of 85 people per year, injures hundreds of people per year, and causes property damage in the millions of dollars. For example, in 1989 over $72 million in civilian property damage was caused directly by lightning.

The vast majority of lightning discharges jump from cloud to cloud, and are abbreviated on observations as LGTCC. A smaller number of discharges appear to occur entirely within a single cloud (LTGIC) or from a cloud to the surrounding clear air (LTGCA). Only a small percentage of lightning discharges occur from cloud to ground (LTGCG) (fig. 1-28). Cloud-to-ground lightning may strike up to 12 miles from the rainfall area in a thunderstorm (fig. 1-29).

Other rarer forms of lightning are ball lightning or St. Elmo's Fire, and lace lightning. Ball lightning appears as ball-shaped parcels of brightly glowing electricity; it may be stationary on a sharp object such as an antenna, mast, or the ridge of a roof; or it may be falling, rolling across a surface, or even bouncing over the ground. Different reports state ball lightning may penetrate windows or wooden walls with little or no trace of passage, while others attribute great damage to ball lightning contact. Ball lightning has been reported to explode in a shower of sparks upon contact with stationary objects.

Lace lightning is occasionally seen moving across the sky through a heavy cirrus spissatus cloud layer well downstream of larger thunderstorms. It appears to move across the sky in pulses forming a fine weblike or lacelike network through the cirrus cloud.

In surface aviation weather observations, thunderstorms are considered to have begun when the first thunder is heard or when overhead lightning is observed, and the local noise level is high enough as might prevent the observer from hearing thunder. Direction, estimated distance to the leading edge of the storm, and direction of movement should be noted if possible. As the storm gets closer, the types of the lightning discharges should be noted, along with the frequency of the lightning discharges, such as occasional lightning or frequent lightning. The Lightning Detection and Tracking System (LDATS) equipment in use at several Navy and Marine Corps weather offices will assist the observer in determining distance, direction, and speed of movement of thunderstorms. Many stations have requirements to set thunderstorm conditions to warn base personnel of anticipated or impending thunderstorm activity. The observer's input as to the existence of thunderstorms, their location, and movement is critical to the thunderstorm warning system.

Figure 1-28.—Cloud-to-ground lightning under a cumulonimbus cloud base.

Figure 1-29.—Cloud-to-ground lightning from the side of cumulonimbus calvus cloud to ground several miles from the cloud base.

Auroras

Auroras are luminous phenomena that appear in the high atmosphere in the form of arcs, bands, draperies, or curtains. These phenomena are usually white but may have other colors. The lower edges of the arcs or curtains are usually well defined; the upper edges are diffuse. Polar auroras are caused by electrically charged particles ejected from the sun that act on the rarified (select) gases of the upper atmosphere. The particles are channeled by earth's magnetic field, so auroras are concentratednear the magnetic poles. In the Northern Hemisphere, they are known as the Aurora Borealis, while in the Southern Hemisphere they are known as the Aurora Australis. Another form of the aurora is *airglow.* Airglow is fainter and lacks definition, but may be seen in the low and middle latitudes as a faint glow in the sky. Unless remarkably intense or vivid, auroras are not reported in surface aviation observations. Shipboard observers will only report auroras when located north of 45" north latitude or south of 45" south latitude.

PHOTOMETEORS

Photometeors consist of a number of atmospheric phenomena attributed to the reflection or refraction of visible light in the sky by liquid water droplets, ice crystals, by the air itself, or by solid particles in the air, such as volcanic ash or dust. Several types of photometeor phenomena may be used to assist in the identification of cloud type, such as the halo, corona, irisation, and rainbows. Fogbows are classified as photometeors, as are superior mirages (objects such as buildings, trees, or mountains seen inverted in the sky) and inferior mirages (shimmering wet appearance of hot surfaces such as roads or sand). Other than aiding in the identification of other phenomena, these phenomena are not significant in surface aviation weather observations; therefore, they are not reported.

Of the weather elements we have discussed to this point, most of the identification of the phenomena is based on the observer's knowledge and on what the observer sees directly. Sure, instruments are in use to help the observer determine visual range and cloud height, but the only method to determine cloud type and what type of weather is occurring is by the observer's classification. Many of the remainder of the observable elements for a surface aviation weather observation are directly obtainable from instruments, so our explanation should be somewhat easier for the remainder of the chapter. Let's continue with the next section on observing the pressure.

REVIEW QUESTIONS

Q45. Any liquid that has a temperature below its normal freezing point but still exists in the liquid state is known by what term?

Q46. What distinguishes drizzle from rain?

Q47. What type of hydrometeor is composed of hard grains of ice formed from freezing rain or the refreezing of snowflakes?

Q48. If rain is accumulating at a rate of 0.25 inches per hour, how should the intensity be classified?

Q49. What does the abbreviation LTGCA mean?

Q50. When is a thunderstorm considered to have begun?

PRESSURE

LEARNING OBJECTIVES: Recognize the importance of an accurate pressure observation. Describe atmospheric pressure, barometric pressure, and station pressure. Explain sea-level pressure and altimeter setting. Define pressure tendency.

Pressure is an important weather analysis and forecasting item used by agriculturalists, pilots, and weather forecasters. Many years ago, farmers discovered that falling atmospheric pressure is associated with poor, unsettled weather and that fair weather is associated with rising atmospheric pressure. Today, most farmers rely on scientific forecasts to regulate their activity. Overland, at and below 18,000 feet, pilots fly aircraft at their assigned flight levels based on the altimeter setting provided by local weather-observation stations.

In forecasting, pressure is used to analyze the *isobar* patterns, or lines of equal pressure. From the isobar patterns, analysts can determine wind speeds, centers of high and low pressure, and other critical information. By tracking the movement of high- and low-pressure centers, forecasters may anticipate future movements of the centers, and their associated weather patterns.

For pressure values to be meaningful to pilots, analysts, and forecasters, the reported readings must be accurate. An error in a reported sea-level pressure may cause an analysis to be in error, especially over data-sparse areas, such as the oceans. But an error in an altimeter setting can be disastrous for a pilot. The responsibility for observing, calculating, and reporting pressures accurately rests solely on you, the weather observer.

In this section, we cover the different types of pressure that must be observed, associated pressure terms, and the pressure values that must be calculated for an observation.

The standard units used to measure and report pressure values are inches of mercury and hectopascals. The term *hectopascals (hPa)* replaced the term *millibars (mb)* several years ago. A hectopascal is exactly equal to one millibar. See Appendix II for conversions between inches of mercury and hectopascals.

ATMOSPHERIC PRESSURE

Atmospheric pressure refers to the pressure exerted by the column of air on any point on the earth's surface. The term is not specific as to where the point in question is located. The vagueness of the term causes some confusion in military weather because the observer can never be sure if the person asking for atmospheric pressure wants station pressure, sea-level pressure, or even an altimeter setting.

BAROMETRIC PRESSURE

Barometric pressure is the pressure read directly from a precision aneroid barometer or a tactical aneroid barometer. On the ML-448/UM precision aneroid barometer, this value may be read in inches or in millibars. Readings in millibars can be converted directly to hectopascals; for example, 978.7 millibars equals 978.7 hPa.

STATION PRESSURE

Station Pressure is the pressure value read on the barometer (barometric pressure in inches or hectopascals) corrected for the difference between the height of the barometer and the station elevation. The correction that is added to the barometric pressure may be an instrument correction, a removal correction, and a temperature correction.

- The *station elevation* is the height of the highest point on the runway above mean sea level (MSL). This is the height that is found published in the Flight Information Publications. Aboard naval ships, the station elevation is considered to be the height of the barometer above the water line, not the height of the flight deck.

• The instrument correction (if used) is determined by the barometer calibration facility during the required semiannual calibration.

• The *removal correction* is the pressure correction based on the difference in height (in feet) of the barometer and the runway or station elevation. To find the removal correction (inches of mercury), multiply the difference in height in feet by 0.001 inch of mercury per foot. The correction in hectopascals is found by multiplying the difference in feet by 0.036 hectopascals per foot. The removal correction is added to the barometric pressure if the barometer is higher than the runway, and subtracted if the barometer is lower than the runway. Once determined, the same removal correction is always added to the indicated barometric pressure unless the barometer is moved.

Aboard naval ships, since the station elevation is the height of the barometer, no removal correction is added when determining station pressure. Temperature corrections are required only for barometers used outdoors.

Station pressure is calculated to the nearest 0.005 inch, or 0.1 hPa. When requested or given in a radio conversation, station pressure is identified with the Q-signal *QFE.*

SEA-LEVEL PRESSURE

Sea-level pressure is a theoretical pressure at the station if the station were actually at sea level. It is calculated on a CP-402/UM pressure reduction computer by using station pressure and an "r" factor that must be obtained from a table.

The "r" factor is based on station elevation and is determined by station temperature. These "r" factors are based on a complex series of calculations found in the *Manual of Barometry,* NAVWEPS 50-1D-510. Tables of "r" values for each station are available from FNMOD Asheville, North Carolina.

Some Navy and Marine Corps weather stations are authorized to use a *constant additive correction* to reduce station pressure to sea-level pressure. Sea-level pressure is always higher than the station pressure with the exception of stations located below sea level (for example, a station located in Death Valley, California, at 280 feet below sea level). A constant additive correction factor (for example, +0.017 inch) for a particular station would be added to the station pressure (in inches) every time a sea-level pressure is required. Authorized shore stations are assigned a constant additive correction factor by FNMOD Asheville.

Since the height of shipboard barometers changes, depending on the load the ship is carrying, shipboard corrections for sea-level pressure are found by multiplying the height of the barometer above the water line in feet by 0.001 inch of mercury per foot (to obtain a correction for inches of mercury) or by 0.036 hectopascals per foot (to obtain a correction for the millibar or hectopascal scale readings). The corrections are then added to the station pressure.

Commonly abbreviated "SLP," sea-level pressure is identified in radio conversations by the Q-signal *QFF.* Sea-level pressure is normally calculated to the nearest 0.1 hPa.

ALTIMETER SETTING

Altimeter setting is a simplified sea-level pressure in inches that may be "dialed" into an aircraft's altimeter so that the altimeter will indicate the correct elevation above mean sea level of an airfield or flight deck when the aircraft's wheels are on the runway or flight deck.

Commonly abbreviated *ALSTG,* altimeter setting is identified in radio conversations by the Q-signal *QNH.* For example, a pilot requesting altimeter setting over the radio should say "What is QNH?" The answer would be "QNH Three Zero Point Zero Two Inches" if the altimeter setting were 30.02 inches.

Weather observers should not underrate the importance of the altimeter setting. Many aircraft accidents have been caused by faulty settings. Altimeter settings are computed for all surface aviation observations with the exception of single-element specials, and must be determined with extreme care.

Altimeter setting is computed using station pressure and a pressure reduction computer. Unlike sea-level pressure (computed on the opposite side of same instrument), the altimeter setting is computed using only the station elevation and station pressure as arguments, and the setting is read to the nearest 0.01 inch.

Altimeter settings may also be obtained from a Digital Altimeter Setting Indicator or an Automated Surface Observing System (ASOS), as you will see in the chapter on equipment.

NOTE: Many years ago, altimeter settings were calculated from the runway elevation (station elevation) plus 10 feet, to compensate for the average height of the altimeter instrument above the wheels of an aircraft.

This practice is no longer followed. Additionally, sea-level pressure should not be converted from hectopascals into inches for use as an altimeter setting, since differences in such calculations could yield false altimeter settings.

Aircraft flying above 18,000 feet overland and on over water flights more than 100 miles offshore routinely use the standard pressure, 29.92 inches, as an altimeter setting. During low-level tactical flights and landings aboard aircraft carriers, however, accurate altimeter settings are required. Now let's consider pressure tendency.

PRESSURE TENDENCY

The *pressure tendency* is the net change in the barometric pressure during a period of time and the trend or characteristic of the change. Normally the pressure tendency is observed for 3-hour periods ending at the intermediate synoptic times 0000Z, 0300Z, 0600Z, etc. Pressure tendencies for 12- and 24-hour periods may also be observed, and routinely replace the 3-hour pressure tendencies in observations taken in the tropics.

The net change is determined by taking the difference in the station pressure between the current observation and the station pressure 3, 12, or 24 hours ago. The trend or characteristic is determined from the barograph trace, or the actual recorded station pressures during the period. The general trends of pressure "higher," the "same," or "lower" than at the beginning of the period are further described in both NAVMET-OCCOMINST 3 141.2 and NAVMETOCCOMINST 3144.1 for reporting purposes.

We will discuss some related pressure calculations on pressure altitude and density altitude later in this chapter in a section on aircraft performance indicators, but first we must cover temperature and moisture observations, which are necessary for those calculations.

REVIEW QUESTIONS

Q51. Ten millibars is equal to how many hectopascals?

Q52. What is meant by the term "removal correction"?

Q53. What would be the "r" factor for a shipboard barometer located 45 feet above the water line?

Q54. What is an altimeter setting used for?

Q55. How is the overall trend or characteristic of pressure tendency determined?

TEMPERATURE

LEARNING OBJECTIVES: Define temperature. Define and describe how to obtain dry-bulb temperature and wet-bulb temperature readings. Define and describe how to calculate dew-point temperature and frost-point temperature. Define sea-surface temperature and describe three methods used to obtain this reading.

Temperature is defined as the amount of sensible heat in a substance or as the measurement of molecular motion in a substance. Molecules in motion cause heat. As energy is added to a substance in the form of light or as infrared radiation (heat energy), the molecules absorb the energy, which increases molecular motion. This increase in molecular motion is measured as an increase in temperature. Higher temperature substances will also give off energy by radiation. Higher temperature substances can also transfer energy from faster moving molecules (warmer) to slower moving (cooler) molecules as the molecules collide. This process is known as conduction.

In surface aviation weather observations, observers take dry-bulb temperature and wet-bulb temperature readings by using sling psychrometers or electric psychrometers, or they obtain the readings from automatic weather station equipment. From these readings, the observer may calculate the dew-point temperature by using the CP-165/UM psychrometric computer, although automatic systems will calculate this important value. Related to the dew-point temperature is the frost-point temperature, which may need to be calculated. Another temperature required for shipboard surface aviation weather observations is the sea-surface temperature. Also, once each day, the observer must obtain a maximum and a minimum temperature reading.

DRY-BULB TEMPERATURE

The *dry-bulb temperature* (also called the ambient air temperature, or simply the air temperature) reflects the amount of heat present in the air. It is read directly from a ventilated thermometer on an electric psychrometer, sling psychrometer, rotor psychrometer, or from automatic measuring equipment. The temperature must be obtained to the nearest 1/10 degree and may be read in either Fahrenheit or Celsius degrees.

WET-BULB TEMPERATURE

The *wet-bulb temperature* is the lowest temperature an object may be cooled to by the process of evaporation. It is read directly from the wet-bulb thermometer on an electric psychrometer, sling psychrometer, or rotor psychrometer. Water evaporating from the moistened wick on the wet-bulb thermometer bulb cools the thermometer bulb and lowers the temperature reading. The cooling effect of the evaporation from the bulb is inversely proportional to the amount of water vapor present in the air: the more water vapor present, the less moisture will evaporate from the moistened wick, and the less cooling of the thermometer bulb will occur. From the dry- and wet-bulb readings, the dew-point temperature and humidity values may be calculated. The automatic weather observation systems do not provide a wet-bulb temperature, but automatically process equivalent measurements to compute dew-point temperature.

DEW-POINT TEMPERATURE

The *dew-point temperature* is the temperature a parcel of air must be cooled to in order to reach saturation. Cooling past the dew-point temperature normally results in condensation or precipitation. Changes in temperature do not alter an air-parcel's dew-point temperature; therefore, dew-point temperature is termed a conservative property. The extraction or addition of moisture, however, from or to an air parcel will respectively decrease or increase the dew-point temperature.

Dew-point temperature is calculated from the dry-bulb temperature and the wet-bulb depression by using the CP-165/UM psychrometric computer. The *wet-bulb depression* is the difference between the dry-bulb temperature and the wet-bulb temperature. Dew-point temperature is automatically calculated by the automatic weather systems.

Many calculations that you will be using call for a dew-point depression as a value. The *dew-point depression* is the difference between the air temperature and the dew-point temperature, expressed as a positive number. For example, if the air temperature is 78°F and the dew-point temperature is 67.5°F, the dew-point depression is 10.5°F.

FROST-POINT TEMPERATURE

The *frost-point temperature* is the temperature, below freezing, that a parcel of air must be cooled to in order to reach saturation. Cooling past the frost-point temperature normally results in sublimation of ice crystals from the air. The frost-point temperature is occasionally referred to as "the dew-point temperature with respect to ice." Calculations using the "low" temperature side of the CP-165/UM psychrometric computer refer to use of the "Ti" scale when the wet-bulb thermometer wick is frozen, and to the "DP" scale if the wet-bulb wick is not frozen. Both scales calculate a dew-point temperature with respect to liquid water, and not a frost-point temperature. The frost-point temperature may be approximated by the following formula:

$$T_F = \frac{9}{10} T_D$$

where T_F is the frost-point temperature, and T_D is the dew-point temperature.

As an observer, you may be asked to calculate frost-point temperatures, especially when working with a Skew T, Log P diagram. Although the frost-point temperature is not usually computed for a surface aviation weather observation, we have introduced it at this point because it is so closely related to the dew-point temperature. Relative humidity, and other humidity computations derived from temperature and dew-point temperature readings are covered later in this chapter.

SEA SURFACE TEMPERATURE

Another temperature reading in shipboard weather observations is the *sea surface temperature.* It is supposed to reflect the temperature of the upper few inches of the sea surface. On some ships with OA divisions, installed sensors automatically measure this value. There are three other acceptable methods for obtaining a sea-surface temperature reading: the bucket temperature method; by expendable bathythermograph; and by use of the seawater injection temperature. The sea-surface temperature reading must be accurate since it is a major input into many undersea warfare (USW) acoustic products.

Bucket Temperature

The *bucket temperature method* is by far the most accurate, yet is also the most work intensive. In this method, a sample of seawater is obtained by casting a lightweight bucket or coffee can with a strong line attached over the side of the ship and retrieving a water sample. This should be done as near to the bow of the ship as possible, since the passage of the ship through the water tends to mix surface water with water from the keel level of the ship. The "bucket" should also be cast ahead of where the observer is standing so that the bucket fills as it drifts by the observer. As the movement of the ship carries the bucket astern of the observer, the bucket should be retrieved. A standard thermometer is then inserted into the water sample, and the water is slowly stirred with the thermometer until the temperature reading stabilizes. The temperature is read to the nearest 1/10 degree Fahrenheit.

Bathythermograph Temperature

The next best method is to obtain a sea surface temperature from an *expendable bathythermograph* sounding. Procedures for conducting a bathythermograph sounding are covered in a later module. Bucket temperatures should be conducted occasionally to verify that the recorded bathythermograph surface temperature is accurate. *Sound* velocimeterreadings may also be used in lieu of a bathythermograph reading.

Seawater Injection Temperature

The least accurate method is the *seawater injection temperature* reading. Seawater injection temperatures are read in the engineering spaces and are usually readily available by shipboard phone from the "main engine room control" watch/operator. Seawater is constantly taken onboard for cooling the engines and for conversion to freshwater. The seawater injection ports are located well below the water line, sometimes as deep as 60 feet on aircraft carriers. Therefore, temperature readings at that point do not accurately reflect a sea surface temperature, but rather a near surface temperature reading.

In tropical waters, the difference between the surface temperature and the near-surface temperature is usually slight. But in certain regions of the mid- and high-latitudes, a strong surface thermocline may exist, which will cause a rapid decrease is temperature from the surface to the injection level. This may cause the difference between the actual surface temperature and the injection temperature to be very large. If injection temperatures are used, they should be routinely checked against bucket temperatures and bathythermograph temperatures, and adjusted if necessary.

REVIEW QUESTIONS

Q56. *Define wet-bulb temperature.*

Q57. *What is the normal result* of *air being cooled to below the dew-point temperature?*

Q58. *Given an air temperature* of *82.5°F and a dew-point temperature* of *70.0°F, calculate the dew-point depression.*

Q59. *What is meant by the term "frost-point temperature"?*

Q60. *List three methods for manually obtaining the sea surface temperature.*

WIND

LEARNING OBJECTIVES: Describe wind speed and wind direction. Define and identify how to determine true-wind direction, relative-wind direction, and magnetic-wind direction. Explain wind character and wind event. Define Foxtrot Corpin.

The atmosphere is essentially an ocean of air surrounding earth. Temperature is unevenly distributed over the earth's surface, varying with latitude and with the seasons. Therefore, all of earth's atmosphere is continuously in a state of fluid motion. Wind is the observed effect of horizontal transport of air masses over earth's surface. Surface winds are the movements of air within 50 feet of the ground. The term *winds* usually refers to both wind speed and direction. Different reporting codes require that observations be made over certain periods of time. Some conventions require a 2-minute observation period, while others require a lo-minute. These time periods are specified in NAVMETOCCOMINST 3141.2 and NAVMET-OCCOMINST 3144.1.

Winds are observed by using the equipment discussed in chapter 2. Automatic observation equipment will immediately report and record winds; other wind-measuring equipment can show a detailed graph of speed and direction over time. Winds are described by wind direction, wind speed, and wind character.

WIND DIRECTION

Wind direction is the average direction from which the wind is blowing during a specified period. Airflow from the north toward the south is referred to as a "North wind." Wind direction always shows minor fluctuations. These minor fluctuations are normally "averaged out" when determining a wind direction.

Several conventions are used to report wind direction. As a weather observer, you must be familiar with the relationship between these direction-reporting conventions.

Wind Direction Conventions

Wind directions are expressed in azimuth bearings or by the 8-point or 16-point compass. In addition, the wind direction may be a true, relative, or a magnetic wind direction. Wind directions are normally observed to the nearest 5° of azimuth, but reported (and forecast) to the nearest 10°.

POINTS OF THE COMPASS.—Points of the compass as represented in figure 1-30 are normally used only to express wind directions in general weather forecasts. They are not used in aviation observations or forecasts. The standard for expressing wind direction in most general public weather forecasts and military forecasts is the 8-point compass. It uses the cardinal points of the compass (north, east, south, and west) as well as the inter-cardinal points (northeast, southeast, southwest, and northwest). General marine area forecasts may also use intermediate compass points, such as North-Northeast, East-Northeast, and East-Southeast. When not specified as "relative" or "magnetic" directions, the points of the compass refer to "true" directions.

When wind direction is critical to the safe conduct of an operation or exercise, such as routine aviation weather operations, parachute operations, or the employment of weapons systems, both observed and forecast wind directions should be provided using azimuth bearings.

AZIMUTH BEARINGS.—In surface aviation weather observations, wind direction is always reported using a 360° azimuth circle with 000°/360° representing True North. Figure 1-30 also shows an azimuth bearing circle. Note that the 0/360 azimuth bearing is aligned with *True North,* the North Pole. Directions are expressed in degrees of azimuth progressing clockwise through 090° representing due East, 180° representing due South, 270° representing due West, and 360° representing due North. In meteorology, an azimuth of 000° is used only when no wind is blowing, while 360° means the wind is from the North.

True Wind Direction

True North is represented on a globe as the North Pole. All directions relative to True North may be called "true bearings." Since the majority of azimuth bearings for wind directions and navigational bearings are required to be oriented to True North, a wind direction or navigational bearing lacking designation is assumed to be a "true" bearing. A *true wind direction* is a wind direction measured with respect to True North. The wind equipment installed at all shore weather stations should be oriented to True North. Therefore, wind direction and wind speed obtained from the equipment is considered to be a true wind direction and a true wind speed.

Relative Wind Direction

A relative bearing uses the current direction that an object (such as a ship or an aircraft) is facing as the 0/360° azimuth alignment. On a ship, a line taken through the centerline of the ship directly over the bow represents the relative bearing 0/360°. The relative azimuth bearings proceed clockwise with directly off the Starboard Beam representing 090°, dead astern representing 180°, directly off the Port Beam representing 270°, and back to the bow at 360°. Wind direction aboard ship is observed by relative bearing and is called a *relative wind direction.* The relative wind direction (and relative wind speed) may be manually converted to a true wind direction (and true wind speed) by using the CP-264/U true wind computer, a maneuvering board, or an aerological plotting board. Do not confuse relative wind with *apparent wind* (the relative wind speed with the wind direction reported using true bearing vice relative bearings). Apparent winds have no application in meteorological observations.

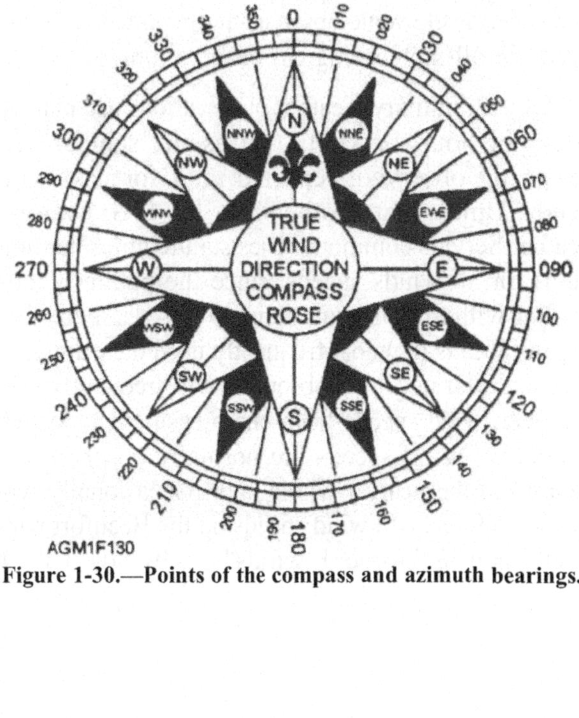

Figure 1-30.—Points of the compass and azimuth bearings.

Magnetic Wind Direction

A *magnetic wind direction* is a direction based on the 360° azimuth circle with the 0/360° azimuth radial aligned with magnetic north. Magnetic bearings are used by tactical weather observers in the field when determining wind directions by using a magnetic compass for reference. The magnetic wind directions thus obtained are converted to true wind directions by adding or subtracting the appropriate magnetic declination for the location. If, for instance, a charted magnetic declination is "7° west," this means that magnetic north is 7° west of actual or True North, and that 7° must be subtracted from the wind direction obtained to convert it to true wind direction. When a location has a declination east of true north, the correction must be added to the magnetic direction. As long as the tactical observer is stationary (not in a moving vehicle), no correction need be applied to the observed wind speed.

WIND SPEED

Wind Speed is the average rate of air motion, or the distance air moves in a specified unit of time. The *instantaneous wind speed* is the speed of the air at any moment. The instantaneous wind speed will usually show minor fluctuations over time. Fluctuations between the highest instantaneous speed and the lowest instantaneous speed are averaged to obtain mean wind speed. *Mean wind speed* is the arithmetic or graphical average wind speed during the period of observation, which is normally 2 minutes. For example, wind speeds on a recorder chart during a 2-minute observation period may constantly vary between 24 and 32 knots. The average, 28 knots, is the mean wind speed. Mean wind speed is the value observed and reported for "wind speed" in all meteorological observations.

All U.S. military weather observations use nautical miles per hour, or knots (kt) as the standard for measuring observed, reported, and forecast wind speeds. Unless stated otherwise, the U.S. National Weather Service commonly uses statute miles per hour (mph) for all winds speeds, since the public is most familiar with that measurement. Overseas, meters per second (m/s) is the most frequently used measurement. Navy and Marine Corps observers will frequently need to convert wind speeds from one measurement system to another. Wind speeds are normally observed and reported to the nearest whole knot. Occasionally, you may see reference to wind speeds on the Beaufort wind scale, such as "force 1 winds" or "winds 3 to 4, becoming 5 by night." *Force* is not always stated, but is assumed. The Beaufort wind scale is included and cross-referenced to standard wind speeds in the table in Appendix V.

Wind speeds aboard ship are affected by ship movement. If the ship is heading into the direction from which the wind is blowing, the observed wind speed across the deck will be greater than the actual wind speed. On the other hand, if the ship is traveling with the wind, the observed wind speed over the deck will be less than the actual wind speed. For this reason, the winds across the deck, as measured on an anemometer, are called *relative wind speeds;* the wind speed is relative to the motion of the ship. Relative wind speed is converted to *true wind speed,* which would be the actual wind speed if measured at a stationary location. You can convert relative wind speed to true wind speed by using the CP-264/U true wind computer (see chapter 2), or a maneuvering board may be used.

Many "descriptive" terms are used to identify wind speed. Some are *light breeze, fresh breeze, gentle breeze, moderate breeze,* or *fresh gale* and *storm.* These terms are part of an accepted scale of nautical wind speeds that may be directly related to wind speed measurements. These descriptive names are included in Appendix V. Others, such as *brisk* or *sultry,* although acceptable in literature, have only a vague relationship to measured wind speeds and should not be used. Only two descriptive terms may be used in military surface weather observations for wind speeds. They are *light,* abbreviated *LGT,* meaning the wind speed is 10 knots or less, and *calm,* meaning there is no detectable motion of the air.

WIND CHARACTER

In addition to wind speed and wind direction, most observations require a determination of wind character. *Wind character* is a description of how the wind (speed or direction) changes during the specified period. Wind speed gusts, the peak wind gust, wind speed squalls, and variable wind direction are all included in wind character, and should be noted during an observation.

Gust

A wind *gust* is a rapid fluctuation in wind speed with a variation between peaks and lulls of 10 knots or more. Gusts are normally observed in the 10-minute period prior to the actual time of observation. Gusts increase the difficulty of controlling aircraft during takeoff and landing. The *gust spread* (the difference in knots between the normal lulls and peaks), if large enough, may cause problems for rotary wing aircraft by initiating rotor chop. Rotor chop is a difficult to control, sometimes hazardous, up and down oscillation of the rotor blades. The *peak wind speed* or *peak gust* is the highest instantaneous wind speed or gust speed greater than 25 knots observed since the last METAR observation.

Variable Winds

Variable winds occur when the wind direction fluctuates by 60° or more. While this condition occurs most frequently when the winds are very light, wind direction fluctuations are most significant when the wind speeds are higher (greater than 6 knots). For observation purposes, the wind direction may be considered variable anytime the observed 2-minute mean wind speed is 6 knots or less.

WIND EVENTS

Certain wind phenomena are included in an observation even though the events did not occur during the 2- or 10-minute period during which the winds were being observed. These phenomena or events may be included in the observation if they occurred within the past hour and were not reported in a previous observation. The events include squalls and wind shifts.

Squalls

A *squall* is a sudden large increase in wind speed (usually accompanied by a change in wind direction) that lasts several minutes and then suddenly dies. For observation purposes, the wind speed must increase by 16 knots or more and the sustained wind speed after the increase must be 22 knots or more for at least 1 minute. Squalls are usually caused by large convective cells, like those that produce strong rain showers and thunderstorms. Squalls may also be produced by dry frontal passages; the presence of precipitation is not a requirement. At sea, strong rain showers at a distance away from the ship are called "squalls" because squall winds are usually present. When lines of thunderstorms form on or move out ahead of a cold front, the line may be called a "squall line" because of the squall winds associated with the thunderstorms.

Wind Shifts

A *wind shift* is any change in wind direction by 45° or more during a 15-minute time period. The change in direction may or may not be accompanied by a change in wind speed. However, wind shifts are only recorded when the mean wind speed is 10 knots or greater during the shift. A wind shift may be very sudden, occurring within a minute or so, or it may occur gradually over the 15-minute period. The most common cause of wind shifts is frontal passage, especially a cold-frontal passage. The onset of a sea breeze may cause a wind shift, as may other locally produced wind conditions.

FOXTROT CORPIN

Foxtrot Corpin is the term used to identify the best course and speed a ship should "come to" to bring the relative wind into the proper window for the launch and recovery of aircraft. For departures and recoveries aboard different classes of ships, the most desirable relative wind speed, acceptable minimum and maximum wind speeds, the most desirable relative wind direction, and the acceptable wind direction variations are specified in the *NATOPS Flight Manuals* for each type of aircraft.

Foxtrot Corpin is routinely provided by Aerographer's Mates aboard aircraft carriers and amphibious assault ships and is computed by using the CP-264/U true wind computer. Detailed instructions for the procedure are printed on the reverse side of the CP-264/U. The procedure uses the true wind and the desired wind to find required ship's course and speed.

REVIEW QUESTIONS

Q61. Explain the difference between True North and Magnetic North.

Q62. Winds blowing directly off the starboard beam are coming from what relative direction?

Q63. How can relative winds be manually converted to true winds?

Q64. How is the mean wind speed determined?

Q65. How would the observed wind speed on a ship be affected by winds blowing from dead astern?

Q66. Define the term "gust."

Q67. Define the term "squall."

SEA AND SWELL WAVES

LEARNING OBJECTIVES: Explain the importance of sea conditions to naval operations. Define duration limited seas and fetch limited seas. Define wave height, wave length, and wave period, and wave direction. Define and distinguish the difference between sea waves and swell waves. Define Romeo Corpin.

Sea conditions are critical to carrier flight operations, replenishment operations, undersea warfare operations, amphibious operations, and search and rescue missions. Your observations of sea conditions are vitally important. They must be accurate so that forecaster and operations personnel may predict the success of planned operations. The majority of waves are disturbances on the surface of the water produced by blowing winds. Although there is some net displacement of water in waves, the majority of the movement of water in a wave is in a circular motion beneath the surface. Waves move across the surface of the water by transferring energy—not matter. Waves move in a sine wave pattern, as shown in figure 1-31.

WAVE PARAMETERS

The success of any operation conducted in the ocean environment may depend on the height of the seas, the direction of the seas, and the wave period. Waves, in general, are described by wave height, wave length, and wave period. Wave direction is another important aspect used to describe waves.

Wave Height

In oceanography, *wave height* (fig. 1-31) is the vertical distance, usually measured in feet, from the *crest* of a wave (the highest portion of a wave) to the *trough* of the wave (the lowest portion of the wave). This differs from the "wave height" or "amplitude" normally used in physics, in which the distance is measured from the "at rest" or midline position to the crests and troughs. When waves are generated by the force of wind acting on the water, the wind speed determines the maximum height of the wave. For a given wind speed, many different wave lengths (frequencies) are produced, and for each wave length (frequency), many different wave heights are developed. Although the general relationship that higher waves tend to have longer wave lengths (lower frequencies) is true, there is no specific relationship between wave height and wave length.

The primary factor that determines the maximum wave height is the wind speed. But the *duration* of the wind (length of time the wind has been blowing at a certain wind speed) and the *fetch,* (distance over the water the winds have been blowing) also limit the maximum wave height. When the highest theoretical wave height based on the wind speed cannot be attained because the winds have not blown for a sufficient period of time, the sea heights are said to be *duration limited.* When the sea heights cannot be attained because the straight line area the winds have been blowing over the water is too short, the sea heights are said to be *fetch limited.* The tables in Appendix V provide a breakdown of wind speeds in relation to wave heights and a wind and sea scale for a fully arisen sea.

The table in Appendix V also refers to the 10 states-of-the-sea, or sea-states. Although not used extensively, sea-states may be mentioned in literature or messages. *Sea-states* refers to general descriptions of wave heights and the appearance of the water surface. They range from "calm" and "sea like a mirror," as in sea-state 0, to "exceptionally high waves" with the "air filled with foam and spray" as in sea-state 9.

Accurate observations of wave height are the most difficult determination in an observation, especially from the catwalk on an aircraft carrier. Typically, observers on large ships, such as an aircraft carrier, significantly underestimate the wave height, while observers on smaller ships and in small boats provide

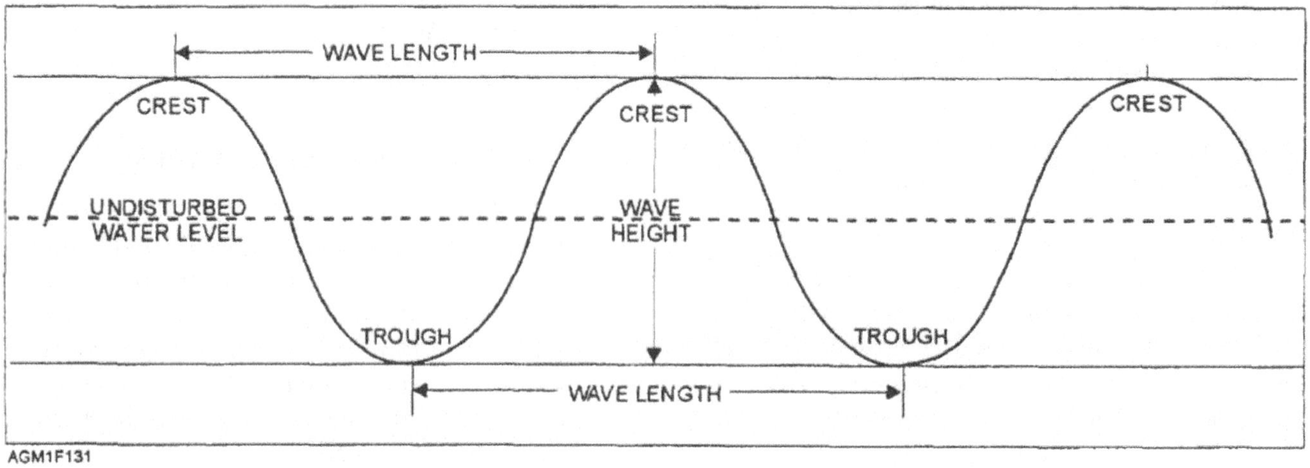

Figure 1-31.—Sine wave pattern and associated parameters in ocean waves.

the most accurate wave height estimations. This is understandable since an observer on a carrier catwalk is some 60 feet above the water line, and the waves look small from that height. Estimation may be improved by observing wave height from the hangar deck, which is only some 30 feet above the water. Until the observer becomes very experienced in observing wave heights, reference objects should be used to judge the wave height.

Good reference objects are ships-in-company or small boats operating alongside. Waves may be compared to the heights of the freeboard along the sides of the ships, or to the size of the small boats. "Load line" markings in feet may be visible on the sides of ships and will assist in wave height determination. Be careful not to observe the waves near the bow of a ship, since the bow-wave is caused by the ship and is not a true representation of the actual wave heights.

The best reference object is something of known size. Some shipboard weather offices keep square 1-, 2-, or 3-foot pieces of cardboard on hand to throw over the side of the ship and use as a reference. Other ships use 1-foot-square pieces of scrap wood as a reference. (Cardboard is preferable, since it will soon become soggy, breakdown, and sink in the water, and it is biodegradable. It is also readily available.) Using a block of wood or cardboard as a reference is known as the *chip* or *block method.*

Wave Length

Wave length is the horizontal distance from one wave crest to the next wave crest, or the distance from one wave trough to the next wave trough. Although difficult to measure at sea, this parameter may be measured on aerial photographs and is directly related to wave period by the approximation $L = 5.12T^2$, where L is the wavelength in feet and T is the wave period in seconds. Wave lengths are not directly observed or reported by observers.

Wave Period

Wave period is the time, usually measured in seconds, that it takes for a complete wave cycle (crest to crest or trough to trough) to pass a given fixed point. Wave period is dependent upon the speed of movement of the wave across the surface. The speed of movement varies with wave length, with shorter wave-length waves moving slower and longer wave-length waves moving faster. This relationship is approximated by $C = 1.34\sqrt{L}$, where C is the wave speed (knots) and L is the wave length. Many calculations dealing with waves use the wave frequency instead of the wave period as a basis for the argument. The wave *frequency* is the number of wave cycles passing a fixed point in 1 second, and it is inversely proportional to wave period. Conversions between wave frequency (f) and wave period(T) are made by the formula $f = 1/T$ or $T = 1/f$.

Wave Direction

Wave direction is the direction, in true degrees of azimuth, that the majority of the waves in a group are coming from. Wave direction is best determined during an observation by sighting along the wave crests and troughs and either adding 90° to or subtracting it from the direction obtained, as shown in figure 1-32. Use the gyroscope repeater on one of the pelorus columns (chapter 2, fig. 2-29) to sight along the wave crests or troughs. Add or subtract 90° to/from the true bearing thus obtained to determine the wave direction. The observer may also sight directly into the oncoming waves, perpendicular to the crests and troughs, to obtain wave direction.

SEA WAVES

Sea waves, often referred to as "seas," are waves generated by the wind in the local area. Light winds

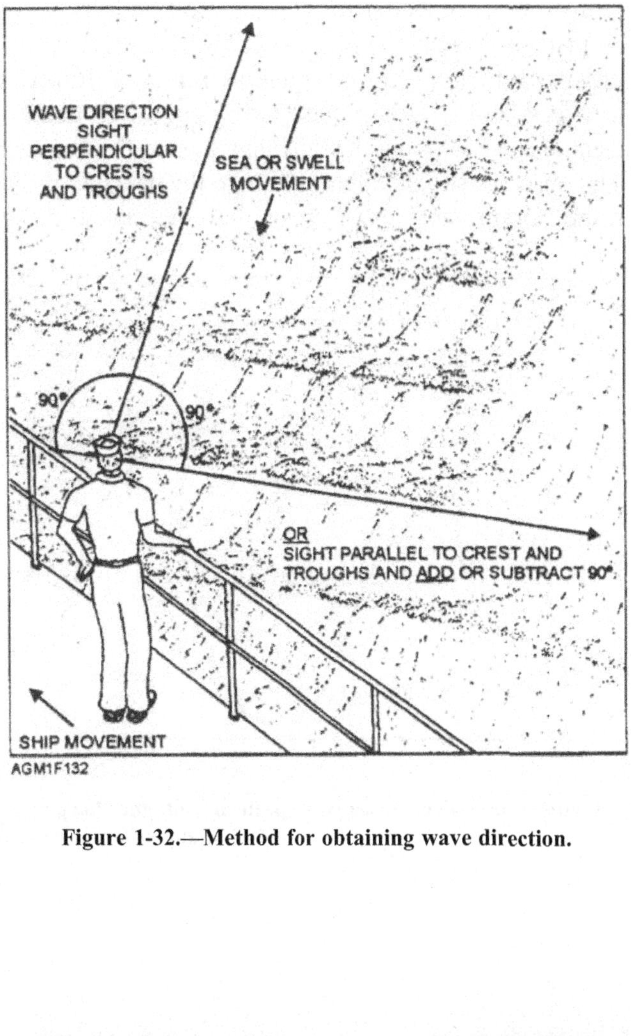

Figure 1-32.—Method for obtaining wave direction.

usually produce seas with small wave heights, small wave lengths, and short periods; higher winds usually produce waves with higher heights, longer wave lengths, and longer periods. When the winds over the water produce sea waves, the wave crests are generally aligned perpendicular to the direction the wind is blowing. The continuing force of the wind on the waves distorts the ideal sine wave pattern, forming sharper crests (fig. 1-33). The waves move in the direction the wind is blowing, with wave crests and troughs perpendicular to the wind direction.

In a given sea condition, many different size waves are present. Observers determine *significant wave height,* or the average wave height of the highest 1/3 of all the waves present. Ideally, the heights of 50 to 100 waves should be recorded on a piece of paper, then the highest 1/3 of the recorded heights should be averaged to obtain significant wave height for the seas. In practice, taking the average height of the "most well defined" waves approximates the significant wave height. Attempt to observe 50 or so waves as a minimum, and then average the height of the "best" 16 or 17 waves.

The average significant wave period in an area of sea waves gives analysts and forecasters a better idea of the total wave energy present in the area than does the observation of the significant wave height. Observations of the average significant period should be made by timing the passage of "well defined" wave crests past a fixed point, such as a buoy, clump of seaweed, wood block, or square piece of cardboard mentioned earlier, and then dividing to find the average. The observer should attempt to time the passage of the same "significant" wave crests that were used in the determination of average significant wave height. If, for example, you timed the passage of 17 "well defined" waves out of 50 waves of various size passing the cardboard square (before you lost sight of the square) in 120 seconds, the average wave period is 120/17, or 7 seconds. The important factor in determining both the average height and the average period for sea waves is that only the highest 1/3 of the waves, the significant waves, are evaluated.

A direction is always determined for sea waves, and the direction found should be in general agreement with the wind direction. If the sea wave direction does not agree within plus/minus 20° of the wind direction, recheck both sea and wind direction. The sea direction is usually not recorded or reported, since it is assumed that the sea direction is nearly the same as the recorded wind direction.

SWELL WAVES

Swell waves are seas that have moved out and away from the area in which they were formed. Because of their different wave lengths and wave speeds, waves move outward from the windy areas where they formed, and separate into groups of waves with distinct wave periods. Since the winds are no longer pushing on the waves, they take on a more typical sine wave pattern with generally equally rounded crests and troughs, and thus are smooth and regular in appearance.

Typically, when only one group of swell waves is present, the wave heights and the wave periods are fairly uniform. Determinations for the average swell wave period, the average swell wave height, and the wave direction may be easily made. When determining

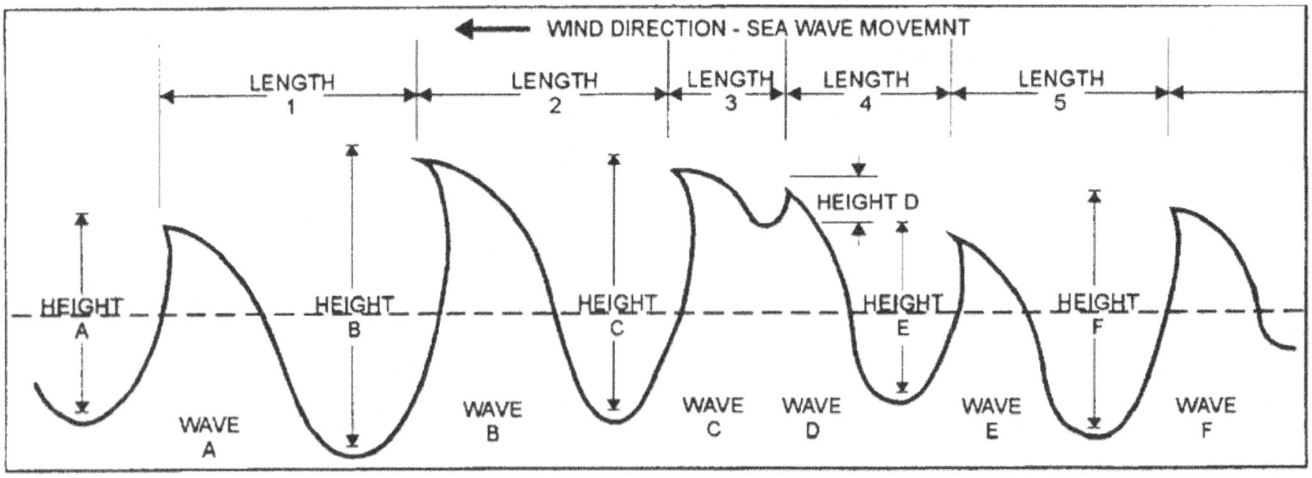

Figure 1-33.—Typical sea wave pattern. Note the sharper crests and the irregular wave pattern caused by the superposition of many different wave length/wave height patterns.

the average swell wave height, use the height of all of the swell waves, not just the highest 1/3, as used for sea waves. Similarly, when determining average period, count and time all of the rounded swell wave crests passing the fixed reference pointed. Make swell wave observations from the side of the ship the waves are approaching from to see the wave pattern better. The swell wave direction should be determined from a relatively high position on the ship so that a larger area of the sea may be observed.

Frequently more than one group of swell waves may be observed, each coming from a different direction. When this happens, you should attempt to determine average height, average period, and direction for each swell wave group. Swell wave groups should differ in direction from each other or from the sea waves by 30° or more to be considered and reported in an observation.

Further information on sea and swell wave observations is contained in NAVMETOCCOMINST 3144.1, as well as H.O. Pub 603, *Practical Methods for Observing and Forecasting Ocean Waves.*

ROMEO CORPIN

Shipboard Aerographer's Mates are occasionally tasked to recommend Romeo Corpin for underway replenishment (UNREP) operations, including connected replenishment (CONREP) and vertical replenishment (VERTREP). *Romeo Corpin* is the best course and speed a ship should "come to" to minimize the effects of the seas and swell on the ship. Wind is also an important consideration. The most desirable course gives the ship the most stable passage to minimize roll, pitch, and yaw of the ship, while limiting the apparent winds across the deck to a safe working level for personnel. During VERTREP operations, Romeo Corpin may be limited by the relative wind requirements for the helicopters involved. General guidance for determination of Romeo Corpin is found in *Underway Replenishment,* NWP 4-01.4. Much of the necessary guidance for the best replenishment course depends on how each individual ship type handles in different sea conditions. This information may be obtained from a qualified Underway Officer of the Deck. All requests for Romeo Corpin should be referred to the Forecaster.

REVIEW QUESTIONS

Q68. Define wave height.

Q69. What factors will determine the maximum height of sea waves for any location?

Q70. What would be the average wave height with a sea state of 5?

Q71. Define wave period.

Q72. How is wave direction determined?

Q73. What is meant by the term "significant wave height"?

Q74. How do swell waves differ from sea waves?

Q7.5. How is swell wave height determined?

ICE ACCRETION

LEARNING OBJECTIVES: Define ice accretion. Describe the characteristics of ice accretion. List elements to be included in the observation of ice accretion.

Ice observations are conducted as part of general shipboard weather observations. *Ice accretion* is the accumulation of clear ice (glaze) or rime ice on the outside structures of a ship. Ice may form on a ship when the air temperature is below freezing and hydrometeors are present. Glaze commonly forms when the air temperature is between 32°F and 25°F (0°C and -4°C) with dense fog, freezing rain or drizzle, or blowing spray. Below 25°F the probability of freezing precipitation drops sharply. As the temperature approaches 14°F, (-10°C), fog and blowing spray form rime ice rather than clear ice, because the freezing occurs too fast for the trapped air bubbles to escape. Ice tends to accumulate first on wires, railings, masts and exposed fittings, and then on flat surfaces receiving little heating from the interior of the ship. Ice accumulates last on decks and bulkheads heated from within the ship. Ice accretion is dangerous not only to personnel who must walk across or work on the weather decks, but to the ship as well. Ice accumulations may break wires and antennas. If the accumulation is heavy enough, the added weight on the superstructure may cause the ship to roll excessively or capsize.

Observations of ice accretion include a determination of the source of the moisture, such as fog, blowing spray, or freezing precipitation, as well as an average measurement of the thickness of the accumulated ice, in centimeters. The observation also requires a determination of the rate of accumulation or melt-off.

REVIEW QUESTIONS

Q76. At approximately what temperature (Fahrenheit) would you expect rime ice to form on a ship, assuming blowing spray is present?

Q77. What elements are included in an ice accretion observation?

ICE IN THE SEA

LEARNING OBJECTIVES: Explain the importance of sea ice to naval operations. Describe the various sea ice classifications, sea ice sizes, and topography of sea ice sheets. Discuss movement of sea ice and ice of land origin. Explain the judgments to be made when observing ice in the sea.

Roughly three percent of the world's water areas are covered in ice. Although small in area, the ice-covered areas of the sea and oceans are important to naval operations because of their proximity to possible hostile forces. Many submarines routinely operate beneath the ice, and surface ships occasionally operate in ice-covered seas or areas frequented by icebergs. The Naval Ice Center in Suitland, Maryland, keeps the Fleet advised of the development, movement, and equatorward limit of the ice edge, as well as of the location and movement of icebergs. Although they make extensive use of satellite imagery to detect and track ice, the ice observations from ships operating near the ice provide valuable input to this critical tracking and forecasting effort. Observations of ice seen floating in the sea are completed as part of each surface weather observation.

There are two main types of ice found floating in the sea: sea ice and ice of land origin.

SEA ICE

Sea ice is ice that forms in the sea. It is, for the most part, frozen seawater. Sea ice accounts for approximately 95% of the ice coverage in the oceans.

For seawater to freeze, the temperatures must be colder for a longer period of time. This is due to the salinity of the water and because of the density changes in the water caused by the salinity. We know that pure water freezes at 0°C (32°F) but the freezing point of seawater varies, depending on the salinity (fig. 1-34). Seawater averages 35‰ or 35 parts per thousand by weight salinity. With this salinity, water begins to freeze at -1.9°C (28°F). Freshwater reaches maximum density at 4°C (39°F). In effect, as freshwater ponds and lakes cool, and the surface waters reach 4°C the water sinks and warmer subsurface water rises to replace it. This slows the process of cooling the surface of the pond below 4°C until the entire body of water is cooled to 4°C. After this point, surface waters cooled to less than 4°C are slightly less dense than the water below the surface, and cooling to the freezing point is rapid. Seawater on the other hand, reaches maximum density at the freezing point. When surface seawater is cooled to the freezing point, but before ice can form, the water sinks and is replaced from below by slightly warmer water. The overturn process continues for a long period of time, even in continued subfreezing air temperatures, until a large column of water can be cooled. Overall, the lower freezing point and greater overturn required makes the freezing process of seawater very slow. The freezing of seawater is further retarded by the mixing action of winds (waves), currents, and tides. Once ice forms, it floats. Ice, even saltwater ice, expands as it freezes, so it is less dense than water at the same temperature.

The formation of sea ice usually begins with the onset of autumn, and the first ice usually appears in the mouths of rivers that empty into shallow seas, such as that off northern Siberia. During the increasingly longer and colder nights of autumn, ice forms along the shorelines (fast ice) and becomes a semipermanent

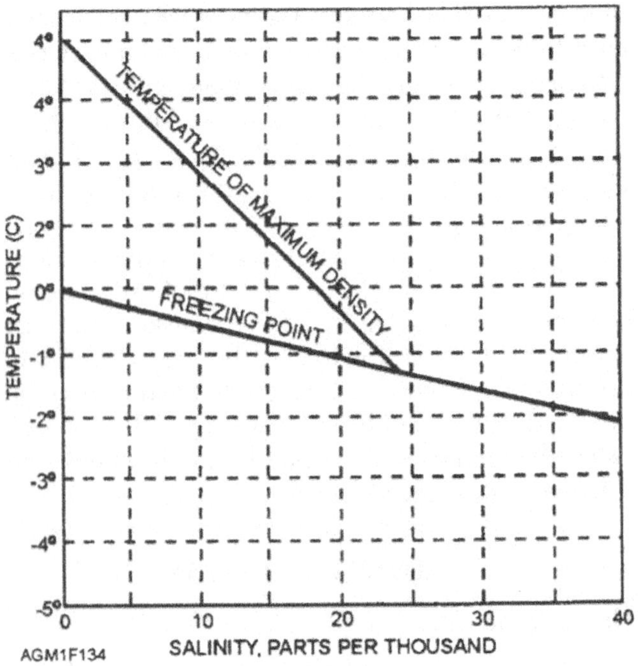

Figure 1-34.—Freezing point and temperature of maximum density versus water salinity.

feature that widens and spreads. When islands are close together, as in the Siberian Sea, fast ice blankets the sea surface, and bridges the waters between all land areas.

On the average in the Northern Hemisphere, sea ice is at a minimum in September, and at a maximum in March. In the Southern Hemisphere, these times are nearly opposite; minimum in March and maximum in September.

The first indication of ice formation is the presence of fine ice crystals on the surface of the water, producing a "slushy" water appearance.

Sea Ice Classification

As sea ice forms and grows, it is generally categorized into one of four groups: newly formed ice, young ice, first-year ice, and old ice.

NEWLY FORMED ICE.—In the open sea, the first sign that the sea surface is freezing is an oily appearance of the water. This is caused by the formation of *spicules* (minute ice needles) and *frazil* crystals (thin plates of ice). As formation continues, the surface attains a thick, soupy consistency termed *grease ice*. Eventually, *slush* and/or *shuga* (spongy white ice clumps) will begin to appear. Next, depending on the wind, waves, and salinity, an elastic or brittle crust forms. The elastic crust *(nilas)* has a matte appearance, while the brittle crust *(ice rind)* is shiny. As the crust thickens, the wind and sea cause the ice to break up into rounded masses known as *pancake ice*. With continued freezing, the pancake ice forms into a continuous sheet.

YOUNG ICE.—This ice sheet forms in 1 year or less, and its thickness ranges from 10 to 30 centimeters (4 to 12 inches). It is further classified as gray ice and gray-white ice.

FIRST-YEAR ICE.—This ice is a more or less unbroken sheet of ice of not more than one winter's growth that starts as young ice. Its thickness is from 30 centimeters to 2 meters (1 foot to 6 1/2 feet). First-year ice may be subdivided into thin first-year ice, medium first-year ice, and thick first-year ice. The latter is more than 4 feet thick.

OLD ICE.—Old ice is extremely heavy sea ice that has survived at least one summer's melt. It occurs primarily in the Arctic and Antarctic polar packs as a mass of converging and dividing ice floes of various ages, sizes and shapes. Old ice may be subdivided into second-year ice and multi-year ice.

Sizes of Sea Ice

Sea ice generally forms in vast sheets frozen solidly to the shores of islands and land masses; such areas of ice are called *fast ice*. The effects of winds and currents may break up fast ice sheets into smaller free-floating pieces of sea ice. Sea ice is categorized into seven different sizes, ranging from "small ice cakes" to "giant ice floes." Refer to figure 1-35 for relative sizes and a comparison to more common features. When the majority of the water area is covered in ice floes, the ice is generally called *pack ice*. A similar term, *the ice puck*, refers to any very large area that is predominately covered in ice.

In the Arctic and Antarctic, some areas of fast ice persist for many years. The frozen layers of seawater collect layers of snow that build up and are compressed into ice. These areas may develop ice sheets many hundreds of feet in thickness. Vast areas of seawater may be covered by these permanent areas of ice, which is called *shelf ice* or an ice shelf. As pieces of shelf ice along the edges of the ice sheet break free, they become icebergs.

Open Water Around Sea Ice

The same forces that separate ice floes from the fast-ice sheet also create various openings of unfrozen water between the ice floes. Naval operations in and around fields of sea ice can be hazardous. The movement of massive floes of ice can cut off ships from open water; worse yet, the ice may close in around a ship, leaving it stranded in a sea of ice. Therefore, changes in the size of open water areas in ice-covered seas becomes very important.

Many water features are associated with sea ice. Some of the more common features are as follows:

- *Fracture*—Any break or crack through the ice sheet
- *Lead*—A long, narrow break or passage through the sea ice sheet or between floes; a navigable fracture. A lead may be open or refrozen
- *Puddle*—A depression in sea ice usually filled with melted water caused by warm weather
- *Thaw hole*—A hole in the ice that is caused by the melting associated with warm weather
- *Polynya*—Any sizable area of seawater enclosed by sea ice. Put simply, a large hole in the ice

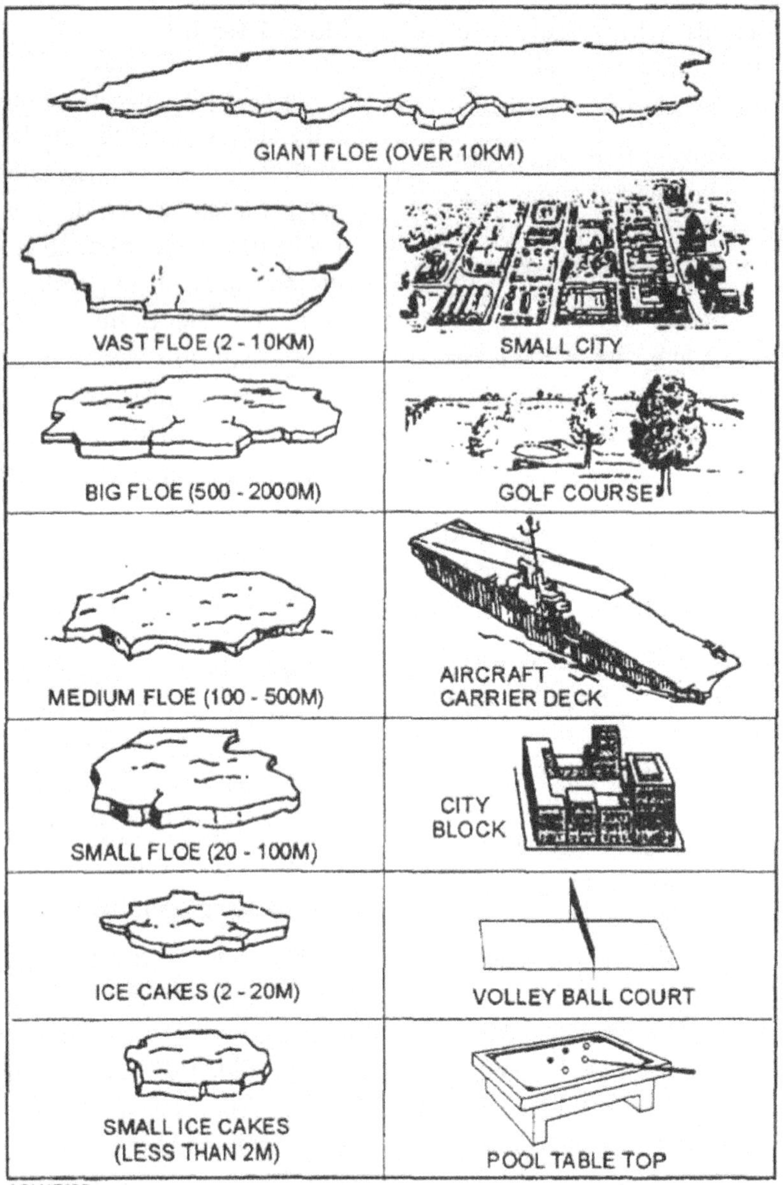

Figure 1-35.—Sizes of sea ice.

Topography of Sea Ice Sheets

In addition to separating pieces of sea ice from the fast ice by pulling it apart, the winds and currents may also push on fast-ice sheets, causing stress budges or stress ruptures in the surface with portions of the sheet overriding other sections. Many of these stress-induced features are identified by specific terms that describe the topography, or configuration of the ice surface. These terms are related to the degree of ice-surface roughness. Figure 1-36 illustrates the types of topography.

RAFTED ICE.—This type of topography occurs when ice cakes override one another. Rafting occurs when wind forces ice cracks or ice floes together. It is associated with young and first-year ice. When the rafting process is occurring, there is great compression within the ice sheet. Ships should avoid operations in any opening in the ice in an area where rafting is occurring, because it indicates the ice is closing in rapidly.

RIDGED ICE.—Ridged ice is much rougher than rafted ice and occurs with first-year ice. Wind and weather eventually smooth the surface of the ridges.

HUMMOCKED ICE.—Hummocking occurs with old ice. It is defined as ice piled haphazardly into mounds or hillocks. At the time of its formation, hummocked ice is similar to rafted ice; the major difference is that hummocked ice, because of its thickness, requires a greater degree of pressure and heaping.

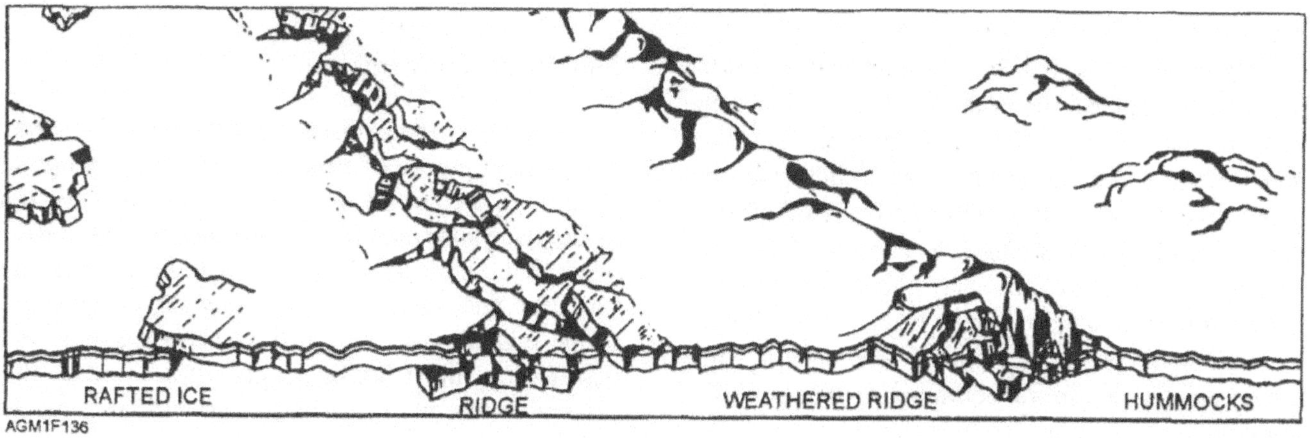

Figure 1-36.—Various types of ice topography caused by pressure.

Movement of Sea Ice

As floes break free of the fast-ice sheet, they move under the effects of the winds and currents. Any movement of ice floes is called "drift." Drift refers to sea ice, as well as broken off portions (outer edges) of fast ice that moves along with wind, tides, and currents. Drift may, therefore, be more specifically defined by direction and speed of movement.

Pack ice usually drifts to the right of the true wind in the Northern Hemisphere (left in the Southern Hemisphere). Observations show that the actual drift is about 30° from the wind direction, or very nearly parallel to the isobars on a weather map. The drift more closely follows the wind in winter than in summer. In summer, the tides play a bigger role in the movement of the ice.

A close estimate of the speed of drifting pack ice is possible by using wind speed. On the average, the drift of ice in the Northern Hemisphere ranges from 1.4% of the wind speed in April to 2.4% of the wind speed in September. Although wind is the primary driving force, the presence or absence of open water in the direction of the drift greatly influences the speed of drift. Ice-free water in the direction of the drift, no matter how distant, permits the pack ice to drift freely in that direction. Ice-clogged water, on the other hand, slows the forward movement of ice.

ICE OF LAND ORIGIN

Ice of land origin is ice that forms on land, usually as glaciers, and moves to the sea. Composed by large accumulations of compacted (freshwater) snow on land areas, the weight of the accumulated snow/ice forces the ice sheets to move as glaciers. If a glacier is located along a coastline and reaches the sea, the leading edge of the glacier may break off (calve) and fall into the sea. This ice then drifts to sea as an *iceberg*.

Since 86% of the world's glaciers occur in Antarctica, most icebergs originate around that continent. Most of the remainder of the world's glaciers are located in Greenland. Greenland is the main source of icebergs in the Northern Hemisphere (about 90%). Nearly 70% of Greenland's icebergs originate along the western coast near 68°N.

ICEBERG CLASSIFICATION

Icebergs are classified by shape and by size. As classified by shape, icebergs are either pinnacled or tabular. *Pinnacled bergs* are generally cone-shaped, but may be very irregular. *Tabular bergs* are generally flat-topped and straight-sided. See figure 1-37.

Figure 1-37.—Types of icebergs.

The structure of an iceberg, and to some extent the appearance, depends upon the ice that produced the berg. Pinnacled bergs come from glaciers that plow across uneven ground on their way to tidewater, but may also develop as fragments that break off of shelf ice. Tabular bergs are produced as portions of sheet ice separate and float free, but may also be produced as large portions of shelf ice break free. Pinnacled bergs are more common in the Northern Hemisphere, while tabular bergs are more common in the Southern Hemisphere.

Icebergs are simply called "icebergs" when they are large or massive. Smaller pieces of ice are called "bergy bits" and "growlers." Like icebergs, bergy bits and growlers originate from glaciers or shelf ice. They may also form as larger icebergs disintegrate. A *bergy bit* is a medium-size fragment of glacial ice, or about the size of a small cottage. A growler is a small fragment of ice about the size of a truck. It is usually of glacial origin, and generally greenish.

Origin Characteristics

Icebergs originating in Greenland average 70 meters in height and 280 to 450 meters in length when first formed. The largest ones may exceed 400 feet in height and several miles in length. The tabular bergs of Antarctica average 30 to 40 meters in height, but their horizontal dimensions greatly surpass the bergs of the Northern Hemisphere. For example, one iceberg observed near Scott Island in 1956 measured 60 miles by 208 miles.

The portion of an iceberg that is visible above the water is dependent upon the type of the berg and the density differences between the seawater and the ice. The type of berg (pinnacled or tabular) determines the height of the ice above the water. In the case of the tabular berg, the depth below the surface is about 7 times the height above the water line. In the case of the pinnacled berg, the depth below the surface averages about 5 times that above the water line.

With regard to density, seawater with a temperature of -1°C and a salinity of 35% produces a density condition that allows nearly 90 percent of the ice to be submerged.

Pinnacled icebergs often have rams (protrusions of ice beneath the surface). These rams can be a great hazard to vessels that might pass close by bergs of this type. The *S.S. Titanic* sank in the North Atlantic with a great loss of life after striking this type of iceberg.

Movement of Icebergs

While the general direction of the drift of icebergs over a long period of time is known, it may not be possible to predict the drift of an individual berg at a given place and time. Bergs lying close together have been observed to move in different directions. The reason for this is that icebergs move under the influence of the prevailing current at the iceberg's submerged depth. The subsurface currents often opposes the existing winds and near-surface currents.

OBSERVING ICE IN THE SEA

When making an observation of ice in the sea, either sea ice or icebergs, you must determine the type of ice present. If the ice is sea ice, you must determine the size of the area covered by ice, the arrangement of ice, and the stage of development. If icebergs are present, you must determine the number of bergs present. If the ship is approaching an ice field, the direction of the ice edge relative to the ship's position must also be determined. Finally, a determination of the ship's ability to operate in the ice and the trend of any changes in the ice conditions must be made. Although these determinations sound complex, they are relatively easy. NAVMETOCCOMINST 3144.1 provides a detailed list of the determinations that must be made. Additional information on ice is available in the *Ice Observation Handbook,* published by the Naval Ice Center, Suitland.

In this section, we have discussed the various procedures used to observe the different elements in surface weather observations and surface aviation weather observations. We have also introduced many technical terms used to discuss and accurately describe the weather elements. Later, we will cover the observation recording methods and the specific code forms used to report surface weather observations. However, before we discuss that information, we must discuss some calculations that the observer is routinely asked to determine from data measured during the weather observation.

REVIEW QUESTIONS

Q78. Why does it take seawater longer to freeze than freshwater?

Q79. When does sea ice reach a maximum in the Northern Hemisphere?

Q50. What is meant by the term "fast ice"?

Q51. Why should ships avoid operating in ice areas where rafting is occurring?

Q82. How does pack ice normally drift in the Northern Hemisphere during winter?

Q83. Where do most of the world's icebergs originate?

Q84. What is the most important influence us to the movement of *icebergs?*

Q85. What publication provides detailed instructions for *reporting ice in the sea?*

COMPUTATION OF PHYSIOLOGICAL INDICATORS FROM OBSERVED DATA

LEARNING OBJECTIVES: Define heat stress. Identify the signs of heat exhaustion and heat stroke. Define relative humidity (RH) and identify how relative humidity and the General Heat Stress Index (GHSI) relate to heat stress. Describe the procedure used to compute GHSI. Identify the difference between the GHSI computed by weather personnel and the Wet-bulb Globe Temperature (WBGT) index used by some military personnel. Explain the effects of cold on the body. Define wind chill temperature and describe the procedure used to determine wind chill temperature. Explain seawater immersion survivability.

In the Navy and Marine Corps, all personnel routinely vary their activity and must wear clothing appropriate for the activity they expect to engage in. How "hot" or "cold" the weather is plays an important part in both operational and training activities, and is especially important in physical readiness training. The temperature is also a factor in off-duty recreational activities. Ashore and aboard ship, weather office observers are routinely asked for various readings used as indicators for the effects of temperature on the human body. In this section, we discuss heat stress and the effects of cold on the human body, as well as the values and indicators operational planners use to avoid exposing personnel to these hazards.

HEAT STRESS

Heat stress is the effect of excessive heat on the body, and the inability of the body to get rid of excess heat fast enough to maintain an internal temperature balance. Sweating is a sign that the body is functioning normally to maintain its heat level. Now let's consider two types of heat stress: heat exhaustion and heat stroke.

Signs of heat exhaustion include profuse sweating with a pale skin color, drowsiness, headache, nausea, vision disturbances, or muscular cramps. Heat exhaustion is a dangerous condition and should be promptly treated. Heat stroke is indicated by a lack of sweating with a hot, dry, red skin, dizziness, restlessness, confusion, or unconsciousness. Heat stroke is a potentially fatal condition requiring immediate medical assistance.

Once a body has been heat stressed, the body's tolerance to heat decreases to a certain degree. Since heat stress will result in permanent physiological changes and may result in death, heat stress should be avoided. Two heat-stress-related indicators, relative humidity and the General Heat Stress Index (GHSI) are routinely computed by weather observers to serve as guidelines for exposure to high heat situations. An additional indicator, the wet-bulb globe temperature, is frequently used as a heat stress indicator in certain situations, but is not routinely computed by weather observers.

Relative Humidity

Relative humidity, commonly abbreviated "RH," is the ratio of how much water vapor is in the air compared to the amount of water vapor, at the current temperature and pressure, that the air can possibly hold, expressed as a percentage. Without adding water vapor to or extracting it from the air, the relative humidity will fall as the temperature rises during the day; and as the temperature falls at night, the relative humidity will rise. The measurements that provide us with the value of water vapor the air can possibly hold are the observed air temperature and the observed pressure. The measurement that yields a value of water vapor actually held by the air is the dew-point temperature, which is calculated from the dry- and wet-bulb temperatures.

Relative humidity is also the most requested indicator of heat effects on the body because it has been in use longer than any other, Just about everyone with an elementary education realizes that when relative humidity is high, the air feels hotter, and when the relative humidity is low, the air feels cooler. when the humidity is high, moisture does not evaporate efficiently from the skin, and the body temperature rises. But when the humidity is low, moisture on the skin evaporates rapidly and provides very efficient cooling of the body. Because of its widespread usage, relative humidity is routinely computed at many shore stations whenever the air temperature and wet-bulb temperatures are measured.

While the effects of moisture in the air affect how people react to temperature by altering the evaporation rate of moisture on the skin, the effects of wind may also alter the evaporation rate of moisture from the skin and affect the temperature of the human body.

General Heat Stress Index

The *General Heat Stress Index* (GHSI), also referred to as *the apparent temperature,* is a measure of how hot the air "feels" to an average person based on the temperature and the humidity. It does not take into account direct sunshine, wind, or the type of clothing a person is wearing. The GHSI value is the "apparent temperature" in National Weather Service public forecasts. The GHSI may be computed from the relative humidity and the air temperature by using figure 1-38.

As an example, with an air temperature of 90°F and a relative humidity of 60%, read the apparent temperature at the intersection of the horizontal air temperature line and the vertical relative humidity line. In this case the GHSI or the apparent temperature is 100°F. Interpolate between the lines as necessary. While interpretation of what a given apparent temperature feels like may vary from person to person, the differences among various apparent temperatures are objective and based on physiological research.

The General Heat Stress Index uses four categories to relate "apparent temperature" to the probable occurrence of heat-stress-related injury as follows:

- Apparent Temperature 130°F and up—Category I, EXTREME DANGER—Heatstroke <u>imminent</u>

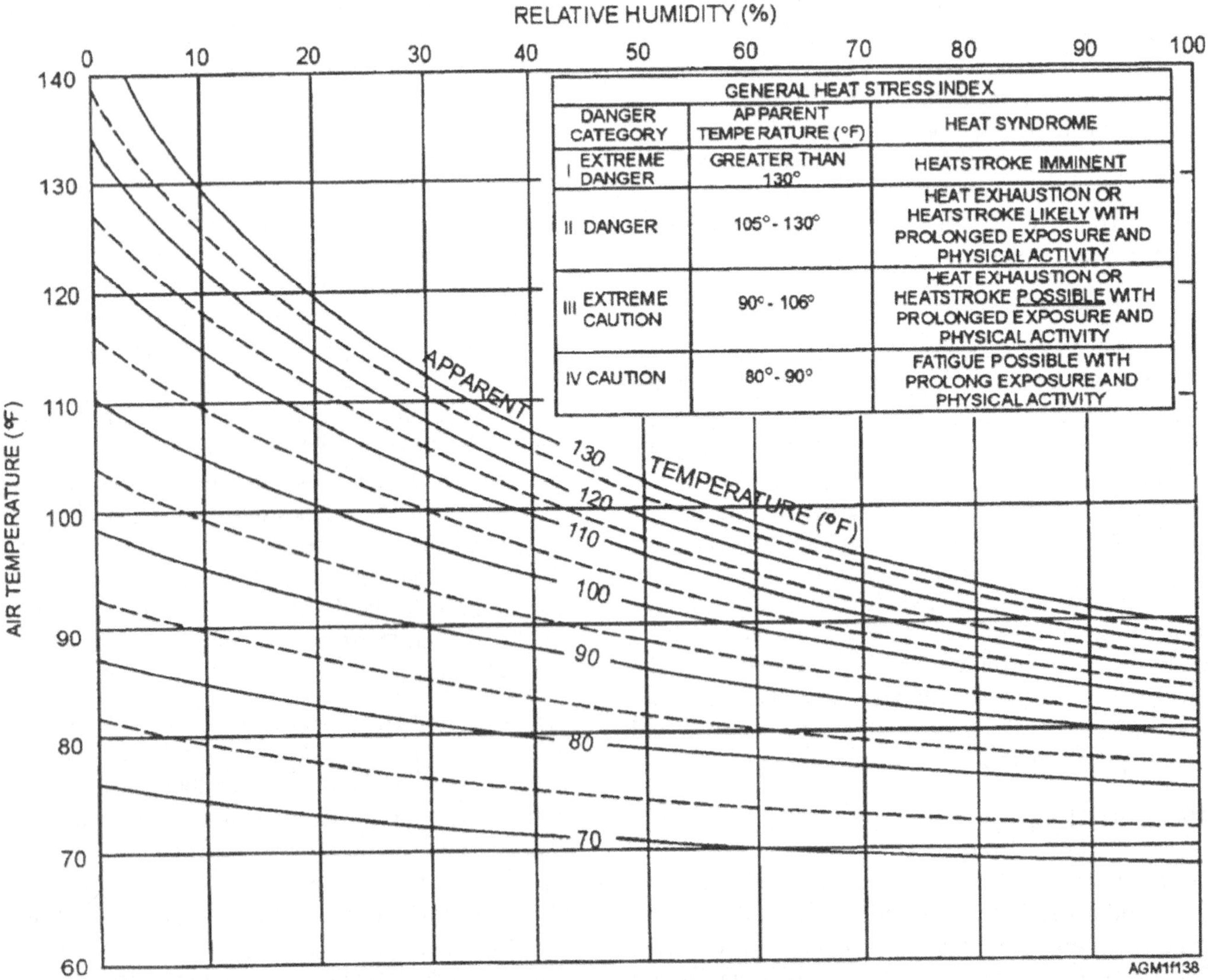

Figure 1-38.—General Heat Stress Index nomogram.

- Apparent Temperature 105°F to 130°F—Category II, DANGER—Heat exhaustion or heatstroke likely with prolonged exposure and physical activity
- Apparent Temperature 90° to 105°F—Category III, EXTREME CAUTION—Heat exhaustion or heatstroke possible with prolonged exposure and physical activity
- Apparent Temperature 80°F to 90°F—Category IV, CAUTION—Fatigue possible with prolonged exposure and physical activity

Since this index is based on the dry-bulb temperature, which by definition and practice is a shielded (in the shade) temperature, the presence of direct sunlight increases the danger.

Wet-bulb Globe Temperature Index

The *Wet-bulb Globe Temperature Index* is a heat stress indicator that considers the effects of temperature, humidity, and radiant energy. The required inputs for the index are measured by a wet-bulb globe temperature meter. The standard wet-bulb globe temperature meter in use by the Navy, called the "Navy Heat Stress Meter," gives a digital readout of dry- and wet-bulb temperatures and globe temperature, and computes a wet-bulb globe temperature (WBGT) index. Although similar to an electric psychrometer in that it has shielded, fan-ventilated dry- and wet-bulb thermometers, this device also has a globe temperature meter (a flat-black metal sphere that measures radiant energy).

Aboard ship, medical Corpsmen usually monitor WBGT readings. Chapter B2 of *the Navy Occupational Safety and Health (NAVOSH) Program Manual for Forces Afloat,* OPNAVINST 5100.19, provides additional information on the shipboard heat stress control program. Since about 1980, the WBGT index has been gaining in usage as a regulatory guide for outdoor work situations; operational activities; and general training, such as field exercises, marching, and physical training. The WBGT index is used in conjunction with a Physiological Heat Exposure Limits (PHEL) chart to determine maximum exposure time for personnel working in high-heat interior environments. When applied to outdoor functions in hot climates, the PHEL charts are used to set activity limits, but a simplified guideline based strictly on the WBGT index is available for certain applications. When used in an outdoor environment, the WBGT index is the only index that compensates for the heating caused by direct or reflected sunshine. A variation of the computation, although not routinely used, also compensates for the cooling effect of the wind.

Various color codes are used outdoors to indicate the varying danger levels of heat stress. Some commands fly appropriate color-coded flags to indicate the heat-stress danger level. WBGT index readings less than 82°F generally mean little threat of heat stress, while readings between 82°F and 89.9°F indicate increasing danger levels. OPNAVMST 6110.1, *Physical Readiness Program,* recommends no physical readiness testing or training be done if the WBGT index is ≥85°F. WBGT index readings ≥90°F indicate a great heat-stress danger, and all strenuous outdoor activity should be avoided. The "danger" level is decreased to ≥80°F when heavy clothing, NBC gear, or body armor is worn. The WBGT index is calculated by the formula

$$W = 0.7WB + 0.2GT + 0.1DB,$$

where

W = the WBGT index, in degrees Fahrenheit;

WB = the wet-bulb temperature (°F);

GT = the globe temperature (°F); and

DB = the dry-bulb temperature (°F).

Additional information on WBGT index readings, PHEL charts, heat-stress dangers, and danger-level color codes may be found in in NAVMED P-5010-3, *Manual* of *Preventative Medicine.*

EFFECTS OF COLD

Just as the temperature alone is not a reliable indicator for how hot a person feels, the temperature of the air is not always a reliable indicator of how cold a person feels. Increased wind speeds may increase the rate of evaporation of moisture from exposed skin areas. This not only will make a person "feel" cooler, but will actually lower the skin temperature, and consequently, body temperature. While the type of clothing a person wears can provide protection from the chilling effects of the wind, the person's state of health and metabolism may affect his or her ability to produce heat. These factors all affect how cold a person will feel. Generally, coldness is related to the actual lowering of internal body temperature by the loss of heat from exposed flesh.

The two primary dangers to people exposed to the cold are frostbite and hypothermia. *Frostbite* is the freezing of the skin, which damages the skin and underlying flesh. *Hypothermia* is the lowering of the internal body temperature due to prolonged exposure to cold air or immersion in cold water.

Frostbite may cause only localized tissue death; but hypothermia, if not reversed, will kill people. When the normal internal body temperature falls below 98.6°F, shivering begins. As internal body temperature approaches 95°F, the body will usually be in uncontrolled, violent-shivering spasms. Lower temperatures cause loss of mental processes, a cessation of shivering, muscle rigidity, unconsciousness, and then death as the body cools below 80°F.

Wind Chill Equivalent Temperature

The wind chill equivalent temperature (also called the *wind chill index,* the *wind chill factor,* or just plain *wind chill*), is the temperature required under no-wind conditions that will equal the cooling effect of the air (the actual air temperature) and the wind on an average size, nude person in the shade. Moisture content of the air, visible moisture on the skin or clothing, presence of sunshine, clothing, and physical activity are not considered.

Wind chill equivalent temperature is found by use of the wind chill nomogram (fig. 1-39). The vertical lines on the nomogram indicate air temperature, in degrees Fahrenheit; the horizontal lines indicate wind speed, in knots; and the curved lines indicate wind chill equivalent temperature. From the intersection of the air temperature line and the wind speed line, follow the curved line downward to the left to read the wind chill. The dashed lines on the nomogram illustrate an example of a wind chill determination with an air temperature of 20°F and a wind speed of 29 knots: the resultant wind chill is -10°F. Interpolate between the lines as necessary. Other tables are used to determine wind chill, but unfortunately, most require wind speed in miles per hour rather than the standard measurement of knots used by the Navy and Marine Corps.

Weather observers should avoid making any recommendations for when and how long people may work outdoors during cold weather. Frostbite and hypothermia depend on the clothing people wear, their level of activity, and the length of time exposed to the cold.

Seawater Immersion Survivability

When a person is immersed in water, the major factor on the length of time the person can survive is the seawater temperature. Some other factors are the sea condition (height and length of the waves), the person's ability to swim, the person's physical condition, and the person's clothing.

When immersed in water, the human body loses heat to the water by conduction. If a person is immersed long enough, internal body temperature falls and unconsciousness or death occurs (fig. 1-40). The survivability assumes that the sea condition is not a factor, and that the person is an average swimmer, in average physical condition, with no special clothing.

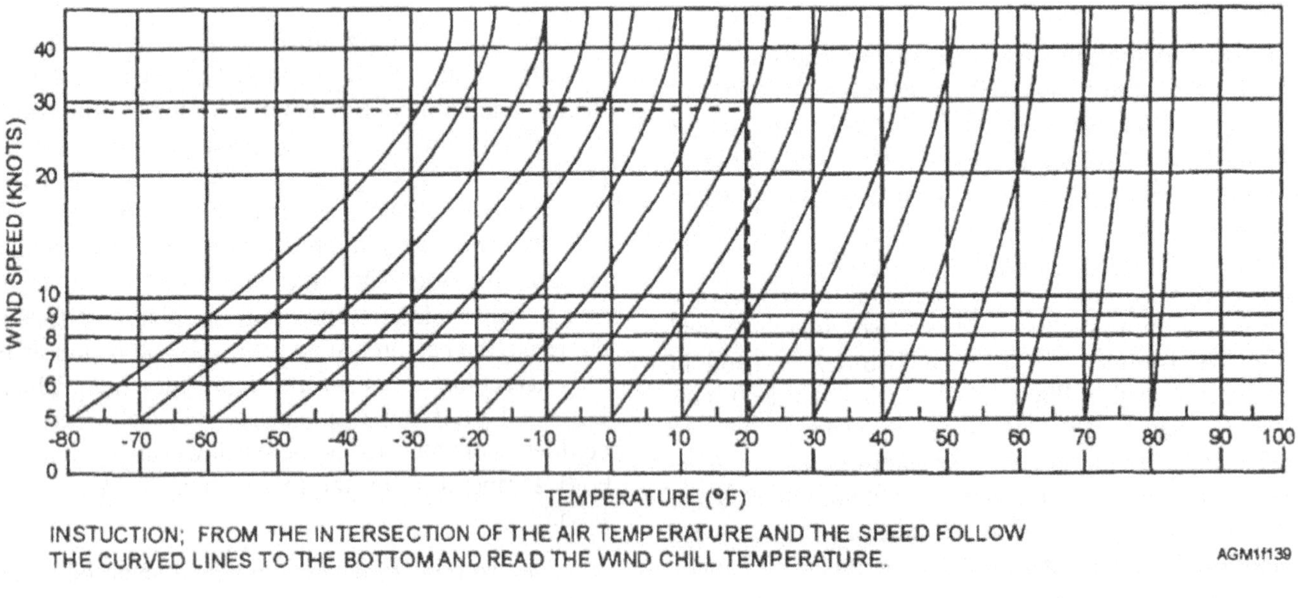

Figure 1-39.—Wind Chill Equivalent Temperature nomogram.

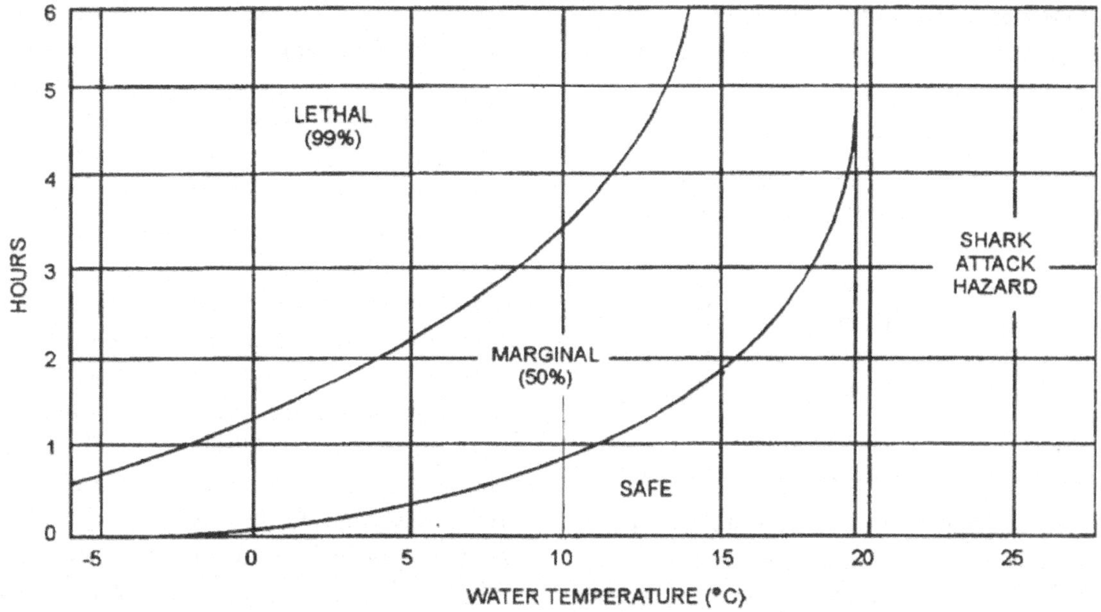

Figure 1-40.—Survivability of a person in water.

Anti-immersion suits are worn by aircrew personnel to retain body heat if the aircrew must ditch in the water. Unfortunately, anti-immersion gear is bulky and hot, and aircrews prefer not to wear the gear. When environmental conditions indicate usage is required, shipboard Aerographers routinely include in flight weather briefings the recommendation that aircrews use anti-immersion suits. This recommendation should also be included in shore-station briefings for over-water flights. Navy and Marine Corps observers usually provide the necessary information to the briefer, based on their observations of the environmental conditions.

Guidance for anti-immersion suit usage by aircrews is given in OPNAVTNST 3710.7, *NATOPS General Flight and Operating Instructions*. Anti-immersion suits or pressure suits with thermal undergarments must be worn when the coldest water temperature in the mission area is ≥50°F or if the coldest wind chill equivalent temperature on the (water) surface during the mission is ≤32°F. When the water temperature is >50°F but ≤60°F, antiexposure undergarments must be worn, but the determination to wear anti-immersion suits is made by the commanding officer. His or her determination should be based on the length of time required to respond with rescue assets to a ditch site, ranging from 1 hour at 50°F to 3 hours at 60°F. With water temperatures above 60°F, anti-immersion gear is not required.

Navy and Marine Corps weather observers routinely provide various values and indicators that are used to gauge the effects of hot and cold environments on personnel. These values are computed from observed measurements made during surface weather observations. All observers should be aware of the effects of heat and cold on the human body, and should be able to calculate the various indicators upon request. Weather affects the performance of equipment as well as personnel. In the next section, we consider aircraft performance in changing weather conditions.

REVIEW QUESTIONS

Q86. *What term describes the inability of the body to get rid of excess heat fast enough to maintain an internal temperature balance?*

Q87. *Define the term "relative humidity."*

Q88. *Given an air temperature of 90°F and a relative humidity of 65%, what is the apparent temperature?*

Q89. *What effects does the Wet-bulb Globe Temperature index (WBGT) take into consideration?*

Q90. *What is meant by the term "wind chill factor"?*

Q91. *What is the survival chance given a seawater temperature of 15°C and an immersion time of 3 hours.*

COMPUTATION OF AIRCRAFT PERFORMANCE INDICATORS FROM OBSERVED DATA

LEARNING OBJECTIVES: Identify three aircraft performance indicators computed from observed data. Define the terms pressure altitude, density altitude, and specific humidity. Describe the procedure used to compute pressure altitude and density altitude. Identify the procedure used to find specific humidity.

Air density and water vapor content of the air have an important effect upon aircraft engine performance and takeoff characteristics. In this section, we describe some of these effects and how they are computed. The three most common elements an Aerographer's Mate must furnish information on are pressure altitude, density altitude, and specific humidity. All ofthesemay be determined by using a Density Altitude Computer, discussed in chapter 2, while pressure altitude and density altitude can be easily obtained from ASOS. Pressure altitude and density altitude are given in feet; while specific humidity is provided in grams per gram or in pounds per pound. Now let's look at pressure altitude.

PRESSURE ALTITUDE

Pressure altitude is defined as the altitude of a given atmospheric pressure in the standard atmosphere. The pressure altitude of a given pressure is usually a fictitious altitude, since it is rarely equal to true altitude. Pressure altitude is equal to true altitude only when pressure at sea level (or the flight-level pressure) corresponds to the pressure of the U.S. Standard Atmosphere. Pressure altitude higher than the actual altitude indicates the air is less dense than normal, and the aircraft may not be able to carry a full (standard) cargo load. Pressure altitude lower than the actual altitude means the air is more dense than normal, and the aircraft may be able to takeoff successfully with a larger cargo load.

Aircraft altimeters are constructed for the pressure-height relationship that exists in the standard atmosphere. Therefore, when the altimeter is set to standard sea-level pressure (29.92 inches of mercury), it indicates pressure altitude and not true altitude. Flight levels-an indicated altitude based on an altimeter setting of 29.92 inches-rather than true altitudes, are flown above 18,000 feet in the United States, and on over-water flights more than 100 miles offshore. The quickest method for approximating the pressure altitude is by using the Pressure Reduction Computer (CP-402/UM), covered in chapter 2. Detailed instructions are listed on the computer. For your own station, you simply dial in the current station pressure and read the pressure altitude on the scale. The solution is more complex when converting forecast altimeter settings to pressure altitude, but the pressure reduction computer may still be used. On occasion, you may find yourself in a situation where this device is not available. Two alternate methods follow that will enable you to calculate approximations of the pressure altitude. Pressure altitude varies directly with the change in pressure multiplied by a complex variable. The variable amount takes into account temperature and station elevation. Both methods simplify the equation but still give fairly close pressure altitude approximations.

The first method uses a set of precalculated pressure altitudes based on pressure differences from standard pressure. These are listed in table 1-5.

Using the table, you may find the pressure altitude value corresponding to your current or forecast altimeter setting or the current or forecast altimeter setting for any other station. This value must be added to your station elevation or the other station's elevation to find the pressure altitude. For example, if your altimeter setting is 29.41 inches and your station elevation is 1,500 feet, you would enter the left side of the table with "29.4" and find the intersection of the column under "0.01" to find 476 feet. Add 476 feet to your station elevation, 1,500 feet, to find the pressure altitude 1,976 feet.

You may also use the table to find pressure altitude by using station pressure. Station elevation should NOT be added to the value when using station pressure.

The second method is useful when you do not have ready access to the table. To calculate pressure altitude, use the formula

$$PA = H_A + PAV,$$

where

PA = pressure altitude,

H_A = station elevation, and

$PA\ V$ = pressure altitude variation approximation (or 29.92 minus the current altimeter setting times 1,000).

Table 1-5.—Pressure Altitude Values

Hundredths →	0.00	0.01	0.02	0.03	0.04	0.05	0.06	0.07	0.08	0.09
PRESSURE inches & ↓tenths↓	**PRESSURE ALTITUDE (FEET)**									
28.0	1824	1814	1805	1795	1785	1776	1766	1756	1746	1737
28.1	1727	1717	1707	1698	1688	1678	1668	1659	1649	1639
28.2	1630	1620	1610	1601	1591	1581	1572	1562	1552	1542
28.3	1533	1523	1513	1504	1494	1484	1475	1465	1456	1446
28.4	1436	1427	1417	1407	1398	1388	1378	1369	1359	1350
28.5	1340	1330	1321	1311	1302	1292	1282	1273	1263	1254
28.6	1244	1234	1225	1215	1206	1196	1186	1177	1167	1138
28.7	1148	1139	1129	1120	1110	1100	1091	1081	1072	1062
28.8	1053	1043	1034	1024	1015	1005	995	986	976	967
28.9	957	948	938	929	919	910	900	891	881	872
29.0	863	853	844	834	825	815	806	796	787	777
29.1	768	758	749	739	730	721	711	702	692	683
29.2	673	664	655	645	636	626	617	607	598	589
29.3	579	570	560	551	542	532	523	514	504	495
29.4	485	476	467	457	448	439	429	420	410	401
29.5	392	382	373	364	354	345	336	326	318	308
29.6	298	289	280	270	261	252	242	233	224	215
29.7	205	196	187	177	168	159	149	140	131	122
29.8	112	103	94	85	75	66	57	47	38	29
29.9	20	10	+1	-8	-17	-26	-36	-45	-54	-63
30.0	-73	-82	-91	-100	-100	-119	-128	-137	-146	-156
30.1	-165	-174	-183	-192	-202	-211	-220	-229	-238	-248
30.2	-257	-266	-275	-284	-293	-303	-312	-321	-330	-339
30.3	-348	-358	-367	-376	-385	-394	-403	-412	-421	-431
30.4	-440	-449	-458	-467	-476	-485	-494	-504	-513	-522
30.5	-531	-540	-549	-558	-567	-576	-585	-594	-604	-613
30.6	-622	-631	-640	-649	-658	-667	-676	-685	-694	-703
30.7	-712	-721	-730	-740	-749	-758	-767	-776	-785	-794
30.8	-803	-812	-821	-830	-839	-848	-857	-866	-875	-884
30.9	-893	-902	-911	-920	-929	-938	-947	-956	-965	-974
31.0	-983	-992	-1001	-1010	-1019	-1028	-1037	-1046	-1055	-1064

INPUT STATION PRESSURE: READ PRESSURE ALTITUDE DIRECTLY FROM TABLE.
INPUT ALTIMETER SETTING: READ VALUE FROM TABLE AND ADD STATION ELEVATION TO FIND PRESSURE ALTITUDE.

For example, using the formula for the same case we just calculated with the table, we find the following:

$PA = H_A + PAV$

$PA = 1{,}500 + 1{,}000(29.92 - 29.41)$

$PA = 1{,}500 + 510$

$PA = 2{,}010$ feet

By comparison, you can see that this value is 34 feet higher than we found by using the table, but it is a close enough approximation when nothing else is available. And, it may be done quickly in your head. With the pressure reduction computer, the same case yields a pressure altitude of 1,979 feet.

Pilots of aircraft, especially rotary wing aircraft, frequently ask for maximum pressure altitude for takeoff and for all destinations. This is calculated using the lowest expected altimeter setting (QNH) for the destination. The forecaster may have to interpret the other station's forecast to determine if the forecast QNH will be valid during the time the aircraft will be in the vicinity. Many rotary wing aircraft have a table in their aircraft technical data that is entered using maximum pressure altitude and maximum temperature to find the maximum permissible load that can be carried. Maximum pressure altitude may be used by the pilot in lieu of density altitude.

DENSITY ALTITUDE

Density altitude is defined as the altitude at which a given air density is found in the standard atmosphere. For a given altitude, density altitude changes with changes in pressure, air temperature, and humidity. An increase in pressure increases air density, so it decreases density altitude. An increase in temperature decreases air density, so it increases density altitude. An increase in humidity decreases air density, so itincreases density altitude. Changes in pressure and temperature have the greatest effect on density altitude, and changes in humidity have the least effect.

If, for example, the pressure at Cheyenne, Wyoming, (elevation 6,140 feet) is equal to the pressure of the standard atmosphere at that elevation, and the temperature is 101°F, the density would be the same as that found at 10,000 feet. Therefore, the air is less dense than normal, and an aircraft on takeoff (at approximately constant weight and power setting) will take longer to get airborne. Air density also affects airspeed. True airspeed and indicated airspeed are equal only when density altitude is zero. True airspeed exceeds indicated airspeed when density altitude increases.

No instrument is available to measure density altitude directly. It must be computed from the pressure (for takeoff, station pressure) and the virtual temperature at the particular altitude under consideration. This may be accomplished by using the Density Altitude Computer (CP-718/UM) or from Table 69, Density Altitude Diagram, of *Smithsonian Meteorological Tables,* NA-50-lB-521. Remember, virtual temperature is used in the computation of density altitude.

The quickest method of calculating density altitude is to use the Density Altitude Computer (CP-718/UM), discussed in chapter 2. Density altitude must be computed from the pressure (for takeoff, station pressure) and the virtual temperature at the particular altitude under consideration. Specific instructions are printed on the device. Density altitude results from the computer may be estimated to the nearest 10 feet between the marked increments of 100 feet. If you are in a situation where you do not have a density altitude computer or the *Smithsonian Meteorological Tables* available, you may ignore the humidity value and calculate density altitude by the formula

$$DA = PA + (120\ V_t),$$

where

DA = density altitude,

PA = pressure altitude at the level you desire density altitude,

120 = a temperature constant (120 feet per 1°C), and

V_t = actual temperature minus standard temperature at the level of the pressure altitude.

For example, let's say the surface temperature is 30°C and your pressure altitude is 2,010 feet. Look at table 1-6 and find the standard temperature corresponding to 2,000 feet. You should find 11°C. Plug these values into the formula to find the following:

$DA = PA + (120\ V_t)$

$DA = 2{,}010 \text{ feet} + [120(3°C - 11°C)]$

$DA = 2{,}010 + 120(19)$

$DA = 2{,}010 + 2{,}280$

$DA = 4{,}290$ feet

Table 1-6.—U.S. Standard Atmosphere Heights and Temperatures

HEIGHTS TO STANDARD PRESSURE AND TEMPERATURE

Altitude, feet	Pressure, hPa	Pressure, inches	Temperature, °C	Temperature, °F	Altitude, feet	Pressure, hPa	Pressure, inches	Temperature, °C	Temperature, °F
0	1013.2	29.92	15.0	59.0	26,000	359.9	10.63	-36.5	-33.7
1,000	977.2	28.86	13.0	55.4	27,000	344.3	10.17	-38.5	-37.3
2,000	942.1	27.82	11.0	51.9	28,000	329.3	9.72	-40.5	-40.9
3,000	908.1	26.82	9.0	48.3	29,000	3 14.8	9.30	-42.5	-44.4
4,000	875.1	25.84	7.1	44.7	30,000	300.9	8.89	-44.4	-48.0
5,000	843.1	24.90	5.1	41.2	31,000	287.4	8.49	-46.4	-51.6
6,000	812.0	23.98	3.1	37.6	32,000	274.5	8.11	-48.4	-55.1
7,000	781.8	23.09	1.1	34.0	33,000	262.0	7.74	-50.4	-58.7
8,000	752.6	22.22	-0.8	30.5	34,000	250.0	7.38	-52.4	-62.2
9,000	724.3	21.39	-2.8	26.9	35,000	238.4	7.04	-54.3	-65.8
10,000	696.8	20.58	-4.8	23.3	36,000	227.3	6.71	-56.3	-69.4
11,000	670.2	19.79	-6.8	19.8	37,000	216.6	6.40	-56.5	-69.7
12,000	644.4	19.03	-8.8	16.2	38,000	206.5	6.10	Constant to	
13,000	619.4	18.29	-10.8	12.6	39,000	196.8	5.81	65,500 feet	
14,000	595.2	17.58	-12.7	9.1	40,000	187.5	5.54		
15,000	571.8	16.89	-14.7	5.5	41,000	178.7	5.28		
16,000	549.2	16.22	-16.7	1.9	42,000	170.4	5.04		
17,000	427.2	15.57	-18.7	-1.6	43,000	162.4	4.79		
18,000	506.0	14.94	-29.7	-5.2	44,000	154.7	4.57		
19,000	484.5	14.34	-22.6	-8.8	45,000	147.5	4.35		
20,000	465.6	13.75	-24.6	-12.3	46,000	140.6	4.15		
21,000	446.4	13.18	-26.6	-15.9	47,000	134.0	3.96		
22,000	427.9	12.64	-28.6	-19.5	48,000	127.7	3.77		
23,000	410.0	12.11	-30.6	-23.9	49,000	121.7	3.59		
24,000	392.7	11.60	-32.5	-26.6	50,000	116.0	3.42		
25,000	376.0	11.10	-34.5	-30.2					

STANDARD PRESSURE SURFACES

Pressure surface, hPa	Height, feet	Height, meters	Temperature °C	Temperature °F	Pressure surface, hPa	Height, feet	Height, meters	Temperature, °C	Temperature, °F
1000	364	111	14.2	57.8	200	38,662	11,784	-56.7	-70.1
925	2,512	766	10.0	50.0	150	44,647	13,608	-56.7	-70.1
850	4,781	1,457	5.5	41.9	100	53,083	16,180	-56.7	-70.1
700	9,882	3,012	-4.7	23.5	70	60,504	18,442	-56.7	-70.1
500	18,289	5,574	-21.3	-6.3	50	67,507	20,576	-55.8	-68.4
400	23,574	7,185	-31.7	-25.1	30	78,244	23,849	-52.7	-62.9
300	30,065	9,164	-44.7	-48.5	20	86,881	26,481	-49.9	-57.8
250	33,999	10,363	-52.4	-62.3	10	101,885	31,055	-45.4	-49.7

For an acceptable result with slightly less precision, you may use the density altitude diagram (fig. 1-41) to obtain density altitude to the nearest 200 feet. This diagram also ignores the effect of humidity on density altitude. Enter the bottom of the diagram with your air temperature and proceed vertically to the intersection of the pressure altitude line, then horizontally to the left side of the diagram to find the density altitude. The light dashed line shows an example using 22°C and a pressure altitude of 10 feet, resulting in a density altitude of about 1,000 feet.

You may interpolate for more precise values, but this precision isn't often necessary for most density altitude calculations. (A quick method of determining standard temperatures in degrees Celsius for all levels up to 35,000 feet is to double the altitude in thousands of feet, subtract 15, and change the sign.)

SPECIFIC HUMIDITY

Specific humidity is the mass of water vapor present in a unit mass of air. Where temperatures are high and rainfall is excessive, the specific humidity of the air reaches high proportions. Accurate information is required to determine the proper amount of horsepower needed for the takeoff roll.

Fog and humidity affect the performance of aircraft. During takeoff, two things are done to compensate for their effect on takeoff performance. First, since humid air is less dense than dry air, the allowable takeoff gross weight is generally reduced for operations in areas that are consistently humid. Second, because power output is decreased by humidity, pilots must compensate for the power loss. Your main responsibility as an Aerographer's Mate is to ensure that the pilot has accurate information. Pilots may request humidity values as either relative humidity (discussed in the previous section) or specific humidity.

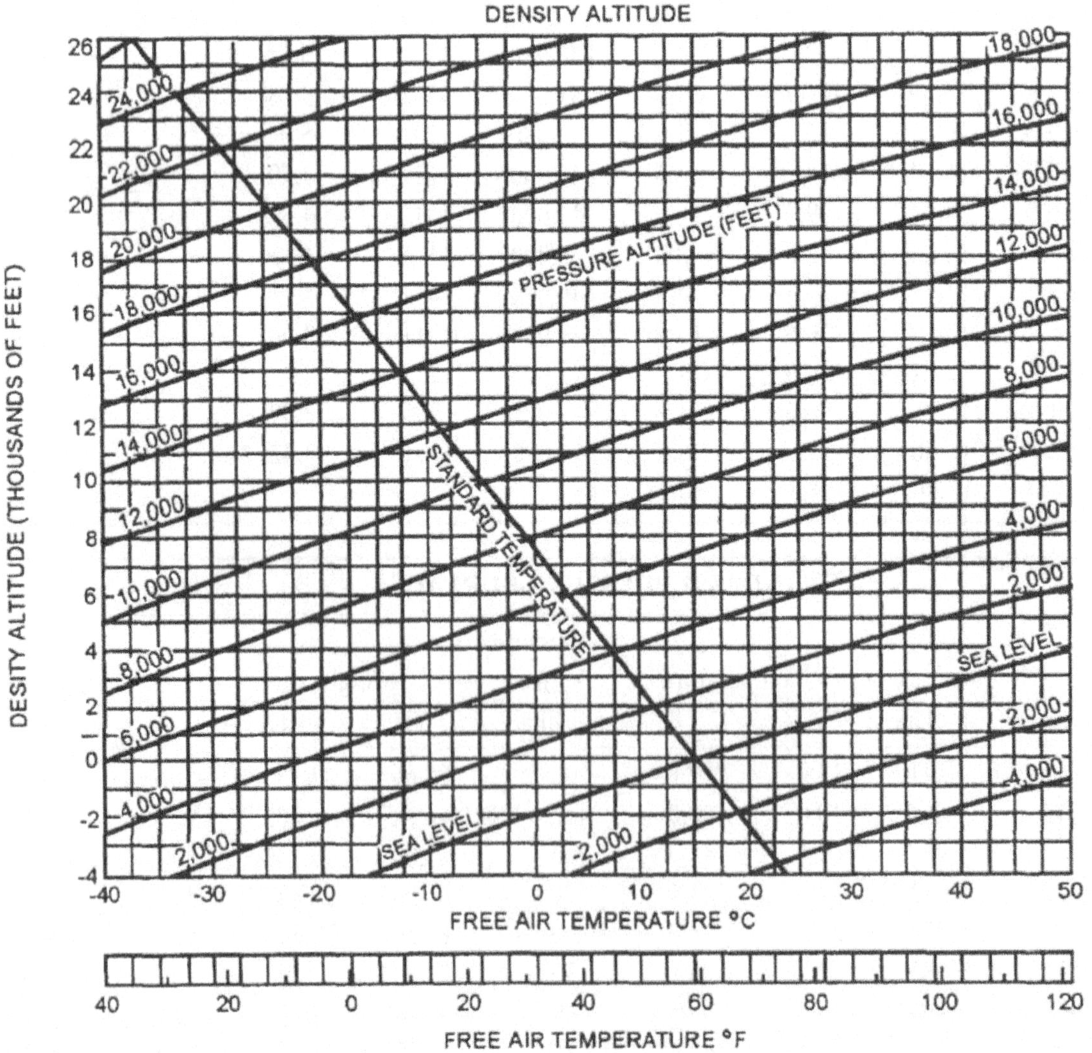

Figure 1-41.—Density altitude diagram.

Specific humidity can be determined from the CP-718/UM density altitude computer following instructions printed on the computer. The air temperature, dew-point temperature, and pressure from the observation are used as arguments.

Proper performance of aircraft engines and the lift produced by the aircraft's wings depend on the density of the air. Air density is affected by pressure, temperature, and water content of the air. Calculated values of pressure altitude, density altitude, and specific humidity are routinely done to provide pilots with specific quantities on which they may base computations for the thrust, lift, and maximum loads for their aircraft.

REVIEW QUESTIONS

Q92. What does a pressure altitude lower than the actual altitude mean?

Q93. What happens to density altitude when the air temperature increases?

Q94. How may density altitude be calculated?

Q95. How does humidity affect air density?

SUMMARY

In this chapter, we have discussed the terms and procedures used when evaluating and measuring surface weather elements, and some of the routinely computed values and indicators associated with surface weather observations. The material has been presented in a manner that, we hope, will guide you to a basic understanding of the subject of "surface weather observations." Additional study of the publications and manuals referenced in the text, especially NAVMETOCCOMINST 3141.2, *Surface METAR Observations User's Manual,* and NAV-METOCCOMINST 3144.1, *United States Navy Manual for Ship's Surface Weather Observations,* will be necessary for a thorough understanding of the subject.

ANSWERS TO REVIEW QUESTIONS

A1. *NAVMETOCCOMINST 3141.2, Surface METAR Observations User's Manual, and NAVMETOCCOMINST 3144.1, United States Navy Manual for Ship's Surface Weather Observations.*

A2. *Fahrenheit, Celsius, and Kelvin.*

A3. *15° of longitude.*

A4. *Coordinated Universal Time.*

AS. *Pressure.*

A6. *27.*

A7. *Cumuliform, stratiform, cirriform.*

A8. *Stratiform.*

A9. *Mechanical lift associated with physical barriers forcing air aloft, convective lift associated with surface heating, convergence resulting from a "piling" of air, and vorticity associated with the rotational motion of molecules in the air and the spinning of the earth.*

A10. *6,500 to 23,000 feet.*

A11. *Variety identifies the specific appearance of the arrangement of elements within a cloud layer, the thickness of the layer, or the presence of multiple layers.*

A12. *Cumulonimbus.*

A13. *The amount of moisture near the surface.*

A14. *If the height of the cumulus congestus cloud appears to be twice the width of the base, it should be classified as towering cumulus.*

A15. *An anvil top.*

A16. *Low-level wind shear and microbursts.*

A17. *The right rear quadrant with respect to CB movement.*

A18. *Nimbostratus.*

A19. *When precipitation begins or when bases drop to less than 6,500 feet.*

A20. *Approaching frontal systems with conditions favorable for thunderstorm activity.*

A21. *Ice crystals.*

A22. *Formed when strong winds moving across mountains set up a wavelike action in the winds downstream from the mountain. The upward moving air in the waves, if moist, is brought to saturation as it rises.*

A23. *Formed when moist air is forced upward by a mountain top and dissipates on the leeward side of the mountain as the moving air descends.*

A24. *In eighths of the sky.*

A25. *Clouds and/or obscuring phenomena aloft either continuous or composed of detached elements that have bases at approximately the same level.*

A26. *The lowest layer that blocks 5/8 ormoreofthe celestial dome from beingseen.*

A27. *Any collection of atmospheric phenomena dense enough to obscure even the portion of the sky directly overhead.*

A28. *Sky cover at any level is equal to the amount the sky cover of the lowest layer plus the additional sky cover present at all successively higher layers (up to and including the layer being considered).*

A29. *12,000 feet.*

A30. *7,500 feet.*

A31. *Prevailing visibility, sector visibility, differing level (or tower) visibility, and runway visual range.*

A32. *The greatest distance that known objects can be seen and identified throughout half or more of the horizon circle.*

A33.*Combat Information Center (CIC).*

A34. *When it differs from the prevailing visibility, and either prevailing visibility or sector visibility is less than 3 miles.*

A35. *When the prevailing visibility is 4 miles or less.*

A36. *Haze.*

A37. *In desert regions on calm, hot, clear afternoons.*

A38. *Blowing sand that reduces visibility to less than 5/16 of a mile.*

A39. *Water vapor bypasses the liquid state and goes directly from a gas to a solid.*

A40. *Moisture, hygroscopic nuclei, and cooling.*

A41. *Four Celsius degrees or less.*

A42. *Advection fog.*

A43. *A fog condition that reduces prevailing visibility to between 5/8 mile and 6 miles.*

A44. *Frost occurs when radiational cooling lowers the temperature of an object below the freezing level and ice crystals form through sublimation.*

A45. *Super-cooled liquid.*

A46. *The slow rate of fall and droplet size (less than 0.02 inch).*

A47. *Ice pellets.*

A48. *Moderate.*

A49. *Lightning cloud-to-air.*

A50. *When the first thunder is heard, or when overhead lightning is observed, and the local noise level is high enough as might prevent the observer from hearing thunder.*

A51. *10 hectopascals.*

A52. *The pressure correction applied to station pressure based on the difference in height of the barometer and the runway or station elevation.*

A53. *1.62 hectopascals or .045 inches of mercury.*

A54. *It is a pressure value used by aircraft to allow correct determinations of height above mean sea level.*

A55. *A barograph trace or the actual recorded pressures during the period.*

A56. *The lowest temperature to which an object may be cooled by the process of evaporation.*

A57. *Condensation and/or precipitation.*

A58. *12.5°F.*

A59. *The temperature, below freezing, that a parcel of air must be cooled to in order to reach saturation.*

A60. *Bucket method, bathythermograph method, and seawater injection method.*

A61. *True North is in reference to the geographic North Pole while Magnetic North is in reference to the magnetic North Pole.*

A62. *090°*

A63. *Using the CP-264/U true wind computer, a maneuvering board, or an aerological plotting chart.*

A64. *By using the arithmetic or graphical average during the 2-minute observation period.*

A65. *It would be less than the actual wind speed.*

A66. *A rapid fluctuation in wind speed with a variation between peaks and lulls of 10 knots or more observed in the lo-minute period prior to the actual time of observation.*

A67. *A sudden increase in wind speed of 16 knots or more and the sustained increase must be 22 knots or more for at least 1 minute.*

A68. *The vertical distance from the crest to the trough of the wave.*

A69. *The wind speed, the length of time the wind has been blowing, and the size of the fetch area.*

A70. *8 to 13 feet.*

A71. *The time it takes for a complete wave cycle to pass a given point.*

A72. *Wavedirection is the direction the majority of waves in a group are coming from.*

A73. *The average height of the highest 1/3 of all the waves present,*

A 74. *Swell waves are sea waves that have moved out of the area of formation and are more smooth and regular in appearance.*

A75. *By using the average height of all the swell waves present.*

A76. *14°F.*

A 77. *The source of the ice accretion, the thickness of the ice, and a determination of the rate of accumulation or melt-off.*

A 78. *It is due to the salinity of seawater and the density changes in seawater caused by salinity. In addition, thefreezingofseawater is slowed because of waves, currents, and tides.*

A 79. *March.*

A80. *Sea ice that is frozen solidly to the shores of islands or land masses.*

A81. *Rafting ice indicates that the ice is closing rapidly.*

A82. *It normally drifts to the right (about 30°) of the wind direction.*

A83. *Antarctica and Greenland.*

A84. *The prevailing sea current at the icebergs submerged depth.*

A85. *NAVMETOCCOMINST 3144.1*

A86. *Heat stress.*

A87. *The ratio of how much water vapor is in the air compared to the amount of water vapor, at the current temperature andpressure, that the air can possibly hold.*

A88. *105°F.*

A89. *Temperature, humidity, and radiant energy.*

A90. *The temperature required under no-wind conditions that will equal the cooling effect of the actual air temperature in conjunction with the wind.*

A91. *Marginal.*

A92. *The air is more dense than normal.*

A93. *It increases density altitude.*

A94. *By using the Density Altitude Computer (CP-718/UM), the Smithsonian Meteorological Tables, the Density Altitude Formula, or the Density Altitude Diagram.*

A95. *It decreases air density.*

CHAPTER 2

SURFACE OBSERVATION EQUIPMENT

INTRODUCTION

As a surface aviation weather observer, you must have a thorough knowledge of the equipment used in the observation process. This chapter will guide you through the various types of equipment used to conduct a surface aviation weather observation, both ashore and aboard ship. Both primary and backup systems are covered.

AUTOMATIC WEATHER STATIONS

LEARNING OBJECTIVE: Describe the major components and characteristics of the Automated Surface Observing System (ASOS), the Shipboard Meteorological and Oceanographic Observing System (SMOOS), and the meteorological buoy.

Automatic weather stations are electronic packages that sample, record, and display or transmit weather information to a collection site or user. In the mid-1970's, several systems were introduced that could measure temperature, wind, pressure, and precipitation. By the early 1980's, sensors were developed that could determine sky cover and visibility. In the mid-1980's remote observation sites were in use, providing full spectrum observation data via satellite and phone lines. In 1988, installation was started on a network of Automatic Meteorological Observation Stations (AMOS) in the Pacific to support the Joint Typhoon Warning Center. In the late 1990's, we will see more of these automatic weather stations installed.

The automated weather stations in use by the Navy and Marine Corps at shore stations are called Automated Surface Observing Systems (ASOS), while the equipment system used for shipboard observations is called the Shipboard Meteorological and Oceanographic Observing System (SMOOS). The widely used meteorological buoys are also a type of fully automated weather stations.

AUTOMATED SURFACE OBSERVING SYSTEM (ASOS)

The Automated Surface Observing System (ASOS) is a configuration of fully automated observation equipment that will replace observation equipment at all shore stations. These automated systems are currently being installed. The AN/GMQ-29 semiautomatic weather station, the AN/GMQ-32 transmissometer system, the AN/GMQ-13 cloud height set, and the AN/UMQ-5 wind-measuring set will be replaced. The ASOS automatically collects, processes, and error checks observation data and formats. In addition, ASOS automatically displays, archives, and reports weather elements included in a surface weather observation.

Display and Control Terminal

The ASOS equipment consists of several components that look very much like a standard desk-top computer. The heart of an ASOS is the acquisition control unit (ACU). The ACU receives data by radio link from up to three data collection packages (DCPs). The DCPs are located near the sensors at the touchdown end of a runway and get raw data from individual sensors via a fiber-optic link.

Inside the meteorological office, a computer keyboard and video monitor serve as an interactive user terminal that allows the observer to both receive data and send commands to the ACU. A computer-style printer prints out selected data and observations. The system is also equipped with an audio alarm and a microphone that is used to record supplements to the voice-produced telephone weather report. Telephone and radio modems are used to link the system with various users, maintenance personnel, and on-line weather reporting circuits. A liquid crystal display (LCD) display maybe located in the control tower to keep the air traffic controllers informed of the latest weather conditions. Software menus guide the user through entry of supplemental observation data, file maintenance, and other procedures.

The ASOS provides several types of data for various requirements. The product outputs are as follows:

- Complete weather observations updated each minute
- Wind speed and direction updated every 5 seconds

- Altimeter settings updated each minute
- An Hourly Surface Aviation Weather Observation report
- Daily weather summary to monitor or printer
- Monthly weather summary to monitor or printer

Sensor Package

One or more ASOS sensor packages (fig. 2-1) are normally located at the touch-down end of main runways. From left to right, figure 2-1 shows the following sensors:

- Visibility sensor-Reports equivalent visibility in l/4-statute-mile increments up to 10 statute miles
- Present weather sensor-Uses an infrared beam to sense precipitation, and reports light, moderate, and heavy rain, snow, or mixed precipitation
- Wind direction and wind speed-Measures speed, in knots, from 0 to 125 knots
- Data collection package (equipment case)-Sends semiprocessed data to the acquisition control unit in the weather office
- Temperature and dew-point sensor-Measures temperature from -80°F to 130°F and dew-point from -30°F to 86°F to the nearest tenth of a degree, and also computes relative humidity
- Heated tipping-bucket rain gauge-Measures precipitation rate from 0 to 10.0 inches per hour, in hundredths of an inch
- Cloud height sensor-A laser detector that reports cloud bases from 100 feet above ground level (AGL) to 12,000 feet AGL

A digital pressure transducer that measures atmospheric pressure from 16.90 inches to 31.50

Figure 2-1.—ASOS sensors: A. Visibility sensor; B. present weather sensor; C. wind direction and wind speed sensor; D. freezing rain sensor; E. DCP antenna and data collection package (equipment case); F. temperature and dew-point sensors; G. rain gauge; and H. cloud height sensor.

inches, a freezing precipitation detector, and a day/night illumination sensor are not shown in figure 2-1. The ASOS will provide station pressure, sea level pressure, pressure altitude (PA), and density altitude (DA) values. Also, additional sensors may be added to the package at a later time.

Operation and Maintenance

Detailed instructions on the operation and maintenance of the sensor package are provided with the manufacturer's User's Manual supplied with each installation.

SHIPBOARD METEOROLOGICAL AND OCEANOGRAPHIC OBSERVING SYSTEM (SMOOS)

The Shipboard Meteorological and Oceanographic Observing System (SMOOS) is an add-on system of supplemental shipboard sensors for the Tactical Environmental Support System (TESS). The "Observer" function in TESS provides the ability to enter local environmental observation data, construct observation report messages, review received messages, and correct erroneous data.

The SMOOS is a suite of environmental sensors that provide continuous automated measurement of meteorological and oceanographic parameters. The sensors automatically send the data to TESS, where it is processed, error-checked, displayed, and distributed. The observer may supplement or override data from the automatic sensors. SMOOS will input data from shipboard wind speed and direction transmitters and from the following sensors:

- Atmospheric pressure sensor
- Temperature/dew-point sensor
- Cloud height detector
- Visibility sensor
- Precipitation sensor
- Seawater temperature sensor

Operation and maintenance manuals are provided with each installation. Now let's briefly discuss the sensors.

Atmospheric Pressure Sensor

The atmospheric pressure sensor assembly is shown in figure 2-2. It is a digital barometer mounted in a weatherproof case with shock protection. It has an accuracy of l 1.O hectopascal (hPa) within a range of 860 to 1,060 hPa.

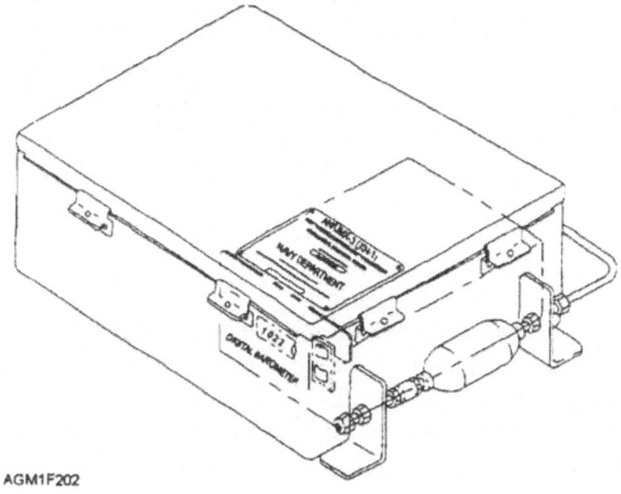

Figure 2-2.—SMOOS atmospheric pressure sensor.

Temperature and Dew-Point Sensor

Figure 2-3 shows the combined temperature/dew-point sensor. The dew-point sensor is a fan-ventilated

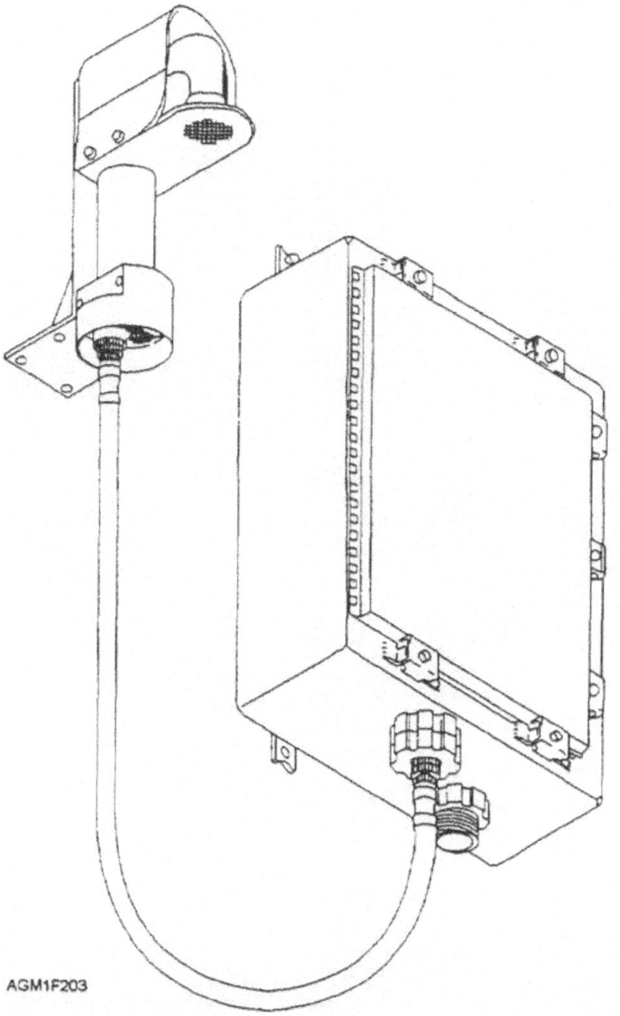

Figure 2-3.—SMOOS temperature/dew-point sensor.

electronic sensor that measures the moisture content of the air. Mounted on the exterior of a ship, the sensor is protected from heat and solar radiation. The temperature measurements are accurate to within ±1.0°F over the range -40°F to 130°F, while the dew-point temperature is accurate to within ±2.0°F over the range -40°F to 100°F.

Cloud-Height Detector

The cloud-height detector is shown in figure 2-4. It is a laser ceilometer that automatically detects cloud layers and provides measurements for up to three cloud-base levels. When the visibility is greater than 3 miles, the detector can measure up to 12,000 feet. It can only measure up to 3,000 feet when the visibility is 1 1/2 to 3 miles or during moderate to heavy rain.

Visibility Sensor

The visibility sensor is shown in figure 2-5. It determines equivalent visibility by measuring forward scattering of an infrared beam by aerosols. The sensor reports equivalent visibility over the range of zero to 10 nautical miles.

Precipitation Sensor

The precipitation sensor (fig. 2-5) uses an infrared beam to detect the droplet size and number of droplets falling through the sensor beam. The sensor reports precipitation rates between 0 and 50 millimeters (mm) per hour and the onset or cessation of precipitation.

Seawater Temperature Sensor

The seawater temperature sensor is usually located near the ship's seawater intake valves, below the water line. It measures the seawater temperature in degrees Fahrenheit andhasarange of 25.0°F to 122.0°F. Due to the sensor being located well below the sea surface, hand-held sea surface temperature measuring instruments may be used in lieu of the SMOOS sensor.

METEOROLOGICAL BUOYS

A third type of automatic weather station is the meteorological buoy. Meteorological buoys may be

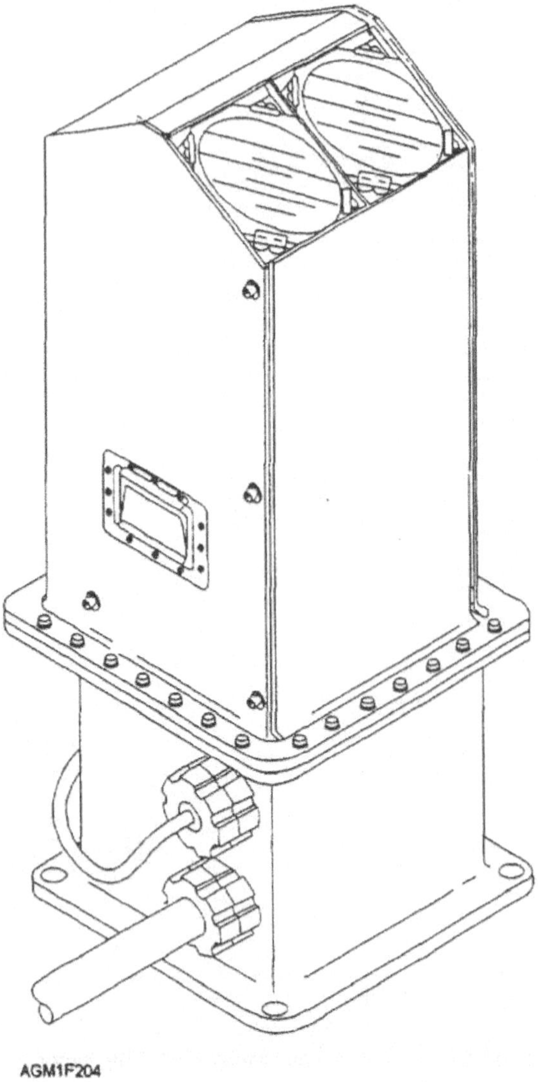

Figure 2-4.—SMOOS cloud-height detector.

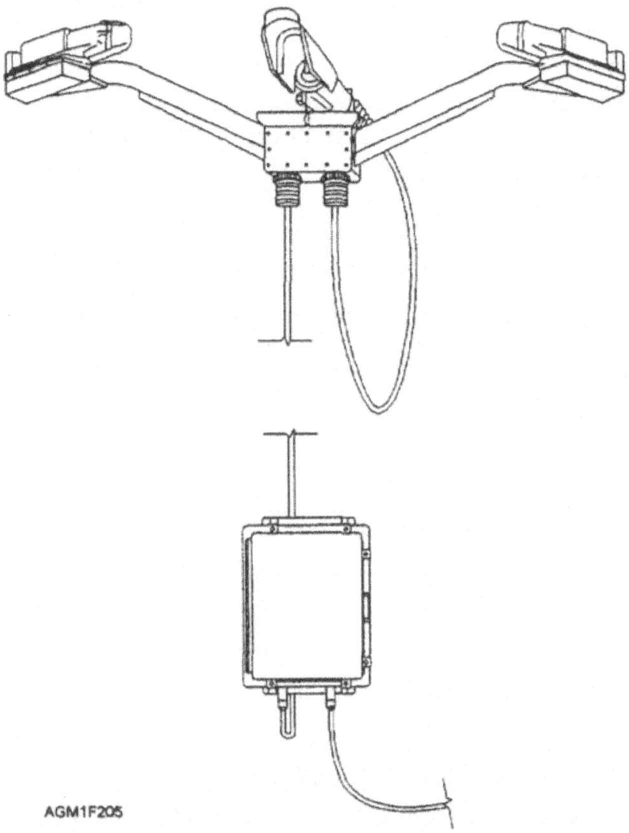

Figure 2-5.—SMOOS visibility sensor (on left and right arms) and SMOOS precipitation sensor (in center).

either moored in a permanent location or drifting buoys. Routinely deployed by aircraft since 1989, drifting meteorological buoys (fig. 2-6) are sonobuoy-size weather stations that provide wind, air pressure, air and sea surface temperature, wave period, and sea temperature/salinity depth profiles to collection points via satellite.

As an observer, you will have little opportunity to see a fully automatic weather station, since no observer support is required. However, you will use observations and data transmitted by these stations. We discuss these products in later modules.

REVIEW QUESTIONS

Q1. How often are ASOS observations updated?

Q2. How does the ASOS system detect precipitation?

Q3. What system does SMOOS interface with?

Q4. The SMOOS can detect clouds at what maximum height?

Q5. How do drifting and moored oceanographic buoys relay information to collection sites?

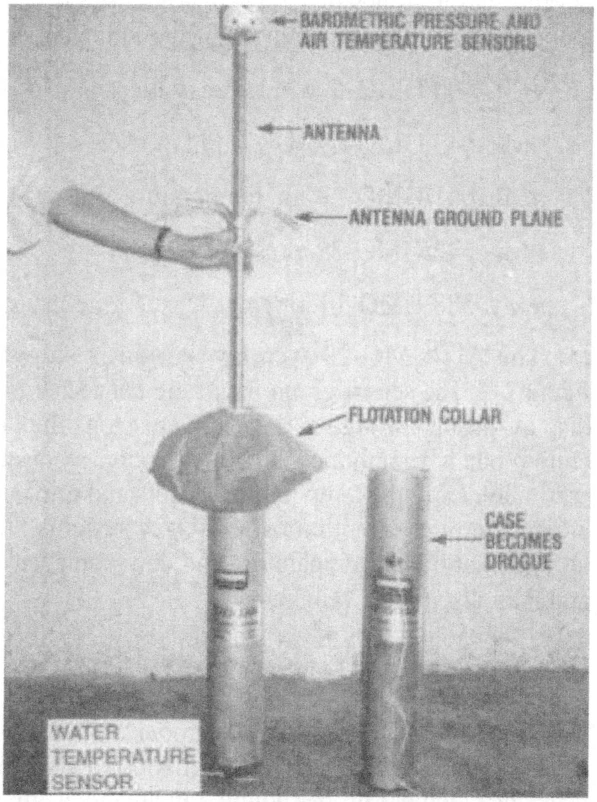

Figure 2-6.—Drifting buoy.

SEMIAUTOMATIC WEATHER STATION

LEARNING OBJECTIVE: Identify the major components and characteristics of the AN/GMQ-29 semiautomatic weather station.

At some observing stations, the primary-use weather observation equipment is still the AN/GMQ-29 automatic weather station. Although called the *Automatic Weather Station* when first introduced, this equipment earned the more popular term *semiautomatic weather station* after truly automatic weather observations were introduced. First operated in 1975, the AN/GMQ-29 uses electronic and electromechanical sensors to measure temperature, pressure, precipitation, and winds. This equipment was installed at all Naval Meteorology and Oceanography detachments, facilities, and centers. The manual that describes operating procedures for this equipment is NAVAIR 50-30 GMQ-29-2, *Handbook of Operation, Service and Overhaul Instructions with Illustrated Parts Breakdown for the Automatic Weather Station (AN/GMQ-29A).* There are two major groups of equipment in the system: the display group and the sensor group. Figure 2-7 shows the display group,

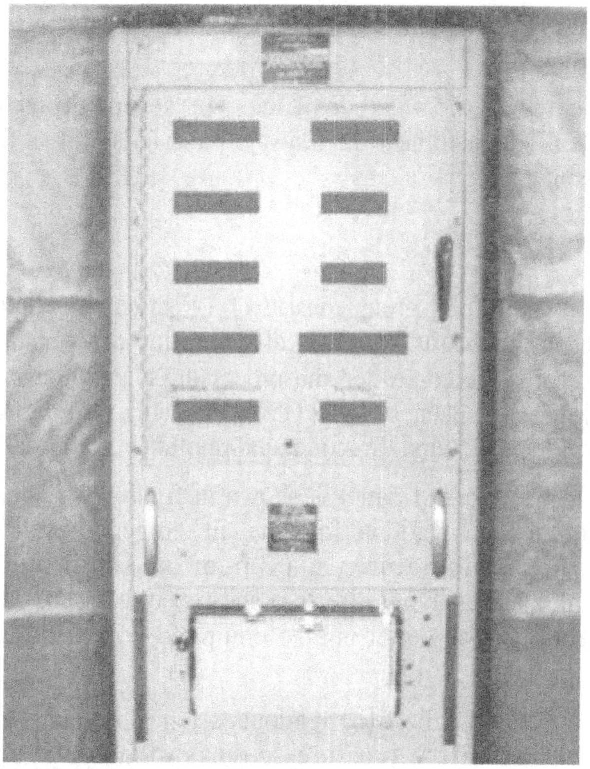

Figure 2-7.—AN/GMQ-29 display group.

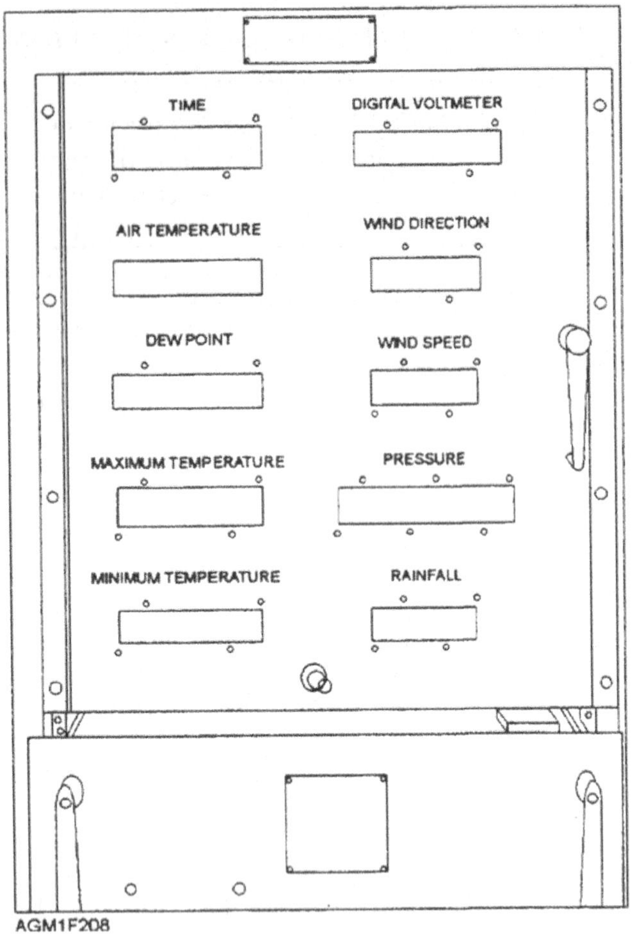

Figure 2-8.—AN/GMQ-29 readout panel.

located in the weather office. The sensor group is usually located near the runway. First, let's look at the display group.

DISPLAY GROW

The display group consists of a light emitting diode (LED) readout panel, a slide-mount drawer that contains the controller and processing electronics and switches, and the RO-447/GMQ-29 analog recorder for wind speed, wind direction, and rainfall.

The readout panel is shown in figure 2-8. It has digital LED readouts for time, air temperature, dew-point temperature, maximum and minimum temperatures, wind direction and speed, pressure, and rainfall. The panel is also equipped with a digital voltmeter.

Some of the LED readouts are not in operation. Because of the pressure sensor system's age and inaccuracy, this feature is no longer used. The wind direction and wind speed display provides a 14- to 16-second averaged wind speed; this 14- to 16-second update is not up to current meteorological standards. The time display, which indicates coordinated universal time (UTC), is not always accurate; therefore, a supplemental clock for time reference may be used.

The air temperature, dew-point temperature, and maximum- and minimum-temperature readouts, provided in degrees Fahrenheit, are reliable and are used as the primary source for these observation readings. Switches in the controller drawer are used to reset the maximum and minimum temperatures.

The accumulated precipitation readout, in inches, tenths, and hundredths of an inch, is also reliable. The precipitation readout is reset to zero by using the reset switch in the front of the controller drawer.

The RO-447/GMQ-29 analog recorder has been disconnected or turned off on most units. The RD-108/UMQ-5 analog recorder is the primary equipment used for evaluation of wind speed and direction. The RD-108/UMQ-5 is discussed later in this chapter.

Now let's look at the AN/GMQ-29 sensor group.

SENSOR GROUP

The AN/GMQ-29 sensor group is located outside the building, usually near the most active runway. Several sensor groups may be used with a single display group. The sensor group consists of the following units:

- ML-400/UMQ-5 wind transmitter
- ML-643/GMQ-29 dew-point sensor
- ML-641/GMQ-29 air temperature sensor
- ML-642/GMQ-29 pressure sensor
- ML-588/GMQ-14 tipping-bucket rain gauge

The AN/GMQ-29 weather station is shown in figure 2-9. The sensor group inputs are connected to the display group through a data transmission line. This equipment is maintained by base ground electronics personnel. Only the temperature sensors and rain gauge are currently used with most GMQ-29 systems. First, let's look at the air temperature and dew-point sensors, and then discuss the rain gauge.

Dew-point and Air Temperature Sensors

The ML-641/GMQ-29 air temperature sensor is shown in figure 2-10. The sensor consists of a resistance element probe mounted in an enclosure. The enclosure shields the sensor from solar radiation and precipitation, yet allows a free flow of air.

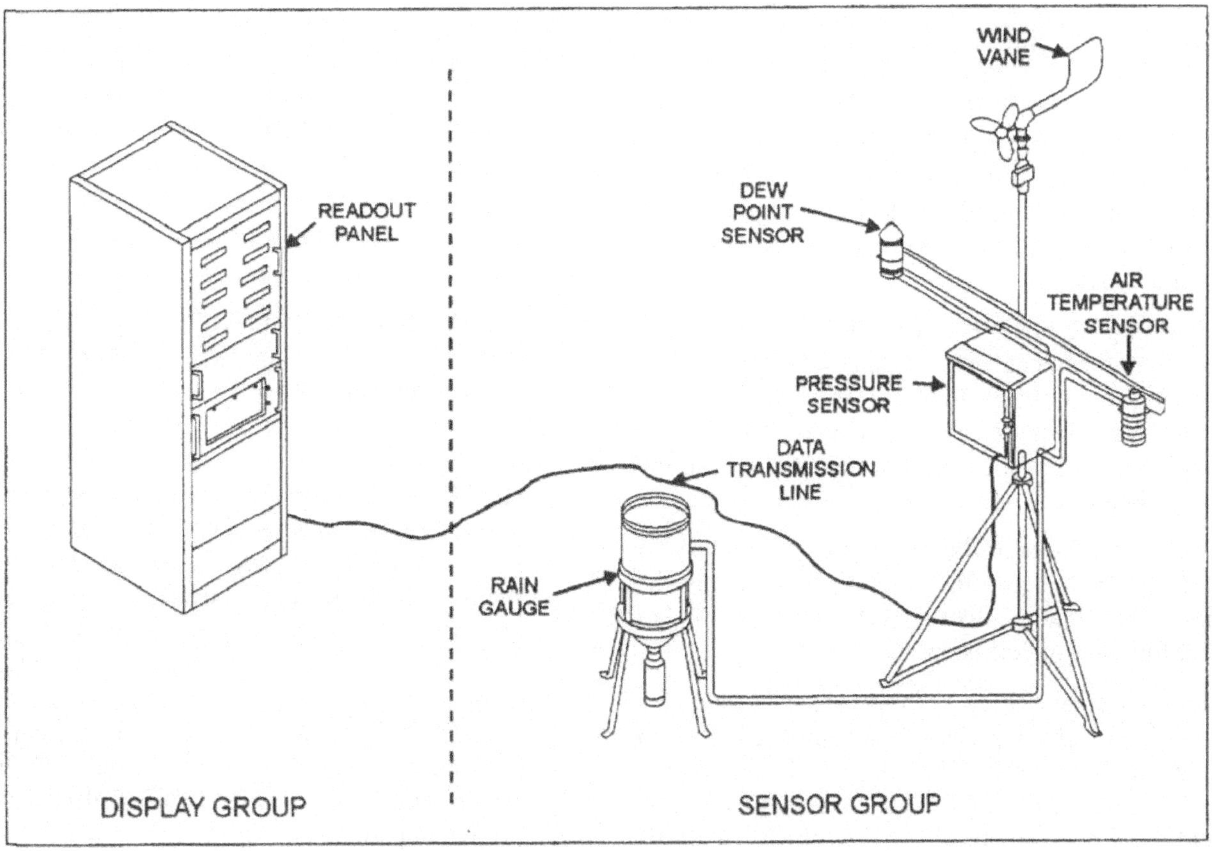

Figure 2-9.—AN/GMQ-29 weather station.

Figure 2-11 shows the ML-643/GMQ-29 dew-point sensor. The sensor consists of a heated resistance probe mounted inside a shield to protect it from rain, wind, and solar radiation.

Tipping-Bucket Rain Gauge

The ML-588/GMQ-14 tipping-bucket rain gauge collects rainwater and funnels it to two small cuplike

Figure 2-10.—ML-641/GMQ-29 air temperature sensor.

Figure 2-11.—ML-643/GMQ-29 dew-point sensor.

"buckets" (see figure 2-12). Each bucket automatically dumps when filled with the equivalent of 0.01 inch of liquid precipitation. The tipping of each bucket activates a mercury switch, which sends a signal to the display controller. The display controller indicates total rainfall on the display and a "step" on the right edge of the RD-108/UMQ-5 analog recorder chart for each "tip" of a bucket.

The tipping buckets normally dump measured water into a second collecting funnel. The funnel has a drain cock at its base, which is used to drain the collected water into a measuring cylinder. The valve is normally left in the open position with the measuring cylinder removed.

You should check the rain gauge frequently to ensure that no foreign objects or dirt is clogging the funnel or the small cuplike buckets. You should also inspect for signs of corrosion.

So far we have discussed only some of the more modern, high-technology electronic equipment used in surface weather observations. Most of the equipment discussed in the following text are not nearly as sophisticated. However, much of the instruments are still retained as back-up equipment because of their simple design and-reliability. The instrument shelter is surface weather observation equipment of this type.

REVIEW QUESTIONS

Q6. What are the two major equipment groups of the AN/GMQ-29?

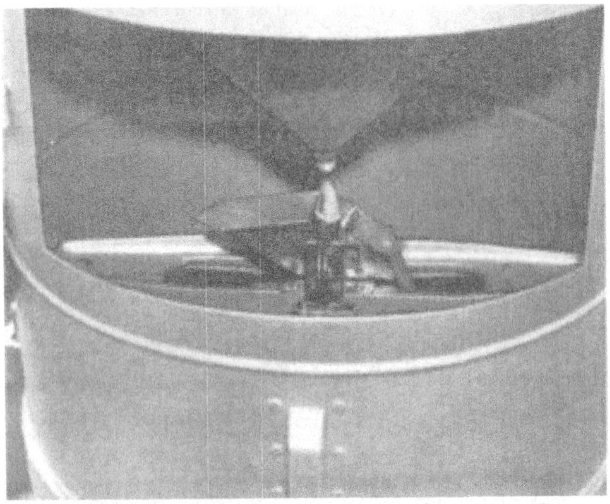

Figure 2-12.—ML-588/GMQ-14 tipping-bucket rain gauge, with the door open showing the twin tipping buckets.

Q7. What information is shown on the AN/GMQ-29 display panel?

Q8. How are sensor group inputs sent to the display group?

INSTRUMENT SHELTER

LEARNING OBJECTIVES: State the purpose of the ML-41 equipment shelter. Describe the routine care required for the shelter.

The ML-41 instrument shelter, shown in figure 2-13, is still in use at many Naval and Marine Corps stations. The shelter is used for the protection and acclimatization of "backup" observation equipment. The shelter is constructed of wood, which is a poor transmitter of heat. It has a louvered door and sides, as well as a double-layered, sloping roof. This type of construction helps keep out water and sunlight, yet allows a free flow of air through the structure. These shelters are always painted white to help reflect sunlight and infrared radiation.

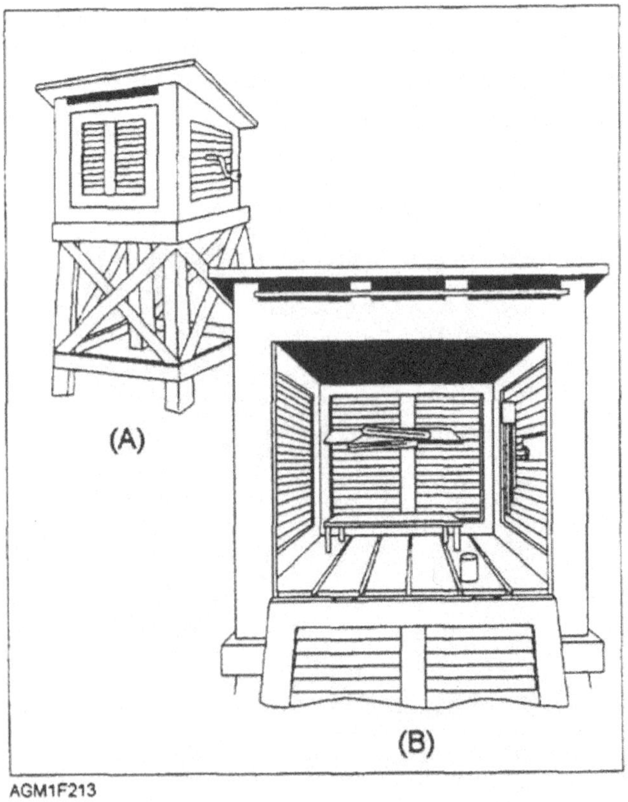

Figure 2-13.—Standard instrument shelter. (A) Construction of support; (B) instrument arrangement inside shelter.

Proper care and maintenance of the shelter is described in NAVAIR 50-30FR-518, *Operation and Maintenance, Standard Air, Maximum and Minimum Thermometers, Townsend Support, Sling and Rotor Psychrometers, and Instrument Shelters.* The routine care of the shelter includes keeping it free of dirt and debris and oiling the door hinges.

Backup equipment usually kept inside the shelter includes the rotor, sling or electric psychrometers, and maximum and minimum thermometers mounted on a Townsend support. When the temperature is above freezing, a bottle of distilled water is also included for the psychrometers.

REVIEW QUESTIONS

Q9. What is the purpose of the ML-41 instrument shelter?

Q10. What tasks are included in the routine care for the instrument shelter?

THERMOMETERS

LEARNING OBJECTIVES: Describe the characteristics of liquid-in-glass thermometers. Explain how to properly read a thermometer. Identify three types of liquid-in-glass thermometers.

Liquid-in-glass thermometers, such as alcohol or mercury thermometers, are found throughout the Navy and Marine Corps in various configurations. Some are simply closed glass tubes mounted on a graduated cardboard, plastic, or metal backing, and others have the graduations etched into the glass. For meteorological and oceanographic readings, calibrated thermometers with the graduations permanently etched into the glass are recommended, since they are considered the most accurate.

In meteorology and oceanography, liquid-in-glass thermometers are used in the rotor psychrometer, the sling psychrometer, electric psychrometers, and as simple thermometers for measuring seawater temperature by the bucket method. The maximum and minimum thermometers found in the instrument shelter are special types of liquid-in-glass thermometers. Both NAVMETOCCOMINST 3141.2 and NAVMETOCCOMINST 3144.1 provide detailed instructions for use and care of the different types of thermometers.

READING A THERMOMETER

Accuracy is the goal when making any temperature reading. To obtain an accurate reading, you must read the thermometer properly. Be sure the thermometer is clean, and allow the instrument to acclimatize prior to reading. Keep the following points in mind when reading temperatures:

- The temperature indicated by a liquid-in-glass thermometer should be read to the nearest 1/10 of a degree at the top of the meniscus, with the eye looking straight at the fluid level (fig. 2-14) to avoid parallax error.

- Avoid obvious scale errors in reading. Errors of 5 or 10 degrees are more common than errors of 1 or 2 degrees.

- Do not allow any moisture to accumulate on a dry-bulb thermometer. Protect the thermometer from spray and precipitation.

- Do not hold a thermometer too close to your body or near a heat source, such as a light bulb or direct sunshine. Always stand downwind of a thermometer.

- When the air temperature is below freezing, wait at least 15 minutes after wetting the wet-bulb wick before using a wet-bulb thermometer. This allows the water to freeze on the wick to prevent false high readings.

ROUTINE CARE

Routine care of thermometers is limited to keeping the glass clean. When using a thermometer to measure wet-bulb temperature, keep the cotton wick clean. A weekly wipe down of glass thermometers with freshwater is usually sufficient. At sea, especially under windy conditions, minute salt crystals tend to accumulate on the thermometers. These crystals readily

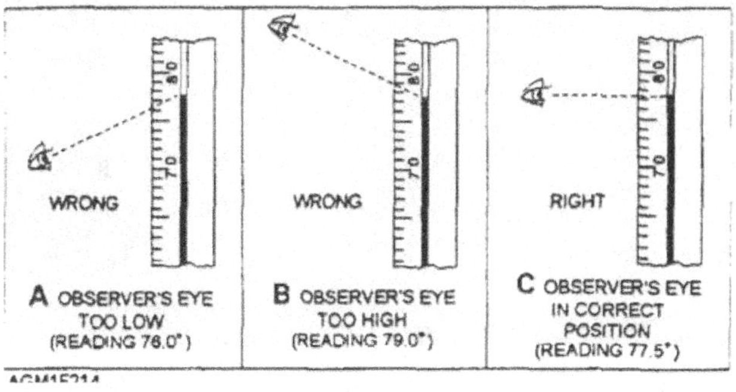

Figure 2-14.—Reading a thermometer to avoid parallax error.

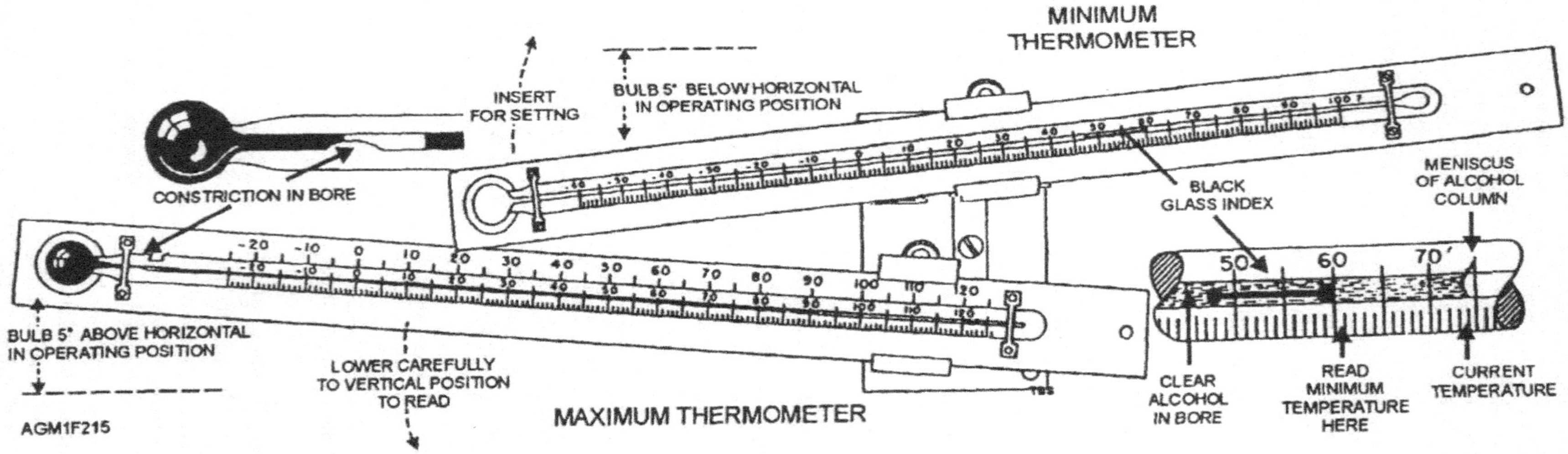

Figure 2-15.—Maximum and minimum thermometers mounted on a Townsend support.

attract moisture, which evaporates when the thermometer is ventilated. Evaporation will yield false low readings. A daily cleaning of the thermometers may be necessary in this situation.

Changing the wick on wet-bulb thermometers should be done weekly aboard ship and monthly ashore. The procedure is detailed in NAVAIR 50-30FR-518.

ALCOHOL THERMOMETERS

Alcohol thermometers may be used to measure temperatures from -115°C (freezing point of alcohol) to 785°C (boiling point of alcohol). The standard thermometer for environmental measurements need only cover the range -20°F to 120°F, or about -30°C to 50°C.

Because alcohol is a volatile fluid, the column in the thermometer frequently becomes separated by mechanical shock. The column may be reunited by dropping the thermometer on a wood surface that is covered with several sheets of paper. Hold the thermometer in a vertical position, bulb end down, 4 to 6 inches over the wood surface covered with paper, and then drop. After the thermometer bulb strikes the paper, catch the thermometer with your other hand to prevent breakage. Small amounts of alcohol may cling to the portion of the capillary tube above the alcohol column. Heating the high-temperature end of the thermometer under an incandescent light may force the alcohol back to the column. Allowing the thermometer to stand in a vertical position overnight will also allow the alcohol to drain down to the column. Do not use a thermometer with a separated fluid column; it will give you inaccurate readings.

MERCURY THERMOMETERS

Recently, several Naval Headquarters have recommended that alcohol thermometers be used in place of mercury thermometers when possible. NAVSEAINST 5100.3, *Mercury, Mercury Compounds, and Components Containing Mercury or Mercury Compounds; Control of,* identifies mercury as a toxic substance that requires special handling and control. Overboard discharge of any amount is prohibited. This instruction also defines mercury-spill decontamination requirements and mercury-handling requirements.

WARNING

Liquid or vaporous mercury is a hazardous material. It is toxic to humans and most forms of marine life, and is highly corrosive to electronic components.

The use of mercury thermometers in sling psychrometers is strongly discouraged, since users of sling psychrometers sometimes strike objects during the spinning operation, breaking the glass and releasing mercury. The release of even a small amount of mercury from a broken thermometer is a "mercury spill," which must be handled in accordance with NAVSEAINST 5100.3. Unbroken mercury thermometers, either in use or in storage, are classified "functional mercury" and are subject to shipboard mercury inventory reporting as directed by NAVSEAINST 5100.3.

Separated mercury columns in thermometers should be rejoined by slowly heating the thermometer bulb under an incandescent lamp. Withdraw the thermometer from the heat source as the mercury approaches the top of the thermometer. Then, carefully control the heating so that the mercury rises slowly in the column and just reaches the end of the thermometer. Overheating at this point will cause the mercury to expand until it ruptures the thermometer, resulting in a spill. Allow the thermometer to cool; the columns of mercury should rejoin as the mercury recedes.

MAXIMUM AND MINIMUM THERMOMETERS

Maximum and minimum thermometers are alcohol-filled thermometers mounted on a Townsend support inside an instrument shelter. These thermometers are strictly backup equipment for the maximum and minimum temperature function of automatic and semiautomatic observation systems. Proper use and care are detailed in NA 50-3OFR-518. Figure 2-15 shows the instruments mounted on the Townsend support, and figure 2-16 shows the detail of the Townsend support.

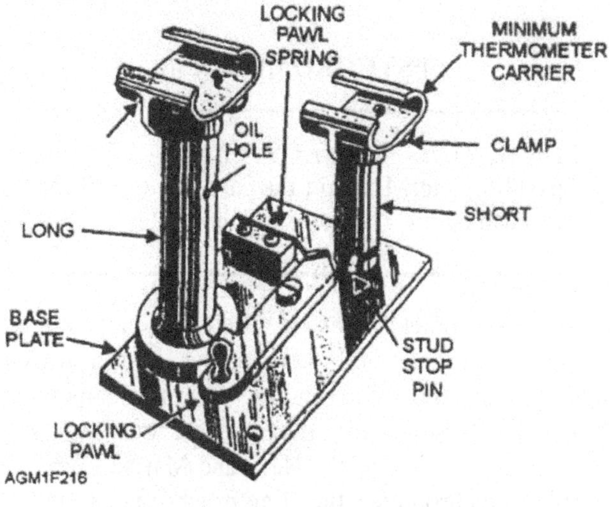

Figure 2-16.—ML-54 Townsend support details.

Figure 2-17.—A typical electric psychrometer.

We have just covered several types of simple thermometers. Most frequently, the thermometers used by Aerographer's Mates are found in psychrometers.

REVIEW QUESTIONS

Q11. What are the two primary liquids found in thermometers?

Q12. Where can you find detailed instructions for use and care of the different types of thermometers?

Q13. How should the temperature be read from a liquid-in-glass thermometer?

Q14. How often should the wick on a wet-bulb thermometer be changed aboard ship?

Q15. What instruction outlines procedures to be followed in the event of a mercury spill?

PSYCHROMETERS

LEARNING OBJECTIVES: Define psychrometer. Explain the care and use of the electric, sling, and rotor psychrometers.

A psychrometer is any device that contains both a dry-bulb and a wet-bulb thermometer used to measure ambient air temperature and wet-bulb air temperature. Currently there are three different types of psychrometers used by the Navy and Marine Corps: the electric psychrometer, the sling psychrometer, and the rotor psychrometer.

ELECTRIC PSYCHROMETER

Electric psychrometers are used aboard some ships as the primary air temperature/wet-bulb temperature-measuring device. They are kept on hand at all shore observation sites as a backup for automatic air temperature and dew-point sensors. There are several slightly different models of electric psychrometers in use: the ML-45O/UM and the ML-450(A)/UM through the ML-45O(D)/UM models, and the "Type III" electric psychrometer. Figure 2-17 shows a typical electric psychrometer. All models contain a dry- and a wet-bulb thermometer, a small, battery-operated electric fan, several batteries, thermometer illumination lights, and a small distilled-water bottle.

Electric psychrometers should be allowed to acclimatize to the ambient conditions if kept inside an office space, especially during extremely cold or hot weather. When the outside air temperature is cooler than 50°F or warmer than 85°F, allow the psychrometer to sit in a sheltered outside area 5 to 10 minutes before using. When the air temperature is below freezing, wet the wick and allow the psychrometer to sit 15 minutes before using, to allow the wick to freeze.

To avoid acclimatization time, many shipboard observers store the psychrometer outside in a sheltered location. You must take care to ensure that the psychrometer is held securely in the storage location to prevent breakage from the ship's motions. Ashore, the electric psychrometers are usually stored outside in the instrument shelter.

To use the electric psychrometer properly, wet the muslin wick with a few drops of distilled water from the bottle. Then, stand on the windward side of the ship or building with the air intake pointed toward the wind, turn the fan motor on, and hold the psychrometer at arm's length. The scale lights should be turned on only when you read the temperatures. After 60 seconds, read the wet-bulb temperature at lo-second intervals until the temperature reads the same for two consecutive readings. Record both the wet- and dry-bulb temperatures to the nearest 1/10 degree. Then, turn the fan motor off.

When issued, each electric psychrometer is supplied with a use, care, and maintenance handbook. Routine care and maintenance of the electric psychrometer include changing the wet-bulb wick, keeping the psychrometer clean and dry, and changing the batteries and light bulbs. Cleaning and lubrication of the electric motor are also required. For more

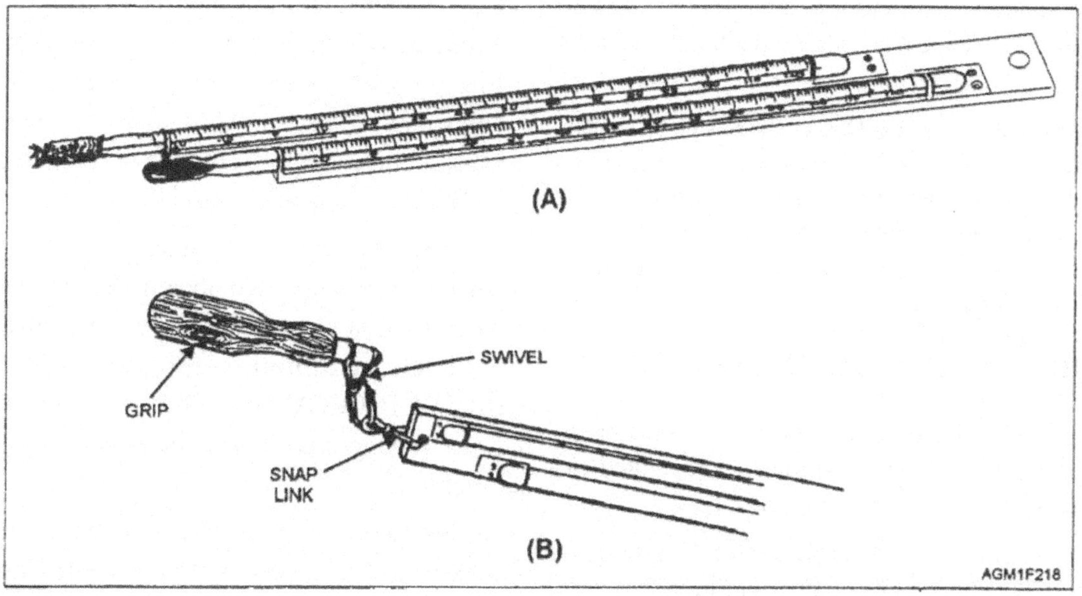

Figure 2-18.—Psychrometer: (A) Standard sling psychrometer; (B) with sling attached.

detailed instructions, refer to the handbook supplied with the equipment.

SLING AND ROTOR PSYCHROMETERS

Sling psychrometers (fig. 2-18) are used to backup the electric psychrometers, both ashore and aboard ship. Because of the widespread use of electric psychrometers, sling psychrometers are rarely used. Rotor psychrometers (fig. 2-19), mounted inside the standard instrument shelter ashore, are also rarely used as backups. Use, care, and maintenance instructions are found in NA 50-30FR-518.

REVIEW QUESTIONS

Q16. Name the three types of psychrometers used by the Navy?

Q17. Explain the proper use of an electric psychrometer.

Q18. What tasks are included in the routine care of electric psychrometers?

PRESSURE INSTRUMENTS

LEARNING OBJECTIVE: Describe the operation of the precision aneroid barometer and the marine barograph.

Pressure-measuring instruments are critical equipment for surface weather observations. The ML-448/UM precision aneroid barometer is still the primary

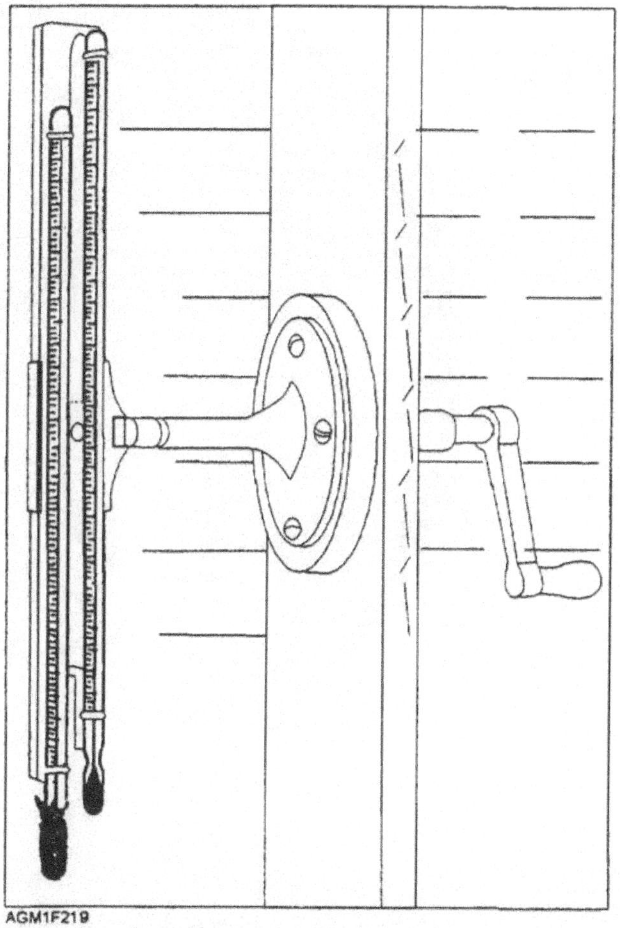

Figure 2-19.—Rotor psychrometer.

pressure-measuring instrument at some stations. Many shore stations also have a Navy digital altimeter setting indicator (NDASI or DASI).

The Automated Surface Observing System (ASOS) equipment also provides pressure readings and becomes the primary pressure-measuring instrument as it is installed. Then, the ML-448/UM aneroid is retained as a backup instrument. Use of the NDASI may be discontinued.

PRECISION ANEROID BAROMETER

The ML-448/UM precision aneroid barometer is shown in figure 2-20. The barometer is mounted in either a brass case or a black, hard, plastic case.

The precision aneroid barometer is designed to accurately indicate atmospheric pressure in inches and hectopascals (hPa). The pressure element is a Sylphon cell, which consists of a sealed, bellows-shaped canister that expands and contracts with changes in air pressure. Gears and linkage arms transfer changes in size of the Sylphon cell into indicator movement. The gears and linkage also correct the movement for changes in temperature. The ML-448/UM has a range from 910 hPa (26.9 inches) to 1,060 hPa (31.3 inches), with an acceptable accuracy of ±1 .0) hPa.

Precision aneroid barometers must be calibrated twice a year when used aboard ship, and once a year when used at shore stations. Basic guidance for the barometer calibration program is provided in NAVMETOCCOMINST 13950.3, *Naval Meteorology and Oceanography Command Barometer Calibration Program.*

Instructions for use of aneroid barometers during a surface aviation observation are provided in both NAVMETOCCOMINST 3141.2 and NAVMETOC-COMINST 3144.1. To obtain a proper pressure reading using any aneroid barometer, use the following steps:

Figure 2-20.—ML-448/UM precision aneroid barometer.

1. Tap the face of the barometer lightly with your finger to reduce the effects of friction.

2. Look straight at the indicator to avoid parallax error. On the ML-448/UM, align the indicator arm directly over the reflection on the mirrored surface of the scale.

3. Read the scale to the nearest 0.005 inch or 0.1 hPa to obtain barometric pressure.

4. Apply any posted correction.

5. Add the appropriate removal correction.

NAVY DIGITAL ALTIMETER SETTING INDICATOR (DASI)

The Navy Digital Altimeter Setting Indicator (DASI), shown in figure 2-21, is located at most Naval meteorology and oceanography detachments and Marine Corps weather stations. The DASI has an electronic pressure-sensing element, and the altimeter values are displayed on a digital LED readout panel. The panels are located both in the observer's spaces and in the control tower. NAVELEX EM-450-AA-OMI-010-DASI, *Operation and Maintenance Instructions, Digital Altimeter Setting Indicator (DASH),* is the operator's manual for this system. It may be held by the air traffic controllers.

The DASI should not be used as a pressure-measuring instrument. Use of the DASI should be discontinued when the altimeter setting indicated by the instrument exceeds the altimeter setting computed from readings from the ML-448/UM by more than ±0.02 inch.

Unless a Space and Naval Warfare Systems Command (SPAWARSYSCOM) certified calibration procedure for the DASI pressure sensor is developed in the near future, the DASI is expected to be taken out of service when ASOS is completely installed.

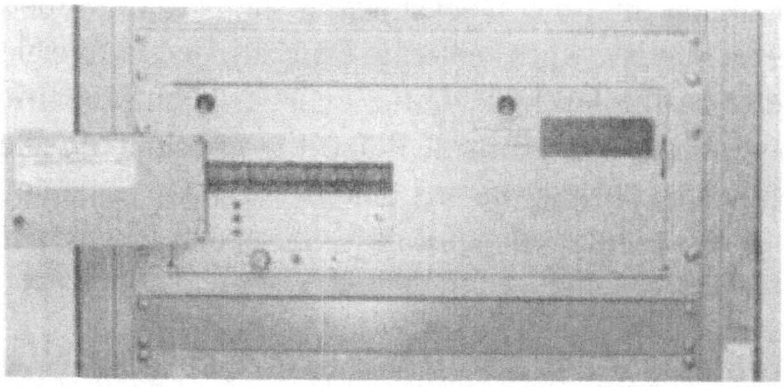

Figure 2-21.—Navy digital altimeter setting indicator (DASI).

MARINE BAROGRAPH

Barographs are instruments used to record or graph atmospheric pressure over a period of time. The marine barograph is the only type of barograph found throughout the Navy and Marine Corps. It is suitable for use aboard ship and at shore stations. The marine barograph has not been assigned a military reference designator.

Barographs are primarily used to determine 3-hour pressure tendencies for surface aviation weather observations. Appendix IV shows general barograph trace patterns used to determine the nine reportable pressure tendencies. Stations not equipped with a barograph may report a simplified tendency based on recorded pressures over a 3-hour period.

The marine barograph (fig. 2-22) is a precision instrument that measures air pressure either in the immediate vicinity or at a remote point outside the ship. The instrument is highly sensitive to changes in pressure and has linkages that accurately compensate for changes in temperature. The instrument maintains its precision aboard ship by the use of an adjustable, grease-filled, dampening cylinder that cancels out shipboard vibrations and the pitch or roll of the ship. Since the air pressure inside a ship may be different from the outside air pressure (especially when material condition Yoke or Zebra is set), shipboard marine barographs should be connected to the outside air. Shipboard marine barographs are connected to the exterior of the ship through use of a flexible rubber hose connected to a fitting on the barograph case.

The marine barograph is equipped with a "magnified scale" for recording pressure. The recording chart scale ranges from 965 to 1,050 hPa, but the instrument has a total usable range from 915 to 1,085 hPa. To prevent the pen arm from recording off the edge of the chart, you should adjust the pen arm upward on the scale by 40 hPa during periods of extremely low pressure, or downward on the scale by 40 hPa during periods of extremely high pressure. The adjustment is made by turning the adjustment knob. Because of the magnified scale, high sensitivity, and accurate temperature compensation, the marine barograph is sometimes called a "microbarograph."

Operation

NA 50-30BIC-1, the *Handbook of Operation, Service, and Overhaul Instructions for the Marine Barograph,* provides detailed instructions for the

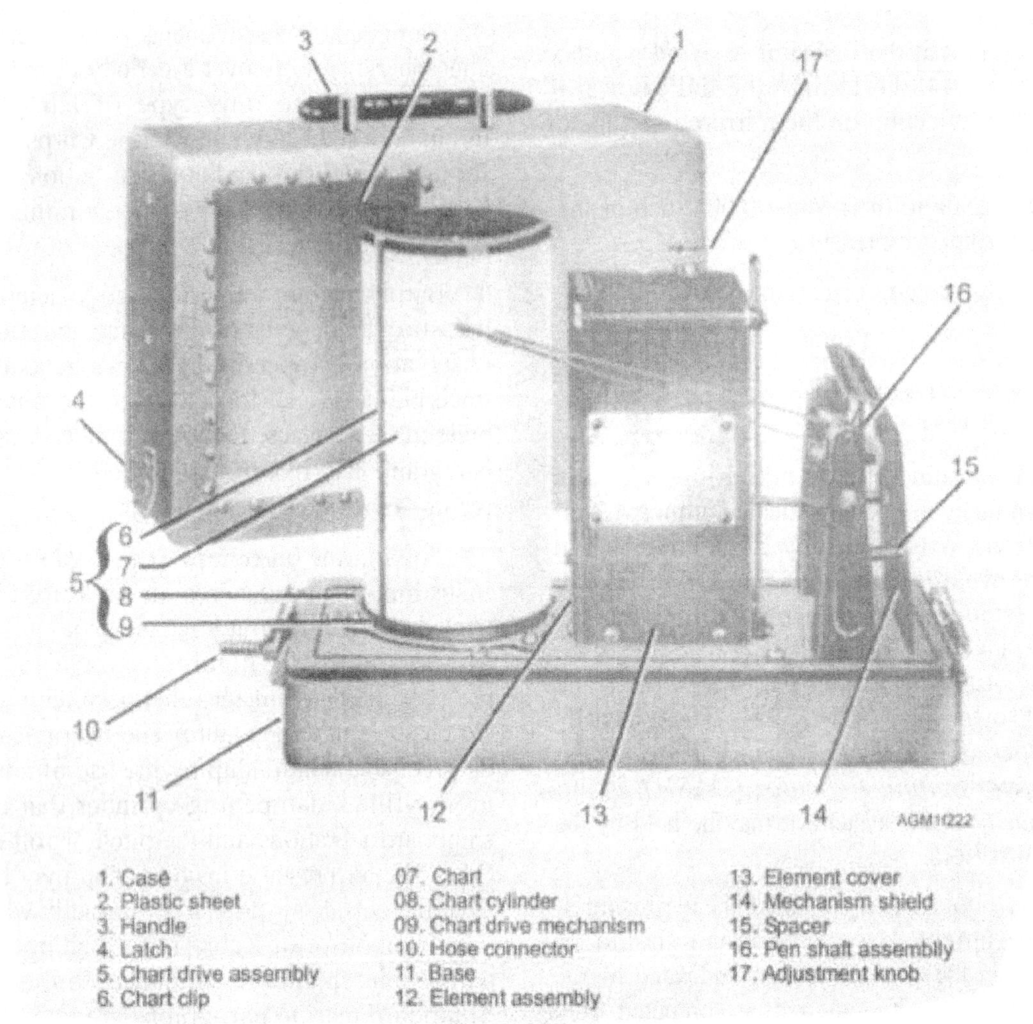

Figure 2-22.—Marine barograph.

proper use and maintenance of the equipment. NAVMETOCCOMINST 3144.1 also provides guidance on replacing charts, rewinding the clock, conducting time and pressure checks, and making other adjustments.

REPLACING THE CHART.—Charts (fig. 2-23) should be replaced every 4 days at 1200 UTC (and on the first day of each month). Fill in as much of the chart identification data as possible before placing the chart on the barograph. All time should be set and entered in UTC. Add the station elevation to the nearest foot if a space is not already provided on the chart. Lift the pen arm from the chart and remove the chart cylinder from the barograph. Remove the old chart, and then place the new chart on the cylinder. The trailing edge of the pressure/time graph should overlap the chart Identification-Data section, and the edge of the paper should be aligned under the metal spring clip. Complete the entries on the old chart for ship's position, date, and (chart) off pressure, date, and time. Rewind the clock mechanism.

REWINDING THE CLOCK.—Although the clock has an 8-day mechanism, it should be rewound every 4th day, when replacing the chart. Seven to eight pulls on the lever inside the cylinder is sufficient to rewind the mechanism. Place the cylinder on the barograph; align the correct time line with the pen point; and lower the pen point onto the chart. Check the pressure.

REINKING THE PEN.—A glass rod is used to transfer a drop of purple recording ink to the pen tip. Very little ink is needed, and the pen is considered adequately inked if the pen tip cavity is half full. The

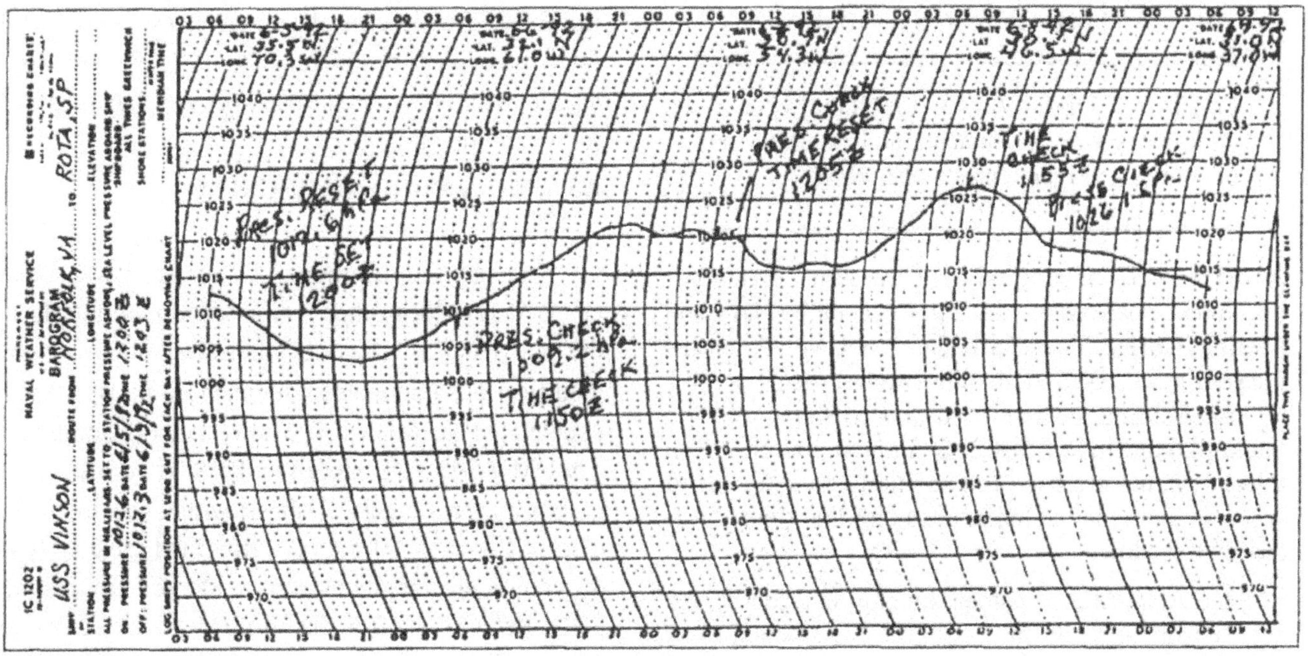

AGM1F223

Figure 2-23.—Marine barograph chart.

ink is hygroscopic and will absorb moisture from the air. When the ink becomes too diluted from absorbed moisture, the trace will appear "washed-out." When this happens, remove the pen tip, rinse it in freshwater, blow it dry, replace it on the pen arm, and reink the pen. A wide pen trace indicates a dirty or damaged pen tip. Clean and reink or replace the tip as appropriate.

Accuracy Checks and Adjustments

Time and pressure checks are completed at each synoptic and intermediate synoptic hour, after reading the 3-hour pressure tendency. At synoptic observation times, touch the pen arm lightly to make a "tic" mark in the pressure trace for time checks. Time adjustments are required when the time is in error by 15 minutes or more, and pressure adjustments are recommended when the pressure is off by 1.5 hPa or more.

TIME ADJUSTMENTS.—Time adjustments are made by rotating the chart cylinder until the pen is marking at the correct time. A time correction entry should be made on the chart, consisting of an arrow pointing to the corrected time, the remark "Time Reset," and the correct UTC time, such as 1155Z. Remarks may be made directly on the mounted chart or entered on a note pad or clipboard near the barograph. If the remarks aren't made directly on the chart, they should be transferred to the chart after it is removed.

PRESSURE ADJUSTMENTS.—Pressure adjustments are made by turning the adjustment knob. Draw an arrow to the correct pressure and enter the remark "Pres. Reset," along with the UTC time, on the chart.

REVIEW QUESTIONS

Q19. *How often should the aneroid barometer be calibrated?*

Q20. *What procedure should you use to obtain a reading from an aneroid barometer?*

Q21. *In what situation should the altimeter setting of the DASI not be used?*

Q22. *What device enables the marine barograph to maintain its precision aboard ship?*

Q23. *When should the chart on the marine barograph be replaced, and what time should it be coordinated with?*

Q24. *Rewinding the clock of the marine barograph requires approximately how many pulls if completed after every fourth day.*

Q25. *What criteria is used for time and pressure adjustments on the marine barograph?*

ANEMOMETER SYSTEMS

LEARNING OBJECTIVES: List three types of wind-measuring instruments and describe the operation and maintenance of each type.

Three types of certified wind-measuring systems or anemometers are widely used throughout the Navy and Marine Corps. These are the AN/UMQ-5 wind-measuring set, used at shore locations; the Type B-3 wind-measuring system, used aboard ships; and the backup, hand-held AN/UMQ-3 anemometer. All of these systems are used to accurately measure wind speed, in knots, and to indicate the direction from which the wind is blowing.

The AN/UMQ-5 system is being replaced by the new sensors and transmitters in the ASOS equipment package. The shipboard Type B-3 system sensors and transmitters will continue to be used although the information will be displayed on the SMOOS display terminal. The reliable AN/PMQ-3 hand-held anemometer is expected to be kept as a backup system for a very long time.

AN/UMQ-5 WIND-MEASURING SYSTEM

The basic equipment of the AN/UMQ-5 has changed very little since its introduction to the fleet in the late 1940's. However, there have been several modifications that added equipment to the system and updated the electronics. The newest modification is the AN/UMQ-5(D), introduced in 1959. A typical system is shown in figure 2-24. A single system may use one or more ML-400/UMQ-5 transmitters, and drive up to six different indicators or recorders. The primary recorder is the RD-108/UMQ-5. ID-300/UMQ-5 indicators and ID-586/UMQ-5 indicators are frequently used with the system.

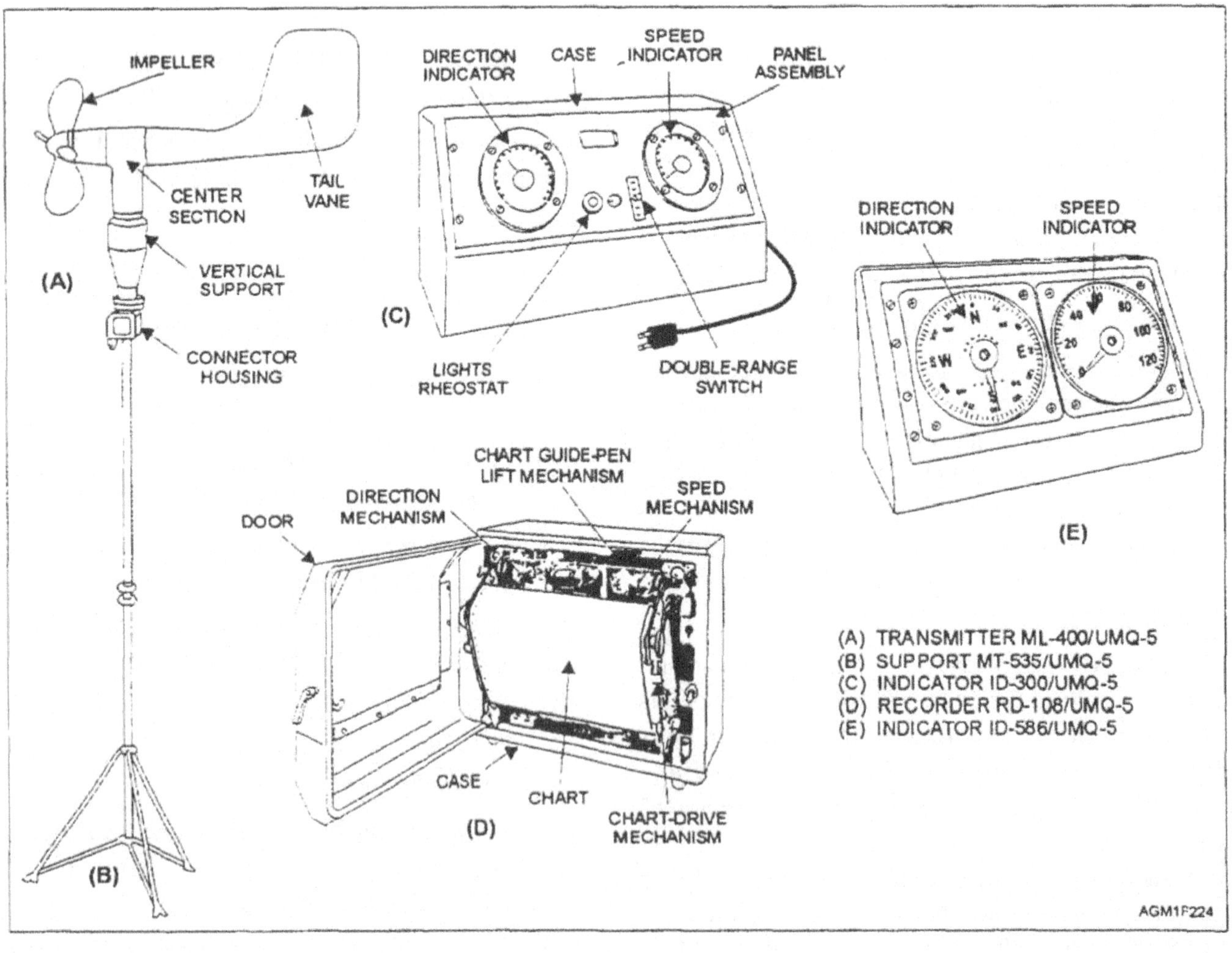

Figure 2-24.—AN/UMQ-5 wind-measuring set.

NOTE: The ID-2447/U indicator (fig. 2-25) is a recent addition to the UMQ-5 system. The indicator is found in weather offices and in Marine Corps radar support spaces, with the primary readout in the Air Traffic Control Center. The indicator provides average, minimum, and maximum wind speeds. For more information on the ID-2447/U indicator, see the operating and maintenance instructions.

Typically, the wind speed and direction indicators and wind speed recorders are installed inside the weather office, while the transmitter is installed on an MT-535/UMQ-5 support mast near the runway or on a rooftop.

Operation and Maintenance

Operation and maintenance instructions for the different components of the AN/UMQ-5 wind-measuring set are detailed in several manuals. Consult the manual for the model installed in your office. NA 50-30FR-525, the *Handbook of Operation and Maintenance Instructions for Wind Measuring Sets AN/UMQ-5C and AN/UMQ-5D,* covers the newer models, including instructions for the later model RD-108(B) chart recorders. NA 50-30FR-501 covers the AN/UMQ-5(A) system, and NA 50-30FR-512 covers the AN/UMQ-5(B) system. Instructions for the older model recorders (RD-108/UMQ-5 and RD-108(A)/UMQ-5) are covered in NA 50-30FR-505, the *Handbookof Operation and Service Instructionsfor the Wind Direction and Speed Recorder RD-108/UMQ-5.*

TRANSMITTERS AND INDICATORS.—Very little operator attention is required for this equipment. The operator's handbook recommends a daily visual check of the transmitters and indicators, and a weekly visual inspection of the transmitter. Jerky movement of the impeller or failure of the wind vane to follow light winds indicates an overhaul is due. Finally, a quarterly friction check is recommended. The transmitter is removed and bench tested with penny and half-dollar weights to check for excessive friction.

Figure 2-25.—ID-2447/U wind speed and direction indicator.

WIND RECORDER.—The analog wind recorder requires somewhat more attention. The recorder pens mark a continuous output of wind direction and wind speed on a continuous, two-column chart (fig. 2-26). The wind direction column, on the left side of the chart, has vertical lines drawn for every 10 degrees; these are labeled every 90 degrees, along with the cardinal compass points. The chart covers 540 degrees to allow for shifting winds. The wind speed column, on the right side, has vertical lines drawn every 2 knots; these are labeled every 20 knots, from 0 to 120 knots. Curving lines across the chart mark the time every 10 minutes and are labeled for every hour, in the center divider strip.

Many recorders have been modified to use preinked pen cartridges, which are simply replaced as a unit when the ink runs dry. Some units still use the older model pens and inkwells. The older pens must be checked frequently; a line that skips on the chart usually indicates a dirty pen.

A standard, 100-foot, wind speed and direction recorder chart, when operated at the standard speed of 3 inches per hour, will run for 15 to 16 days before it requires changing. Instructions on how to replace the chart in the recorder are contained in the handbook on operation and maintenance.

Recorder Charts

The analog recorder charts should be carefully preserved and treated as a permanent official record. Each recorder chart roll should be identified at the start and end of the roll, with the station name and the date and time the recording began and ended. The chart is adjusted so the times marked on the chart correspond to UTC. The chart should be changed at 0000 UTC on the first day of each month and at intermediate times to prevent the loss of data. See NAVMETOCCOMINST 3141.2 for more information.

TIME CHECKS.—Time checks are made on the recorder chart by drawing a short line on the recorder chart where the pen is marking data, and entering the date and time to the nearest minute. At the minimum, time checks are required as follows:

- At the beginning and end of each chart roll

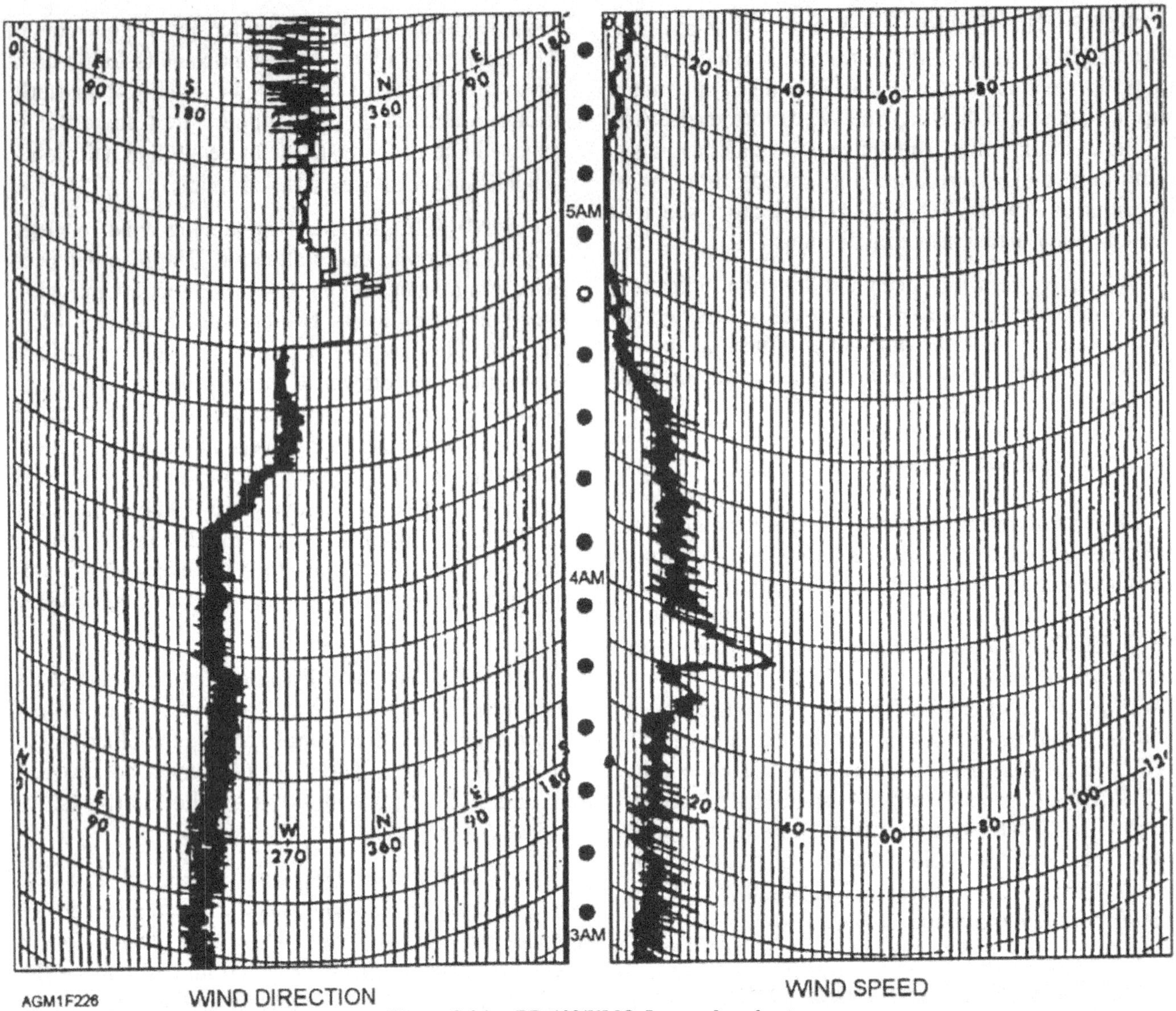

Figure 2-26.—RD-108/UMQ-5 recorder chart

- Near the time of each synoptic hour (0000Z, 06002, 12002, and 1800Z)
- When notified of an aircraft mishap
- For each disruption or discontinuity in the trace; for instance, when returning the equipment to service after a power outage or maintenance downtime
- At the time of the first observation of the day when the station does not operate for 24 hours a day

TIME ADJUSTMENTS.—Time adjustments must be made whenever the time error is more than 5 minutes. Draw an arrow to the adjusted time, and then enter "Time Corrected," followed by the date and UTC time to the nearest minute. If an adjustment is necessary because of a power outage or maintenance, enter the appropriate reason for the gap in the record.

CHART FEED-RATE ADJUSTMENTS.—Whenever the chart feed-rate is changed, enter a time check and an appropriate note, such as "Begin 6 inches per hour" or "Begin 3 inches per hour." Chart speed rates of 1 1/2 inches per hour, 3 inches per hour, and 6 inches per hour are available by manually changing the drive gears. At the standard 3-inch-per-hour rate, a chart will last 15 days in continuous operation.

TYPE B-3 SHIPBOARD ANEMOMETER

The Type B-3 wind-measuring set falls under the cognizance of NAVSEA. It is serviced by the shipboard IC electricians, who maintain the technical manuals on the system. The only responsibility the Aerographer's Mate has in connection with this equipment is to read the indicators and to verify that the indicator is operating properly. If the indicated relative speed or direction appears to be in error, the problem is reported to the IC shop.

All ships have a minimum of two wind transmitters, located on the port and starboard yardarms. The switch used to control the transmitter in use is usually located on the bridge on smaller ships, but may be located in the geophysics office aboard aircraft carriers. Normally, the windward-side transmitter is used. The transmitters are very similar to the ML-400/UMQ-5 transmitters.

The transmitter selected reports wind direction relative to the bow of the ship and wind speed relative to the ship's motion. These winds are called relative winds. Several Type B-3 relative wind indicators (fig. 2-27) are installed throughout each ship in various locations. To convert relative wind to true wind, you must also know the ship's true heading (available from the magnetic compass [fig. 2-28] or gyroscope repeater [fig. 2-29]), as well as the ship's speed (available from the underwater-speed log indicator [fig. 2-30]). If a magnetic compass is used for the ship's course, to obtain a true heading, the magnetic declination must be added to the compass heading before that number is used in a true-wind computation. Gyro repeaters are preferable for conversion, since these readings are true headings. Both gyro repeaters and underwater-speed log indicators are usually located in the geophysics office, as well as on the bridge and at other strategic locations throughout the ship.

Figure 2-28—Shipboard magnetic compass.

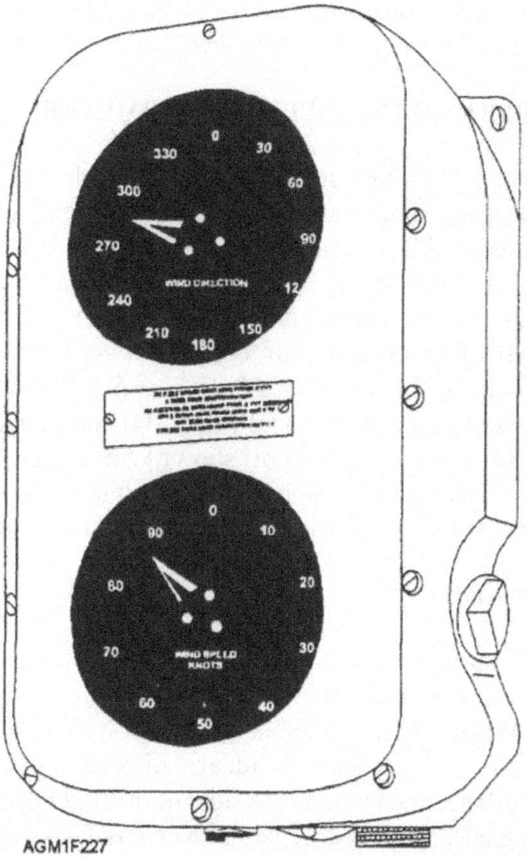

Figure 2-27.—Type B-3 shipboard relative-wind direction and relative-wind speed indicator.

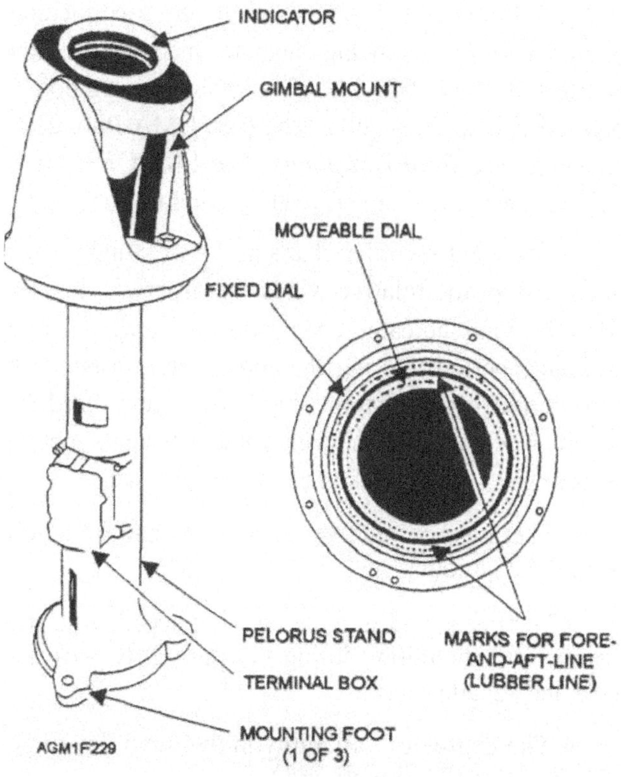

Figure 2-29.—Shipboard pelorus with a gyroscope repeater.

Figure 2-30.—Shipboard underwater-speed log indicator.

NOTE: There are two manual methods used to convert relative wind speed and direction to true wind speed and direction. The most frequently used method is the CP-264/U true-wind-computer method (see later section on the CP-264/U). The maneuvering board method is also frequently used (see DMA Pub. 217, *Maneuvering Board Manual).* The SMOOS system will automatically convert relative wind to true wind.

Since wind recorder charts are not routinely used aboard ship, the relative winds are observed directly from the wind indicator, and winds are averaged over a 2-minute period. During the appropriate observation time period, ensure that the ship's course and speed are steady. Never observe the wind while the ship is turning or changing speed. Make note of the following:

- The average relative-wind speed and direction during the period

- The peaks and lulls in the wind speed, with the appropriate direction, during the period. (Check for gust and squall criteria.)

- The degree of variability in the wind direction. (Check for variability criteria.)

True wind computations should be made for both the average wind speed and direction and the maximum gust speed and direction.

Verify computed true wind speed and direction with a check of the sea direction and sea state. Criteria for estimating wind speed from sea state are contained in Appendix V. The true wind direction is normally the same direction the seas are approaching from, particularly the smaller wavelets. If similar differences between estimated winds and computed true winds occur for several observations, a problem with the wind instruments is indicated. The following points should be reviewed when verifying true winds:

- The true wind direction is always on the same side of the ship as the apparent wind direction, but is further away from the bow (in degrees) than the apparent wind

- When the apparent wind direction is abaft the beam, the true wind speed is greater than the apparent wind speed

- When the apparent wind direction is forward of the beam, the true wind speed is less than the apparent wind speed

Whenever the shipboard anemometer is inoperative or the results are suspect, the winds should be observed by using the AN/PMQ-3 hand-held anemometer

AN/PMQ-3 HAND-HELD ANEMOMETER

The AN/PMQ-3 anemometer is found aboard ship and at many shore stations as backup wind-measuring equipment. All models of the AN/PMQ-3 are nearly identical in design and operation. Detailed guidance for operation and maintenance is found in NA 50-30PMQ3C-1, *Manual of Operation and Service Instructions for the Wind Measuring Set AN/PMQ-3(C).* A standard AN/PMQ-3 (fig. 2-31) comes with a special storage case (not shown) in which the anemometer should be stored when not in use. The case also contains a spare wind vane and spare transmitter.

Operation

To properly use the AN/PMQ-3, you need a location with unobstructed windflow and a reference line. When ashore, a field or ramp area with a known true north reference is ideal. Aboard ship, an unobstructed area (windward side) of the flight deck or the signal bridge is acceptable. Keep as far away as possible from deck edges, since the eddy effects of wind

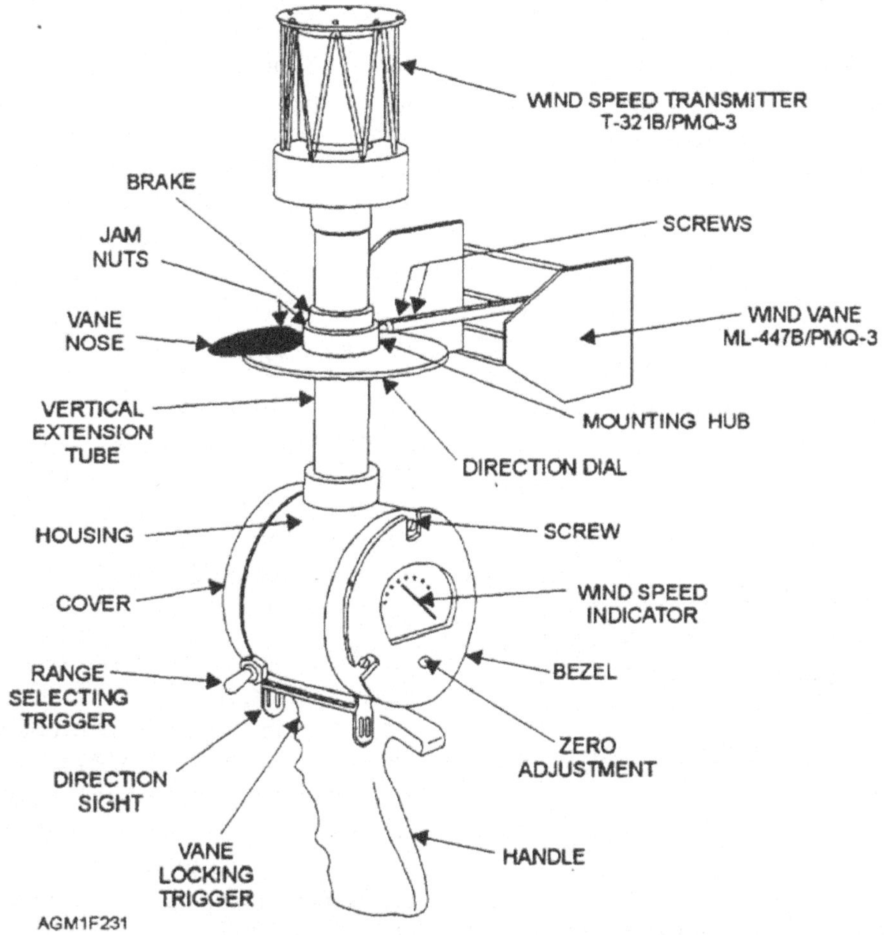

Figure 2-31.—AN/PMQ-3 hand-held anemometer.

flowing over the edge of the deck and wind combings will give inaccurate wind speed and direction. It is best to aim the sights of the AN/PMQ-3 toward the bow on a line parallel to the centerline of the ship.

Hold the instrument at arm's length and in a vertical position, with the indicator at or slightly above eye level, and align the sights with the true north reference or the bow of the ship (fig. 2-32). Depress the vane unlocking trigger to release the wind vane, and, at the same time, observe the indicated wind speed.

CAUTION

Activate the low-speed, range-selecting switch on the side of the casing on the newer models, or on the handle of the older models only if the wind speed is less than 15 knots. The indicator will be damaged if the switch is activated during higher winds.

Release the vane unlocking trigger when the wind vane yields a representative wind direction, and read the wind direction on the wind vane azimuth circle.

Figure 2-32.—Aerographer's Mate using the AN/PMQ-3 on the flight deck during flight operations. Note that hearing protection and a helmet are worn.

If a true north reference line was used at a shore station, no further computations are necessary. Aboard ship, however, you will need to obtain the ship's course and speed during the time of the observation, and compute true wind speed and direction.

Maintenance

The only recommended routine maintenance for the AN/PMQ-3 is a semiannual oiling of the wind-speed transmitter unit. Partial disassembly is required; consult the handbook for instructions. You should also wipe the instrument down with a towel dampened in freshwater to remove any salt deposits. Mild detergent may be used to remove dirt and grime.

REVIEW QUESTIONS

Q26. Other than automated observing systems, name three types of anemometers still used by the Navy?

Q27. When should the analog recorder chart of the AN/UMQ-5 be changed?

Q28. When should a time check be made on the analog recorder of the AN/UMQ-5?

Q29. What type of wind measurement is made by the Type B3 wind indicator?

Q30. At sea, what method should you use to verify true wind speed and direction?

Q31. How does the true wind speed compare to the apparent wind speed when the apparent wind direction is forward of the beam?

Q32. What procedure should you follow when using the AN/PMQ-3 at sea?

Q33. When *should the low-speed, range-selecting switch be activated on the AN/PMQ-3?*

RAIN GAUGES

LEARNING OBJECTIVES Describe the operation and maintenance of standard rain gauges ML-588 and ML-217.

In addition to automated observing systems, the Navy has two other standard rain gauges: the ML-588/UMQ-14 tipping-bucket rain gauge, discussed earlier as part of the AN/GMQ-29 system, and the ML-217 4-inch plastic rain gauge.

The ML-217 rain gauge (fig. 2-33) has a 4-inch funnel opening that collects and dumps precipitation into a small, graduated cylinder. The cylinder collects up to 1 inch of liquid, and measures the liquid to the nearest 1/100 of an inch. A 4-inch-diameter plastic outer casing is used to collect any overflow from the graduated cylinder. The steel support tripod that holds the rain gauge should be securely mounted on a firm, level surface.

OPERATION

The rain gauge is used to measure liquid precipitation or liquid equivalent of frozen precipitation. The amount of liquid precipitation in the graduated cylinder is read directly from the scale. A cylinder full to the brim is exactly 1 inch. After emptying, the graduated cylinder is used to measure water from the overflow cylinder.

Although it is a backup for liquid precipitation measurements, the ML-217 is the primary measuring instrument for frozen precipitation not handled by the ML-588 rain gauge or ASOS/SMOOS. When solid precipitation is expected, the collector funnel and the graduated cylinder should be removed from the 4-inch overflow cylinder to allow the frozen precipitation to accumulate directly into the overflow cylinder. The

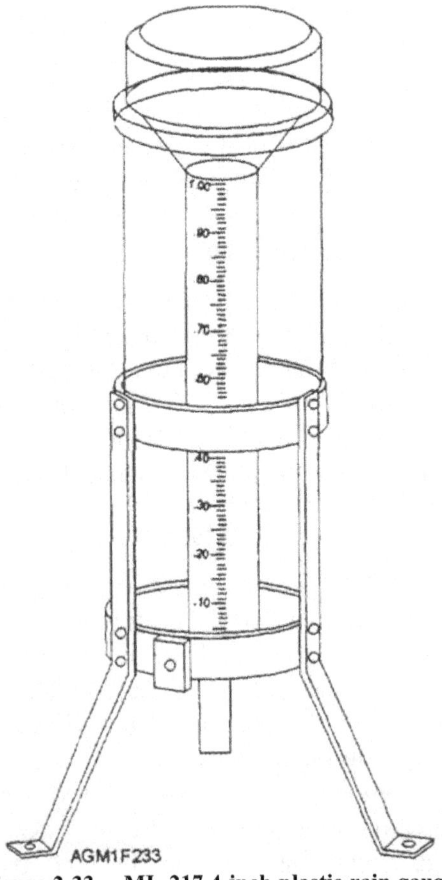

Figure 2-33.—ML-217 4-inch plastic rain gauge.

melted precipitation is measured to determine liquid equivalent. The overflow cylinder may also be used to take a core sample of accumulated frozen precipitation. When the sample is melted, it is measured in the graduated cylinder to obtain liquid equivalent.

MAINTENANCE

The primary maintenance action for the ML-217 rain gauge is keeping the gauge clean. Water and, if necessary, mild detergent may be used to clean the plastic parts. The steel support tripod will need periodic rust removal and repainting.

REVIEW QUESTIONS

Q34. How much precipitation can the graduated cylinder of the 1-inch rain gauge hold?

Q35. Which rain gauge is used primarily for the measurement of frozen precipitation?

AN/GMQ-32 TRANSMISSOMETER

LEARNING OBJECTIVES: Define transmissometer. Describe the operation and maintenance of the major components of the AN/GMQ-32 system. Explain the required AN/GMQ-32 recorder chart notations. Describe the operation and maintenance of the OA-7900/GMQ-10 converter/indicator group.

A transmissometer is an instrument that measures how well light travels through a substance. In meteorology, a transmissometer is a device used to measure the transmissivity of light through air to determine visibility. In addition to the visual-range sensor on the ASOS, there is one system used by Naval and Marine Corps shore activities-the AN/GMQ-32 transmissometer system (formerly called the AN/GMQ-10).

The AN/GMQ-32 transmissometer provides a visual and graphic record of the transmissivity of light between the projector and the detector. This charted value, with some standard adjustments, may be converted to sector visibility or to runway visual range. *Sector visibility* is how far an unlighted object can be seen during the day, while *runway visual range* (RVR) is how far the runway lights can be seen by a pilot. Without any change in weather conditions, the runway visual range may be increased by simply turning up the intensity of the runway lighting.

MAJOR COMPONENTS

The AN/GMQ-32 transmissometer system (fig. 2-34) consists of one or more sets of ML-461/GMQ-10 transmissometer projectors and R-547/GMQ-10 receivers, located on the airfield; their associated electrical equipment and cables; and either the ID-353B/GMQ-10 or the ID-820/GMQ-10 transmissometer indicator/recorder unit. A separate equipment system, used with most AN/GMQ-32

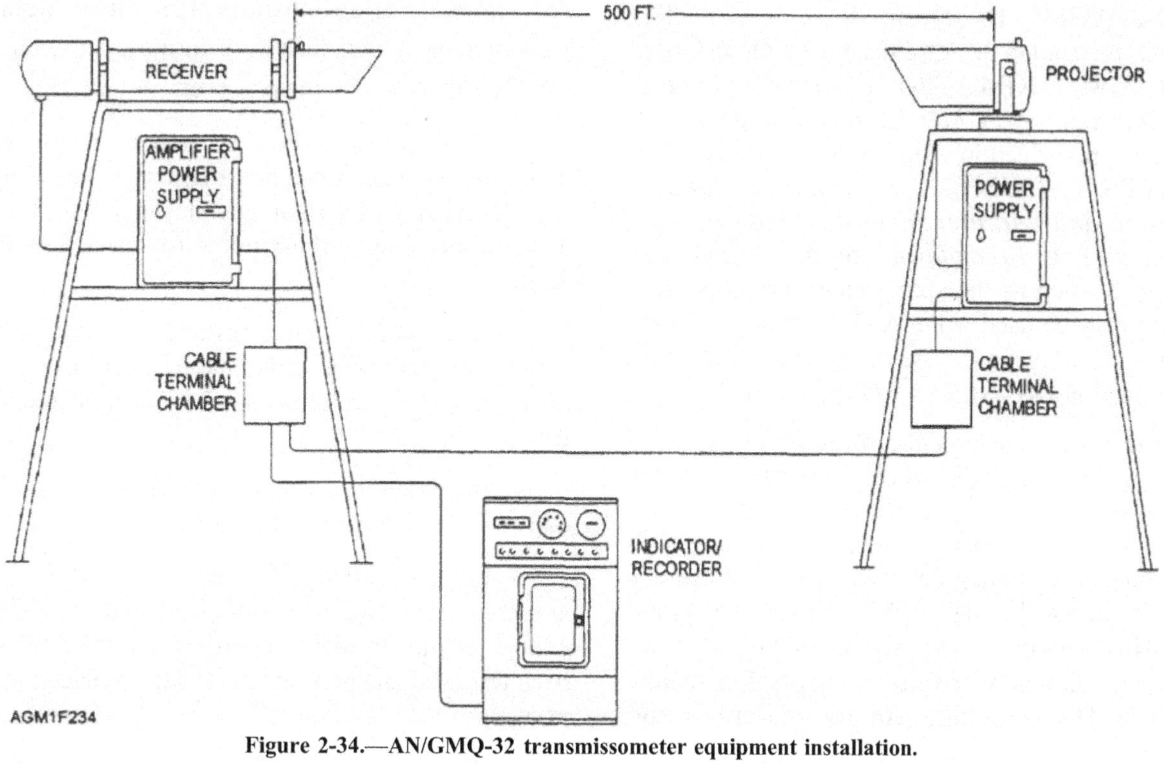

Figure 2-34.—AN/GMQ-32 transmissometer equipment installation.

installations, is the OA-7900/GMQ-10 converter/indicator group. It provides a LED digital display of runway visual range in hundreds of feet.

The AN/GMQ-32 system is an upgrade of the AN/GMQ-10 system. This major change replaced the vacuum-tube electronics of the AN/GMQ-10 with solid-state circuitry. The external appearance of the equipment was not changed. The equipment still retains the original identification plates.

INDICATOR/RECORDER UNIT

The ID-353B/GMQ-10(B) or the ID-820/GMQ-10(C) transmissometer indicator/recorder unit (fig. 2-35) is found in the observer's work area of Naval Meteorology and Oceanography detachments and Marine Corps weather offices. The only visible difference between the two models is that the calibration setting meter is found only on the ID-353B, which was originally designed for use overseas where the local electricity did not conform to the carefully maintained 60-hertz AC found in the United States.

Operation

The transmissometer equipment is normally left on. It is turned off only for maintenance. If the system has been turned off, start-up calibration adjustments are required after a 1-hour warm-up period.

Detailed operational procedures for the upgraded AN/GMQ-32 system, including the ID-353/GMQ-10 and ID-820/GMQ-10 transmissometer indicator/recorder, are found in Air Force Technical Order (AFTO) 31M1-2 GMQ-32-1, *Operation Manual, Transmissometer AN/GMQ-32,* provided with each equipment modification. Instructions provided in NAVWEPS 50-30GMQ-10-2, *Technical Manual, Operation and Maintenance Instructions with Illustrated Parts Breakdown,* for the mechanical components, such as the spring chart drive, are still valid and may be used if the AFTO is not available. Instructions and criteria for use of transmissometer data are contained in NAVMETOCCOMINST 3141.2.

The transmissometer indicator/recorder provides two output readings: the recorder graph and the transmissivity meter. Both outputs provide transmissivity percentages. Two sensitivity settings are used to process incoming data. The indicator/recorder is normally operated in the LOW sensitivity setting, but the HIGH sensitivity setting may be used during low-visibility conditions, when transmissivity falls below 20%. The HIGH setting simply multiplies the indicator/recorder output by 5, which allows the operator to read the output more accurately. Both readings must be corrected for background illumination and converted by using the tables provided in NAVMETOCCOMINST 3141.2 to yield runway visual range or sector visibility.

The background illumination correction is found by placing the Background switch in the TEST position, obtaining a background illumination level, and then subtracting the background illumination level from the transmissivity reading to obtain a corrected transmissivity reading.

The runway visual range (RVR) table in NAVMETOCCOMINST 3141.2 is in two sections: one for use during the day and the other for use at night. Columns are provided for runway light settings (LS) 5, 4, 3, and Other. The Other column is used when runway lighting is either not in operation or at setting 1 or 2. The runway light setting is obtained from the air traffic controllers. Use the transmissivity value in the proper light setting column to find RVR.

Maintenance

General maintenance for the AN/GMQ-32 system is performed by base ground electronics personnel or trained METEM (METeorological Equipment Maintenance) Electronics Technicians. Refer to the care-and-use publications for more detailed information. The following routine maintenance is usually performed by the observer:

- Replacing the recorder paper—Replace the paper by using the reloading diagram exposed in the recorder window. Charts in continuous operation run at 3 inches per hour and are normally changed every 2 weeks

- Rewinding the recorder chart drive spring—Rewind when setting the recorder chart time or once per week. A fully wound drive will operate for 8 days

- Adjusting the recorder zero adjustment—Adjust daily when the recorder is in use

- Refilling/cleaning the inkwell and pens—Empty and clean inkwells, indicator pens, and marker pens once a month. In many recorders, the inkwells have been removed and pen cartridges have replaced the ink pens

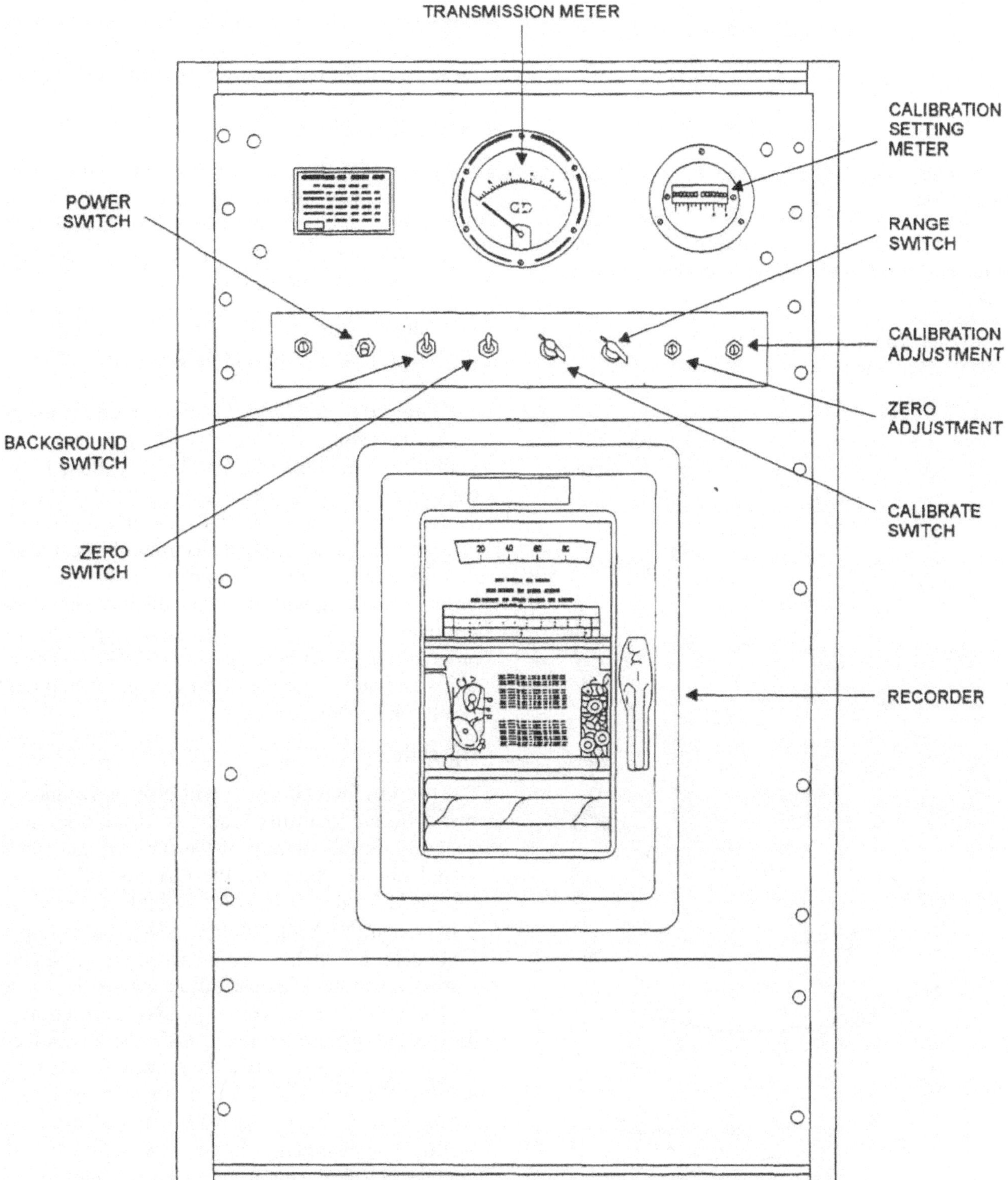

Figure 2-35.—ID-353B/GMQ-10(B) or ID-820/GMQ-10(C) transmissometer indicator/recorder.

Required AN/GMQ-32 Recorder Chart Notations

Transmissometer recorder charts (fig. 2-36) should be set to Coordinated Universal Time (UTC); they must have a time check with a date-time group entered at the following times:

- The beginning and end of each roll
- The actual time of each synoptic hour observation
- The beginning and ending of maintenance shutdowns or other periods when the recorder is not in service
- When you are notified of any aircraft mishap at or near your station

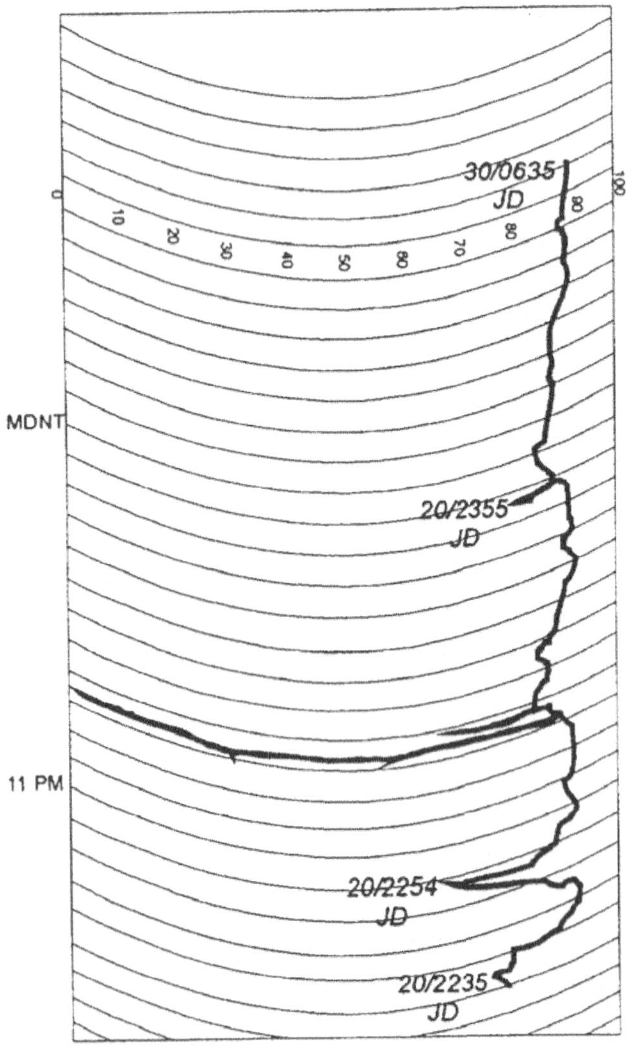

AGM1F236

Figure 2-36.—AN/GMQ-32 transmissometer recorder chart notations.

In addition to the regular time checks with a date-time group, other notations are required on the recorder chart on the following occasions:

- When the chart time differs by more than 5 minutes from the actual time. (Place a date-time group on the position the pen was recording on the chart, correct the chart time position, and enter a new date-time group on the adjusted position.)
- When a chart or any portion of the chart is provided for any special study or investigation. (Enter the station name, runway number, length of the baseline, and month and year on the chart, in addition to the required time checks.)

OA-7900A/GMQ-10 CONVERTER/INDICATOR GROUP

The OA-7900A/GMQ-10 converter/indicator group, consisting of an ID-1939/GMQ-10 digital LED visibility indicator and a CV-3125/GMQ-10 signal converter (fig. 2-37), was added to most AN/GMQ-10 systems in the early 1970's. The original ID-l939 visibility indicators used an electromechanical display dial that was very difficult to maintain and suffered a great deal of downtime. The display dials were removed in the mid-1970's and replaced with programmable-read-only-memory (PROM) computer chips and an LED readout for runway visual range (RVR), in hundreds of feet.

Operation

The OA7900/GMQ-10 will provide readouts for two different visibility readings. It provides sector visibility, in the vicinity of the runway, and runway visual range. Operating the OA7900/GMQ-10 in NORMAL mode yields sector visibility. Operating the converter in DAY or NIGHT mode, and setting the Light-setting switch to the indicated runway-lighting intensity, activates different calculations in the memory to yield runway visual range. No correction for background illumination is required. Check with the air traffic controllers to verify the runway light-intensity setting. Runway lights are frequently operated at setting 2 or 3, but a pilot may request that the level (setting) be changed. Level 1 is dimmest and 5 brightest. Ideally, the setting used is the lowest setting that lights can be seen through any obstruction to visibility. Study the operator's manual NA 50-30GMQ10-7, *Operator and Maintenance Instructions with Illustrated Parts Breakdown, OA-7900A/GMQ-10,* for further details on operating and maintaining this system.

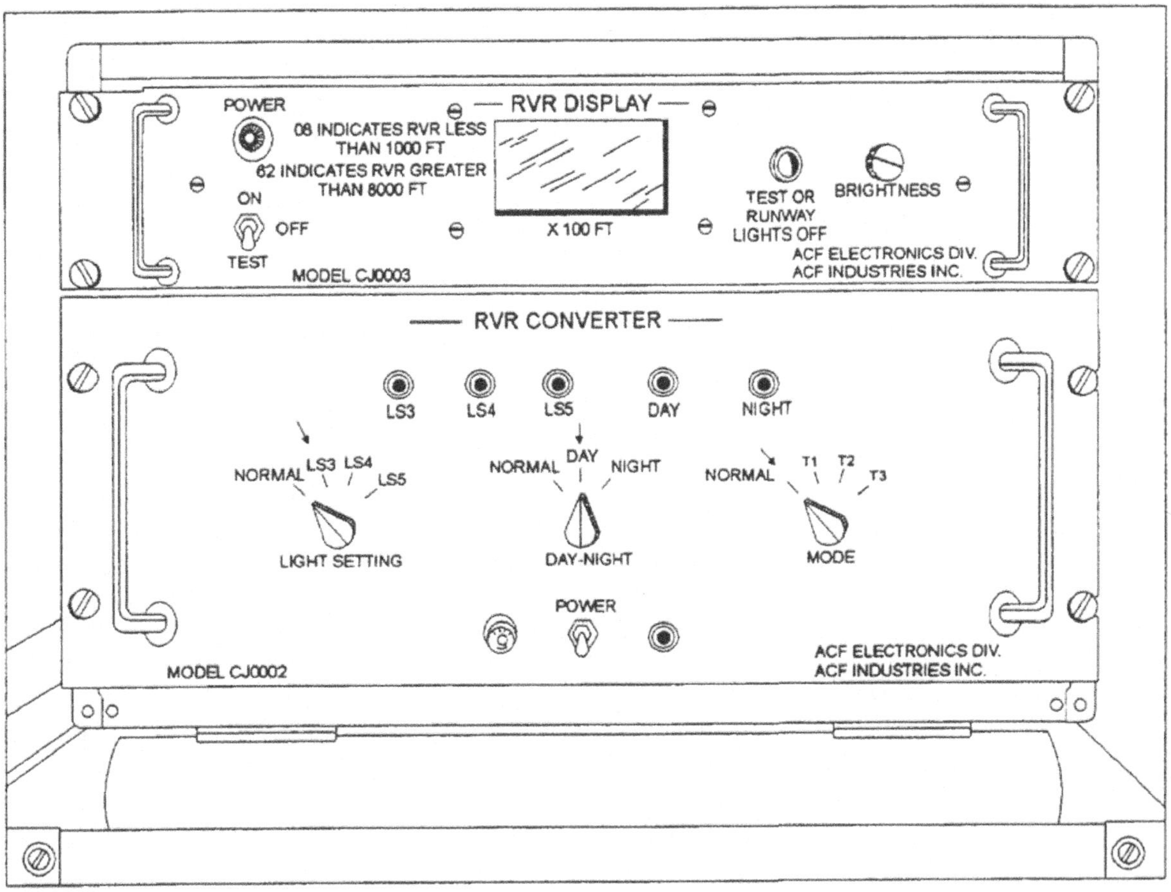

Figure 2-37.—OA-7900A/GMQ-10 Converter/Indicator Group consisting of the ID-1939/GMQ-10 digital LED indicator and the CV-3125/GMQ-10 signal converter.

Maintenance

This equipment requires very little operator maintenance. Maintenance includes cleaning the fan screen on the rear of the converter and keeping external surfaces clean. Refer to the operator's manual for details.

REVIEW QUESTIONS

Q36. What does a meteorological transmissometer actually measure?

Q37. How do the terms "sector visibility" and "runway visual range" differ with respect to the AN/GMQ-32?

Q38. What publication contains criteria and instructions for the use of transmissometers?

Q39. How are transmissivity values presented on the AN/GMQ-32 and how are these values converted?

Q40. What routine maintenance is normally accomplished by the observer with respect to the AN/GMQ-32?

CLOUD HEIGHT EQUIPMENT

LEARNING OBJECTIVES: Describe the operation, recorder trace outputs, and maintenance of the AN/GMQ-13 cloud height set. Explain the operation and maintenance of the ML-121 ceiling height projector and clinometers used ashore and at sea. Describe two types of ceiling balloons, how they are inflated, and how they are used to determine cloud height.

In this section we will discuss some of the more frequently used types of cloud-height measuring equipment. The AN/GMQ-13 cloud height set (also called the rotating beam ceilometer or RBC), is still the primary cloud-height measuring equipment in use at some shore stations. The AN/GMQ-13 is currently being replaced by the ASOS cloud height detector. Most shore stations also have backup cloud-height measuring equipment, which consists of an ML-121 ceiling light projector and ML-119 clinometer. Aboard ship, an ML-591/U shipboard clinometer is used in

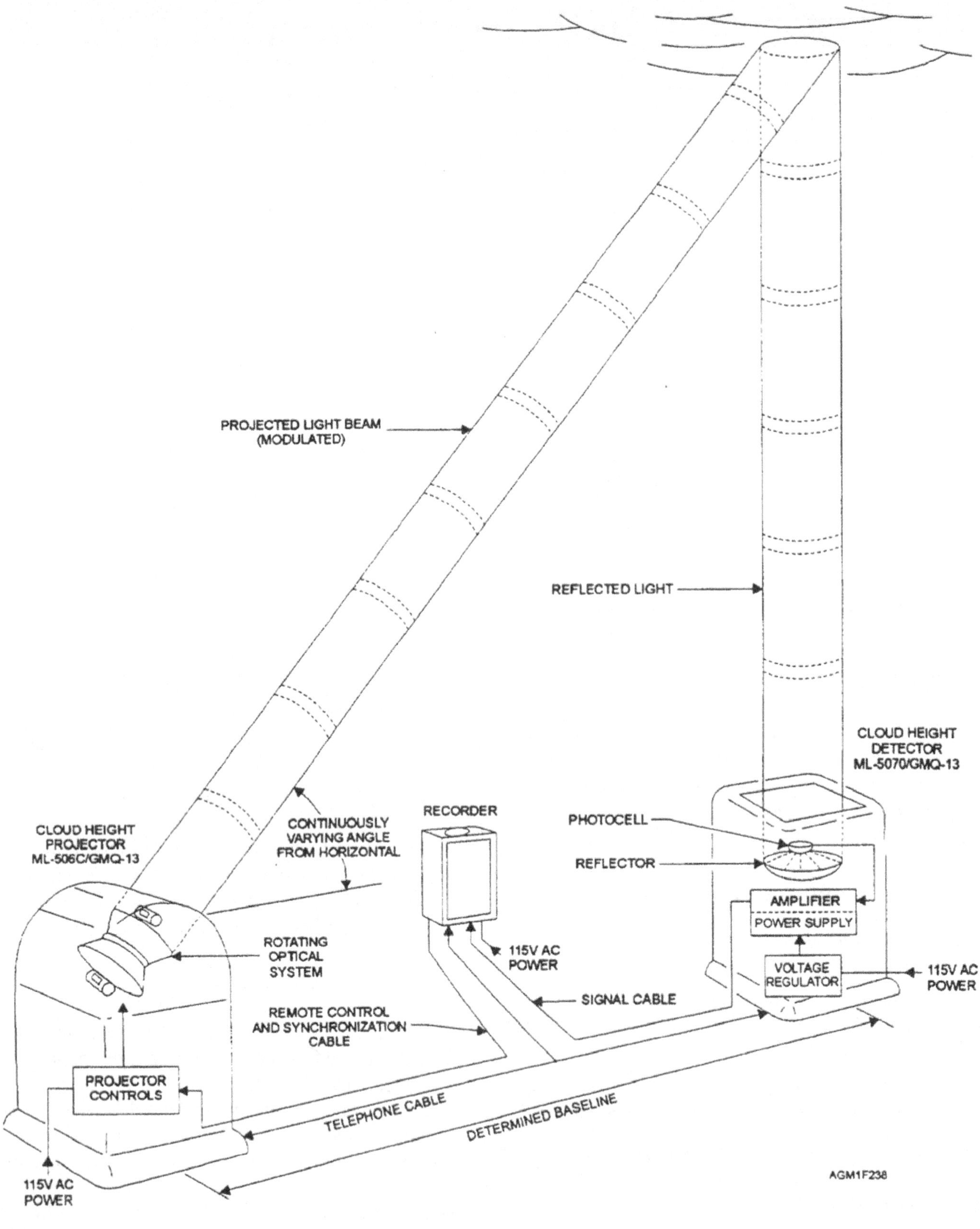

Figure 2-38.—ML-506/GMQ-13 RBC projector and ML-507/GMQ-13 cloud height detector setup.

Figure 2-39.—RO-546/GMQ-13 GIFFT cloud height recorder.

conjunction with the ML-121 ceiling light projector. Ceiling balloons may also be used to determine cloud height, both ashore and at sea.

AN/GMQ-13 CLOUD HEIGHT SET

The AN/GMQ-13 cloud height set was introduced to the fleet from 1955 to 1959. Over the years many modifications have been made to the individual pieces of equipment. Currently, the cloud height set consists of an ML-506/GMQ-13 rotating beam projector, an ML-507/GMQ-13 cloud height detector, and an RO-546/GMQ-13 GIFFT ceiling height recorder. The projector and the detector are permanently installed near the main runway, but are separated by a 400- to 900-foot baseline. See figure 2-38 for a standard setup diagram. The standard baseline is 400 feet. Cloud height is determined by the angle of the projected light beam reflected from the base of the clouds directly overhead the receiver. The GIFFT cloud height recorder, usually rack-mounted in the observer's space (fig. 2-39), records angle-of-reflection on a chart.

Operation

Detailed information on the theory of operation, operating instructions, and maintenance guidance for the projector and receiver is contained in NA 50-30GMQ13-1, *Handbook, Operation and Service Instructions, with Illustrated Parts Breakdown, Cloud Height Set AN/GMQ-13(C).* Operation and maintenance guidance for the RO-546/GMQ-13 GIFFT recorder is published in the manufacturer's instructions provided with the equipment.

Evaluation of the Recorder Trace

The RO-546 recorder chart is marked in hours and tens of minutes. The time marking "234" equates to 2340Z. Time adjustments are covered in the operator's manual.

The procedure used to determine cloud height using the GIFFT recorder has been refined since the manufacturer provided the operator's manual. The recorder elevation angle is evaluated differently for various weather conditions. In some cases, the indicated elevation angle must be corrected to obtain an accurate cloud-base height.

CORRECTION FACTOR.—An equipment correction must be determined for each individual GIFFT recorder before any recorded elevation angles may be used to determine cloud height. The following procedure is used for RO-546 recorders when the ML-506 RBC projector and ML-507 RBC receiver are separated by the standard, 400-foot, baseline.

To obtain the equipment correction, make a recording of a high, solid cloud cover with bases between 10,000 and 20,000 feet. This solid cloud cover should produce a fairly consistent solid band, 4° to 8° wide, on the recorder chart. Solid, well-defined cloud bases will produce a shorter trace, while diffuse or ragged bases will produce a longer trace. See figure 2-40. The occasional horizontal lines marked outside

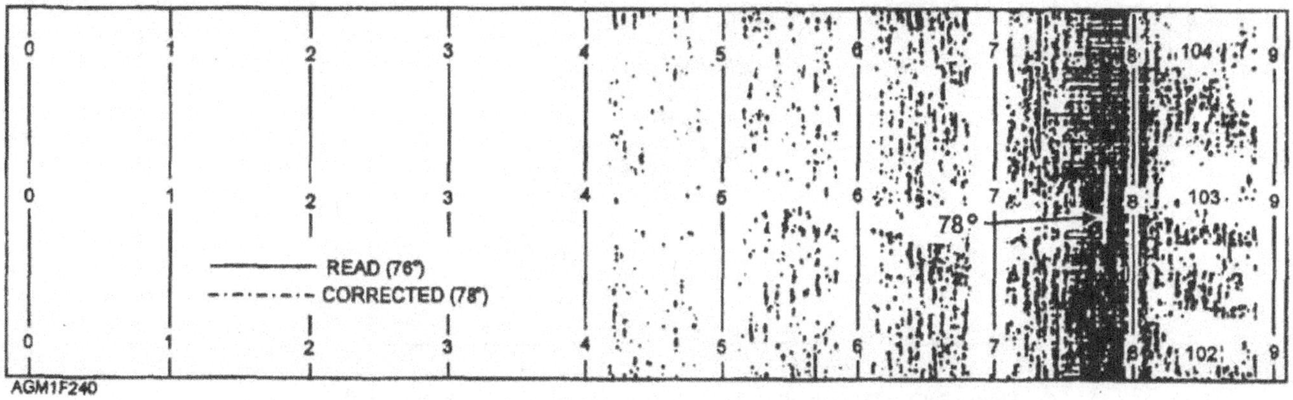

Figure 2-40.—Determining equipment correction from high, solid cloud cover.

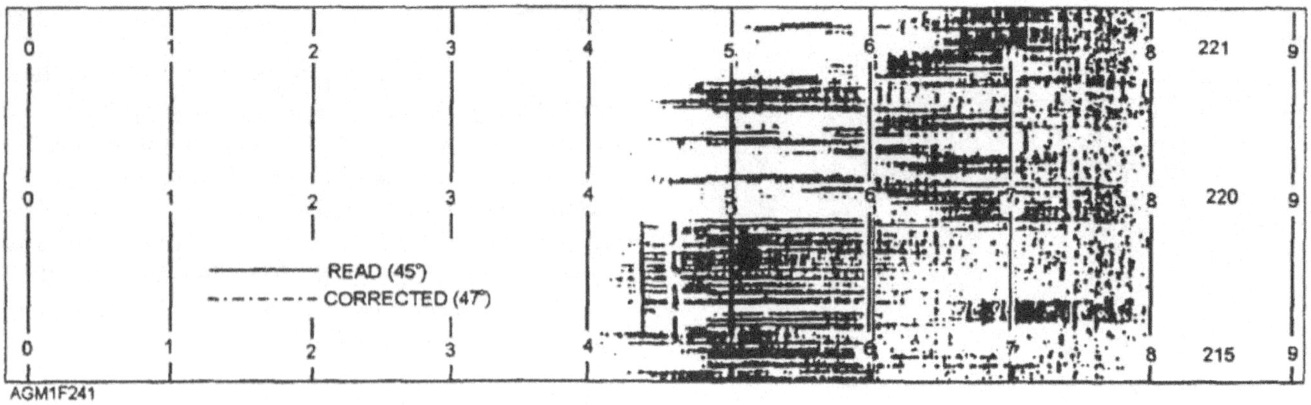

Figure 2-41.—Determining corrected cloud elevation angle with multiple cloud layers.

the band are ignored. The width of the solid band, in degrees, divided by 2 is the equipment correction for the recorder. The correction factor is applied to recorded elevation angles for all solid cloud bases and diffuse stratus bases over 45° elevation.

The length of the thin marking lines is an indication of the reflected light sensed by the receiver. Line length is affected by the focus and output candlepower of each of the two lights in the RBC projector. Replacing the bulbs in the projector, cleaning a dirty projector lens, or refocusing the projector lights will change the characteristic length of the recorder marking lines; and a change in line length will require that anew equipment correction be determined.

SOLID CLOUD COVER.—The cloud elevation angle of solid cloud cover, such as is indicated in figure 2-40, is read at the left edge of the solid band. The equipment correction, 2° in this case, is then added to the indicated elevation angle to find the corrected elevation angle. For solid cloud cover only, the corrected elevation angle is the same as reading the elevation angle at the center of the solid band.

MULTIPLE CLOUD LAYERS.—Read the angle at the left side of the solid band, and then add the correction to obtain the proper cloud angle. For example, in figure 2-41, at time 2154Z, the solid line at 45° elevation was drawn for the left side of the solid portion of the band. The equipment correction, 2°, found from figure 2-40, is added to find the corrected angle of the cloud bases, 47°.

FOG AND LOW STRATUS.—Read the elevation angle at the left side of the solid band. Do not add the correction unless the correction yields an elevation angle greater than or equal to 45°. Figure 2-42 shows a low-stratus situation that became fog on the deck at time mark 000. Between 2320Z and 2330Z, the stratus shows a solid band at 44°. Adding the equipment correction results in a corrected elevation angle of 46°,

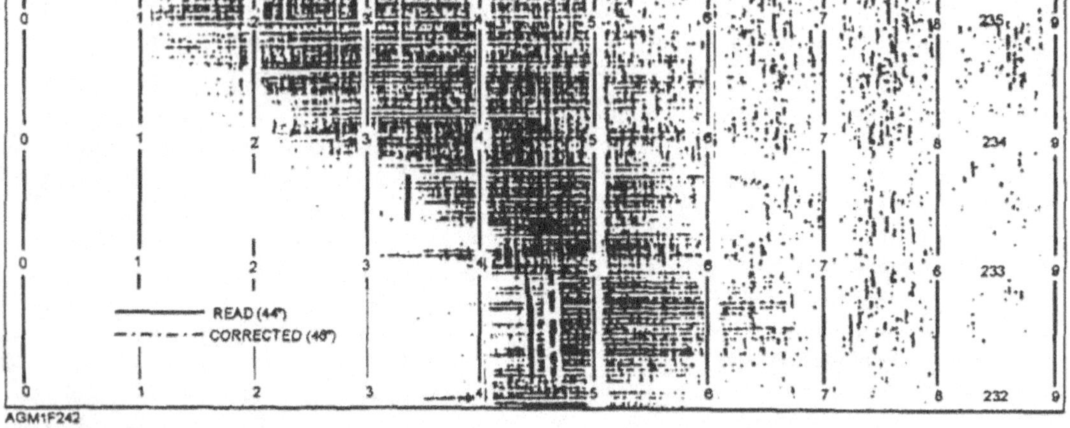

Figure 2-42.—Determining cloud elevation angles for low stratus and fog.

so adding the correction is acceptable. At 2334Z, we read the left side of the solid band at 33°. Adding the equipment correction of 2° would result in only a 35° corrected angle. Since the correction is less than 45°, the correction is not added, and the 33° elevation angle should be used.

CONVERTING ELEVATION ANGLE TO CLOUD HEIGHT.—The operator's manual for the GIFFT recorder contains a table for converting elevation angle to cloud height for units using the standard 400-foot baseline. A table may be constructed for your unit's baseline, if different from the standard, by multiplying the tangent of the elevation angles from 1° through 89° by the length of your baseline, in feet.

Maintenance

Maintenance requirements for the RO-546 recorder are described in the operator's manual. All calibrations and electrical checks are performed by base ground electronics personnel. Aerographer's Mates are responsible for changing the recorder chart paper and replacing the chart-marking stylus when the old stylus becomes worn, broken, or bent. Pilot-reported cloud heights that are consistently different from RO-546 indicated heights, or other indications of malfunction, should be reported to ground electronics. Backup equipment, such as the ML-121 ceiling light projector and the ML-119 clinometer, should be used when the AN/GMQ-13 is out of service.

ML-121 CEILING LIGHT PROJECTOR

The standard light projector used by the Navy and Marine Corps is the ML-121 ceiling light projector, shown in figure 2-43. This equipment projects a narrow, concentrated beam of light vertically onto cloud bases up to 10,000 feet. An observer, sighting at the illuminated spot on a cloud base, uses clinometer elevation angle and baseline distance to find cloud height. The light projector is effective only at night.

Operation

The ceiling light projector is activated by a switch located in the weather office or outside near the observation point. The projector must be activated only when conducting a measurement. The high-intensity light may confuse pilots, both ashore and aboard ship; therefore, you must request permission from your supervisor to activate the projector.

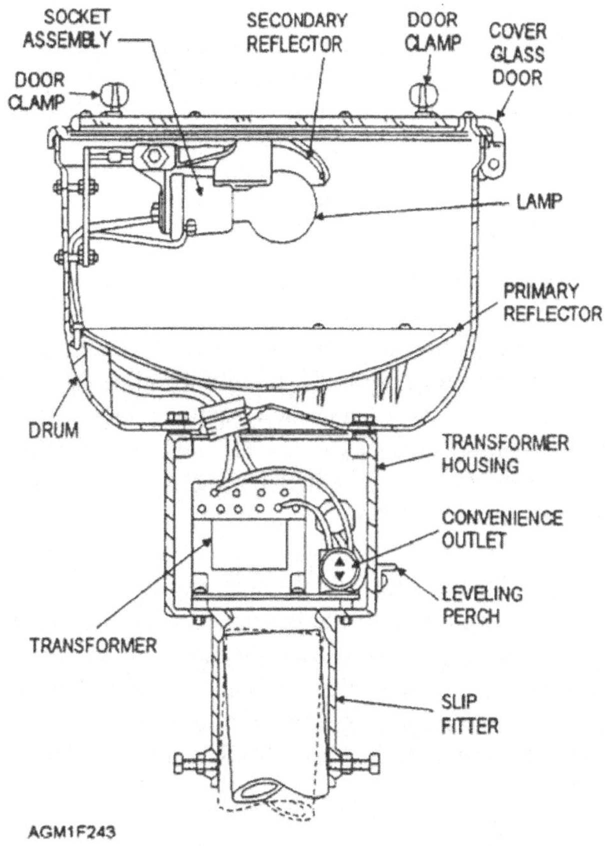

Figure 2-43.—ML-121 ceiling light projector.

Maintenance

Ground electronics personnel should provide maintenance support for the projector. Maintenance procedures are described in NA 50-30FR-521, *Handbook, Operation and Maintenance Instructions, Ceiling Light Projector ML-121 and Clinometer ML-119*. Recommended operator maintenance includes the following:

- Weekly cleaning of glass cover plate and inspection and cleaning of drainage/ventilation holes in the projector housing
- Replacing lamps that are burned out or have blackened or sagging filaments
- Quarterly checking of lamp alignment and focus, and inspecting the projector to make sure it is level

CLINOMETER

Clinometers are used to measure the elevation angle of the projected light spot on a cloud base. Two different types of clinometers are used: the ML-119 clinometer, found only at shore stations, and the ML-591/U shipboard clinometer, used both aboard ship and ashore. The ML-119 clinometer is shown in figure 2-44. It is hand-held, whereas the shipboard clinometer is mounted on a bracket and supplied with a weatherproof cover; operation is similar.

Operation

To operate the clinometer, loosen the clutch screw to allow the elevation scale to swing freely. After sighting the light spot on the cloud base, bring the cross hairs to bear on the spot. Then, with the cross hairs on the spot, tighten the clutch screw, and read the elevation angle from the scale to the nearest whole degree. Cloud height is calculated by multiplying the baseline distance by the tangent of the elevation angle. Normally, a minimum of three readings are used to determine an average. Aboard ship, add the height of the clinometer above the waterline to the calculated distance. Where established premeasured baselines are used, tables for converting measured elevation angle to cloud height may have been previously calculated and may be available. NAVMETOCCOMINST 3144.1 provides summarized procedures for using the shipboard clinometer.

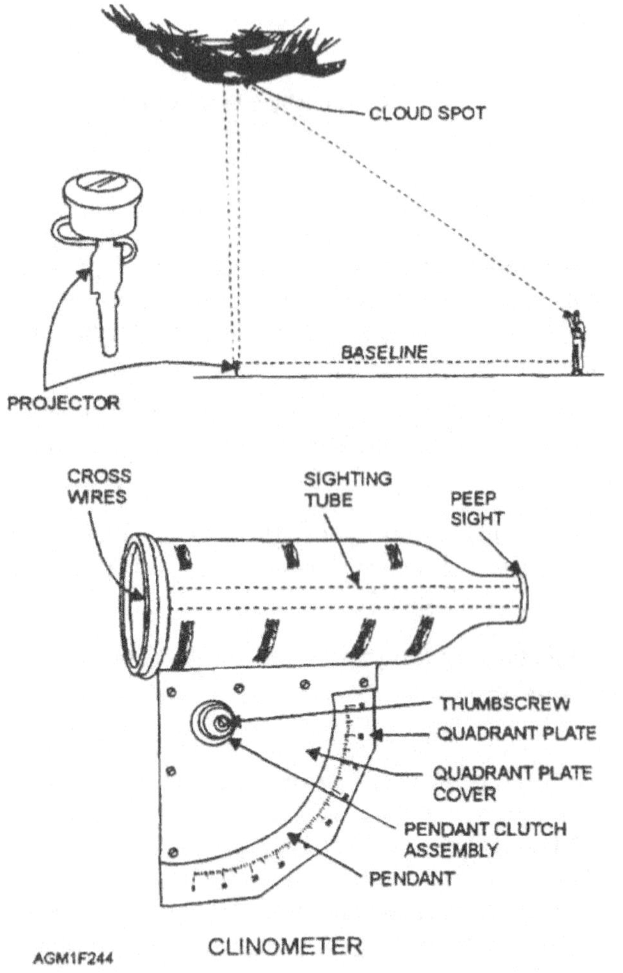

Figure 2-44.—ML-119 clinometer.

Maintenance

Maintenance procedures for the ML-119 clinometer are described in NA 50-30FR-521 and for the ML-591/U clinometer in NAVWEPS 50-30FR13, *Technical Manual, Operation, Maintenance, and Overhaul Instructions with Illustrated Parts Breakdown, (Shipboard) Elevator Clinometer ML-591/U.* Monthly cleaning is recommended, along with inspection to make sure the parts move freely. Aerographer's Mates are normally responsible for all maintenance on clinometers.

CEILING BALLOONS

Ceiling balloon measurements are still a valid procedure to determine cloud height when other methods fail. Although not routinely accomplished at most Naval meteorology and oceanography detachments, ceiling balloon observations are common aboard ship and at most Marine Corps weather stations. Ceiling balloons are primarily used as a backup method to determine cloud height during the day when ceiling lights cannot be used.

Ceiling balloons provide a reasonable height determination up to 2,000 feet. Since heavy precipitation will slow the ascension rate enough to invalidate any measurement, avoid balloon measurement during such conditions.

The equipment required for a ceiling balloon observation includes the balloons, balloon weight set, a stopwatch, helium, and a helium regulator with a balloon-inflation nozzle. In this section we discuss the balloons, the balloon-inflation procedure with the weight set, and the method used to determine cloud or ceiling height.

Balloon Types

Several different sizes and colors of balloons may be used for ceiling determination. The balloons are available in both red and black; the red balloons are used for thin cloud cover or on hazy days, while the black balloons are used for all other cloud conditions. Although the standard ceiling balloons are the lo-gram balloons, 30-gram balloons provide a faster ascension rate and are preferable for use during high-wind conditions. Use of the slower ascending, lo-gram balloon during high winds may result in the balloon being carried out of sight horizontally before it reaches the cloud layer or ceiling.

Balloon Inflation

Balloons should be inflated slowly, with helium. A 10-gram balloon should take 3/4 to 1 minute to fill, and a 30-gram balloon should take 2 1/2 to 3 minutes to fill. This slow inflation rate allows the balloons to stretch properly during filling.

The balloons are inflated by using the ML-575/UM universal balloon balance set (fig. 2-45). Refer to the technical data supplied with the set for detailed information. The 10-gram balloons, attached to the inflation nozzle weighted to 45 grams, are inflated to neutral buoyancy. This yields 45-grams of free lift. The 30-gram balloons are inflated to 139-gram free lift. Chemical lights and balloon-tying cords are added to the inflation nozzle during inflation, but are not calculated in the weighting figures.

Determination of Cloud Height

You must always request permission from the Officer of the Deck (OOD) and the Air Boss aboard a carrier, or from the air traffic controller at an air station, to release a ceiling balloon. After clearance is received, the ceiling balloon is released and the ascension is timed. Cloud height is determined by the elapsed time, in minutes and seconds, that the balloon takes to enter into the cloud base. Consider the point of entry as midway between the time the balloon first begins to fade and the time of complete disappearance for layers aloft. Use the point at which the balloon disappears to estimate vertical visibility. Use the standard ascension rate tables in either NAVMETOCCOMINST 3141.2 or NAVMETOCCOMINST 3144.1 to determine the height.

So far, we have covered most of the equipment used in surface aviation weather observations. We have mentioned a few of the calculators used to compute observational data, such as the psychrometric computer and the true wind computer. Now let's discuss the various types of calculators used in surface observations.

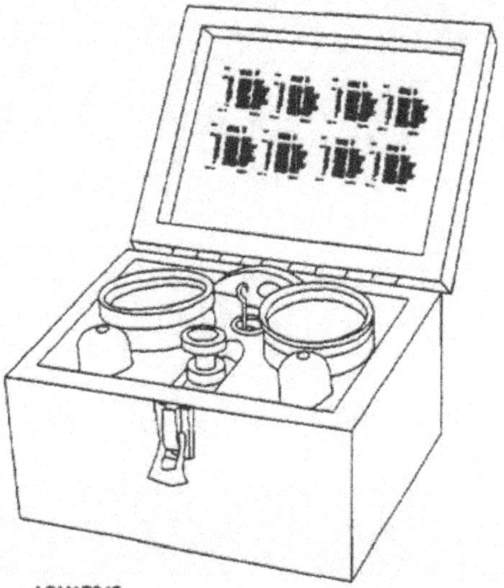

Figure 2-45.—ML-575/UM universal balloon balance set.

REVIEW QUESTIONS

Q41. What are the three main parts of the ML-507/GMQ-13 Cloud Height Set?

Q42. When you use the ML-507/GMQ-13, how are cloud heights determined?

Q43. What data must be computedfor each individual RO-546/GMQ-13 GIFFT recorder?

Q44. When is the correction factor of the GIFFT recorder not applied?

Q45. What instrument is used in conjunction with the ML-121 Ceiling Light Projector to obtain cloud heights?

Q46. What publication contains instructions for using the shipboard clinometer?

Q47. How are cloud heights determined when using a clinometer?

Q48. Under what conditions should red ceiling balloons be used to measure cloud heights?

Q49. Why should larger, 30-gram balloons be used in high wind conditions vice 10-gram balloons?

Q50. How is cloud height determined when you use ceiling balloons?

CALCULATORS

LEARNING OBJECTIVE: Explain the care and use of four types of hand-held calculators.

In the 1930's and 1940's, the military first made serious attempts to study the weather and apply it operationally. Much of the data, such as dew-point temperature and relative humidity, had to be determined from printed tables (such as those available in the *Smithsonian Meteorological Tables)* or calculated manually. In the ensuing decades, as meteorology became more of a precise science, many paper nomograms and metal or plastic calculators were developed to make the calculations easier. These nomograms and calculators helped free the user from bulky printed tables. Although the built-in computational power of the ASOS and SMOOS perform many of these calculations, we include this section on the proper use and care of the most frequently used calculators, because they will continue to be used as backup equipment. The calculators used for computations of surface observational data are the pressure reduction computer, the psychrometric computer, the density altitude computer, and the true wind computer.

Proper care of all plastic and metal calculators is important to keeping these devices functional. The plastic and metal surfaces should be kept free of any sand or abrasive particles. Clean the space between adjacent plastic or metal disks by drawing a piece of blotter paper through the space. Wash the calculator in freshwater, using a mild detergent if necessary. Dry the calculator thoroughly. Use ordinary desk blotter paper to dry the space between disks. Do not use solvents to clean calculators; solvents will dissolve or soften the plastic.

Store plastic and metal calculators in a cool, dry location. If the storage temperature exceeds 140°F, the

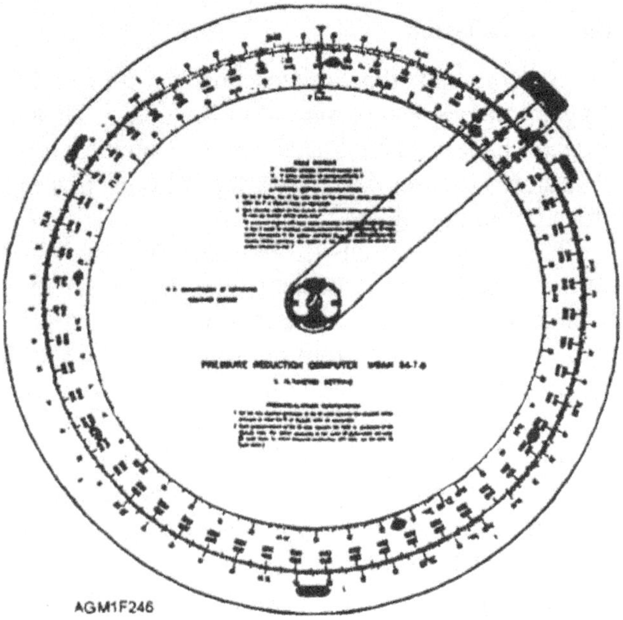

AGM1F246

Figure 2-46.—CP-402/UM pressure reduction computer.

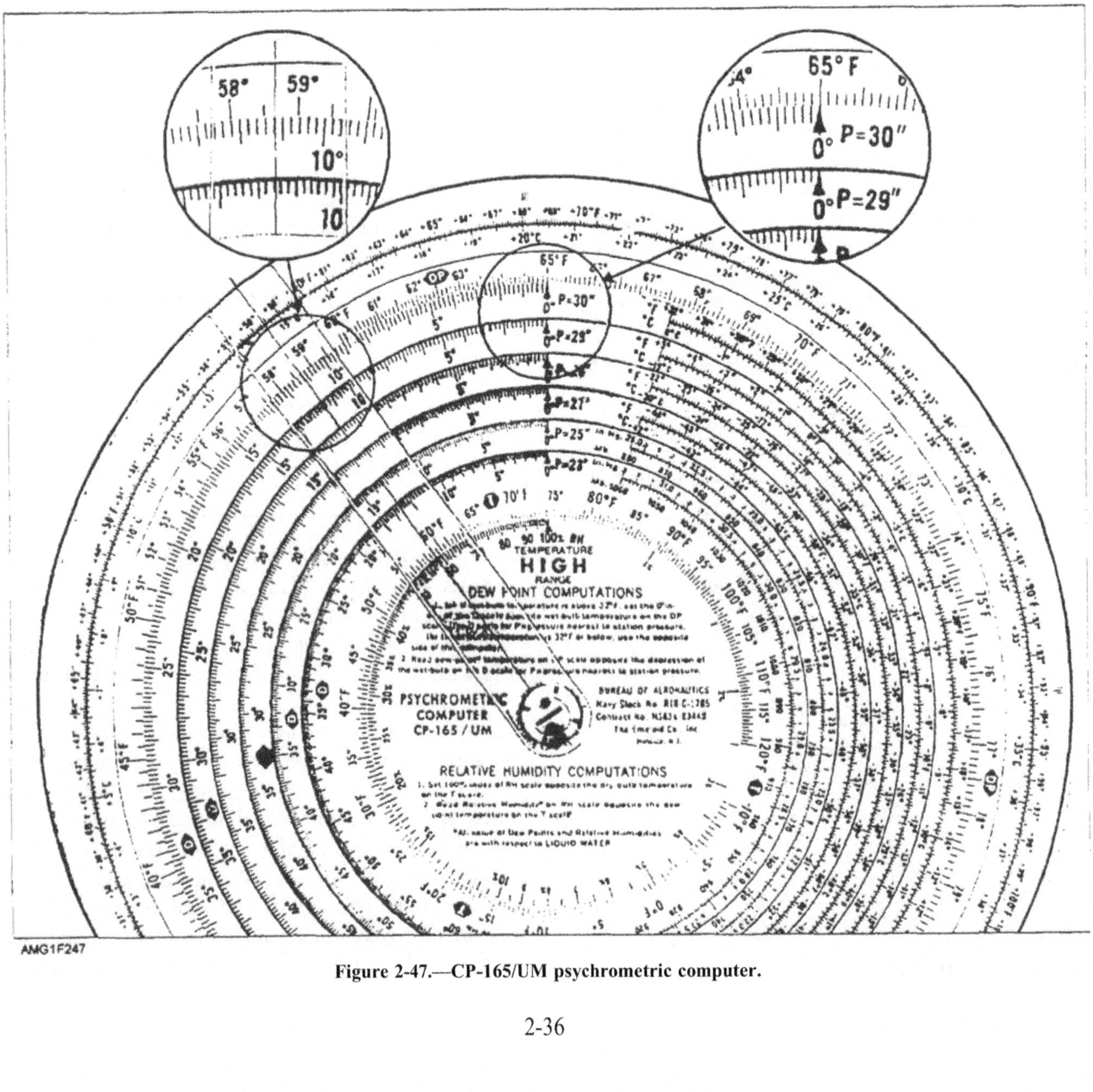

AMG1F247

Figure 2-47.—CP-165/UM psychrometric computer.

plastic parts will warp, melt, or shrink. Aboard ship, stow plastic calculators in an accessible but secure place. The shock of falling off a plotting table or desktop in high seas may crack the plastic disks.

CP-402/UM PRESSURE REDUCTION COMPUTER

The CP-402/UM pressure reduction computer (fig. 2-46) is a two-sided calculator used to compute sea-level pressure, altimeter setting, and pressure altitude from station pressure. Detailed instructions for each process are printed on the calculator.

CP-165/UM PSYCHROMETRIC COMPUTER

The CP-165/UM psychrometric computer (fig. 2-47) is a very colorful, two-sided calculator used to compute dew-point temperature from air temperature, wet-bulb temperature, and pressure. One side is used when the air temperature is above freezing (32°F), while the other side is used for computations when the temperature is at or below freezing. The psychrometric computer may also be used to calculate relative humidity from air and dew-point temperature, and to convert between the Fahrenheit and Celsius scales. Detailed instructions for use are printed on the calculator and are also published in NA 50-30FR-523, *Handbook, Operation and Care of Aerological Calculators, Computers, and Evaluators.*

CP-718/UM DENSITY ALTITUDE COMPUTER

The CP-718/UM density altitude computer is shown in figure 2-48. It is primarily used to compute

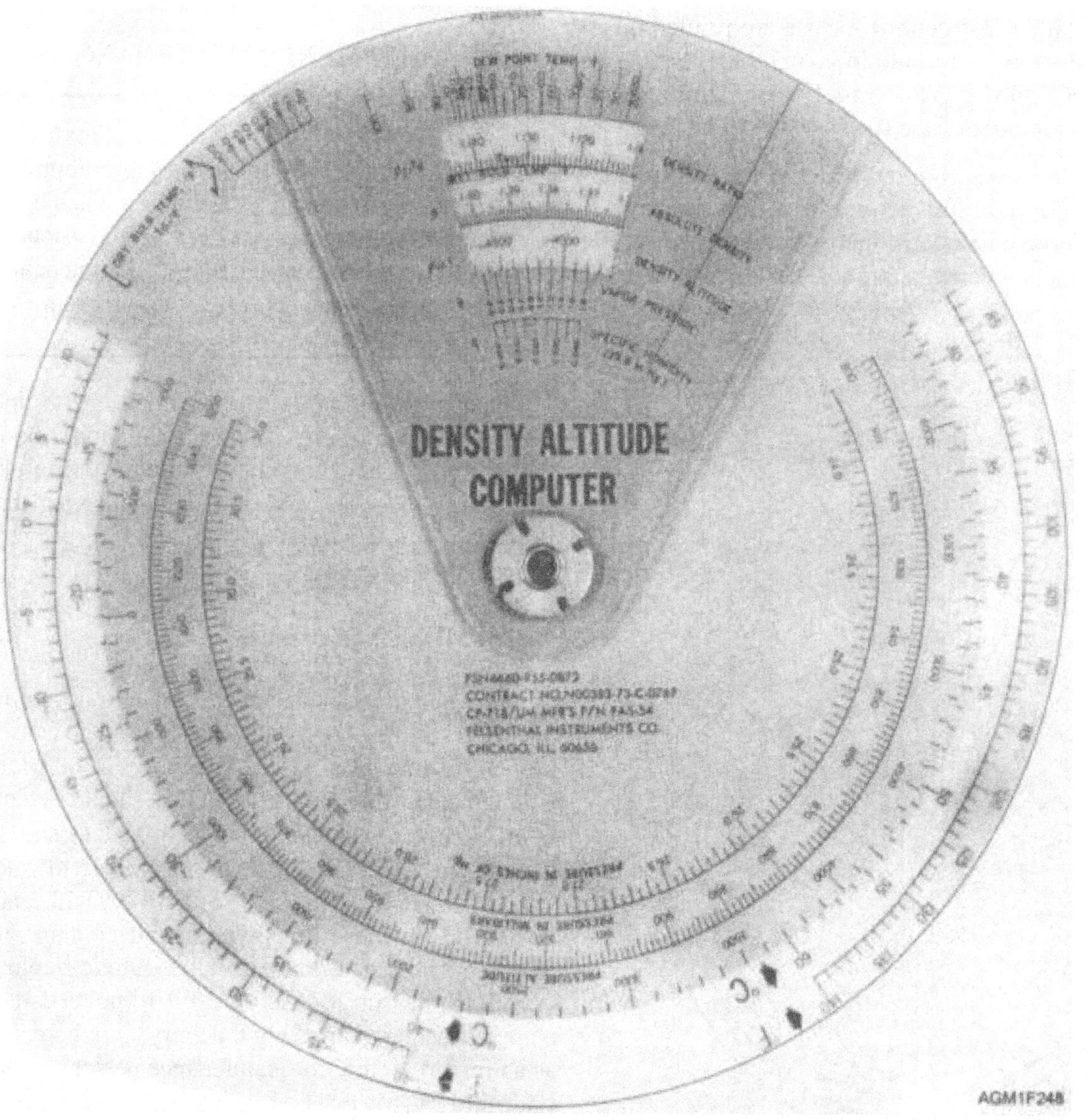

Figure 2-48.—CP-718/UM density altitude computer.

pressure altitude from pressure and air temperature, and to compute density altitude using the pressure, air temperature, and dew-point temperature. It may also be used to compute density ratio, absolute density, vapor pressure, and specific humidity. Detailed instructions for the computation of pressure altitude and density altitude are printed on the back of the calculator.

CP-264/U TRUE WIND COMPUTER

The CP-264/U true wind computer (fig. 2-49) is used to convert relative wind direction and speed to true wind direction and speed, using the ship's heading and speed. It may also be used to compute ship's course and speed, necessary to obtain a relative wind direction and speed across the deck for aircraft operations. The inputs for this calculation are the actual (true) wind speed and direction, and the required relative wind speed and direction. Detailed instructions for use are printed on the back of the computer and in NA 50-30FR-523. NAVMETOCCOMINST 3144.1 also contains information on the use of the CP-264/U calculator.

Throughout this chapter we have discussed "backup" equipment for primary-use observation equipment. You will occasionally need to use the backup equipment when your primary equipment fails.

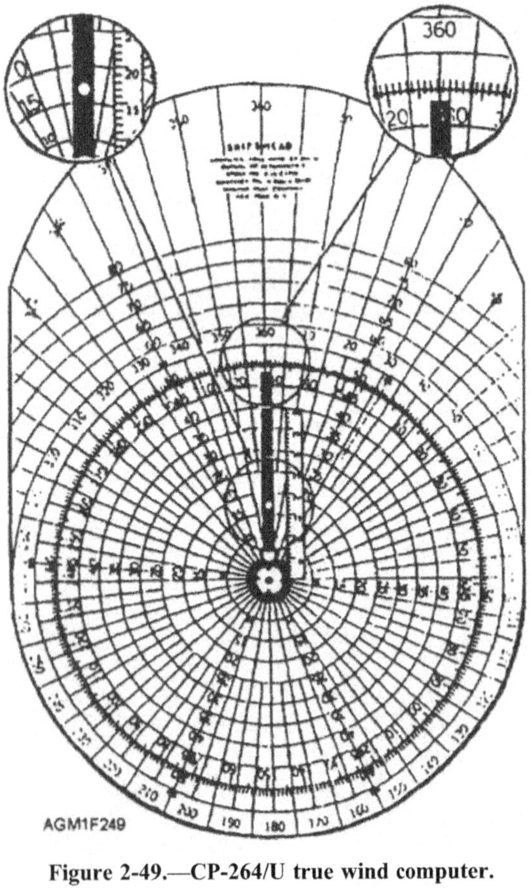

Figure 2-49.—CP-264/U true wind computer.

You will then need to begin repair of the primary system that has become inoperative. In the next section, we discuss equipment maintenance and repair.

REVIEW QUESTIONS

Q51. *What is the purpose of the CP-402/UM calculator?*

Q52. *What values may be computed by using the CP-165/UM calculator?*

Q53. *Which calculator is used to compute vapor pressure and specific humidity?*

Q54. *Besides true wind direction and speed calculations, what additional information can be obtained from the CP-264/U True Wind Computer?*

EQUIPMENT MAINTENANCE

LEARNING OBJECTIVES: Describe procedures to be followed during equipment outages. Recognize the resources available for equipment maintenance and repair assistance. Identify the basics of the Navy's Maintenance and Material Management (3-M) program.

Equipment outages will occur from time to time. In this section, we briefly discuss what should be done when equipment becomes inoperative, and the technicians that will provide maintenance and repair. We also briefly discuss the Navy's preventive maintenance program.

EQUIPMENT OUTAGES

An equipment outage occurs when a device is not available for use. Equipment outages can be caused by power failure, equipment breakdown, or planned maintenance. All equipment outages should be entered in an equipment outage logbook. Logbooks should be maintained for each major equipment system, such as ASOS, SMOOS, AN/GMQ-32, AN/GMQ-13, and the AN/UMQ-5. These logbooks should contain, at the minimum, the date and time the outage began and ended, the reason for the outage (if known), and the name of the person making the entry. Logbooks serve as an important tool for maintenance personnel. Notify your supervisor or section leader of all equipment outages. Your supervisor may then ask you to contact maintenance personnel.

MAINTENANCE RESOURCES

Repair of meteorological or oceanographic equipment can be accomplished through a variety of sources, depending on the severity of the problem. Meteorological and oceanographic equipment is usually maintained and repaired by electronics technicians that have completed Meteorological Equipment Maintenance (METEM) training. These technicians are stationed at most sea and shore commands with weather support facilities. Equipment problems beyond the capability of local technicians may be referred to civilian Field Technical Representatives (FTRs) based at major fleet concentration centers, such as Norfolk, Virginia, and San Diego, California. The contractor for a specific piece of equipment may also provide technical assistance.

3-M SYSTEM

The 3-M System provides support for all shipboard equipment, including all meteorological and oceanographic equipment. Equipment used ashore that is identical to shipboard installed equipment with Ships' 3-M Planned Maintenance System (PMS) support will be maintained using existing PMS coverage. The 3-M System includes both the PMS and the Maintenance Data System (MDS).

The PMS uses maintenance requirements cards (MRCs) to provide detailed step-by-step instructions for preventive maintenance on a particular equipment system. The main purpose of the Planned Maintenance System is to ensure that routine maintenance is completed as required and that maximum equipment operational readiness is achieved. This will help prevent major equipment breakdowns. When breakdowns do occur, shipboard personnel are called first to repair the equipment. All repair actions are documented in the 3-M System with formatted reports. Further information on the shipboard 3-M System is provided in OPNAVINST 4790.4, *Ships' Maintenance and Material Management (3-M) Manual.* NAVMETOCCOMINST 4790.2, *Maintenance and Material Management (3M) Systems Policies and Procedures,* provides detailed guidance on PMS and MDS unique to Naval Meteorology and Oceanography Commmand (NAVMETOCCOM) activities.

REVIEW QUESTIONS

Q55. What information should be entered in an equipment outage logbook?

Q56. Which personnel are primarily responsible for routine maintenance and repair of meteorological and oceanographic equipment?

Q57. What is the main purpose of the 3-M system?

SUMMARY

In this chapter we have discussed the equipment commonly used to conduct a surface aviation weather observation. Although your unit may not have some of this equipment, you should be able to identify the purpose of each piece of equipment. More importantly, you should be able to identify the reference manuals that provide the necessary instructions on the operation and maintenance of each equipment system. We strongly recommend that you review the operator manuals for each equipment system or major component in use at your command.

ANSWERS TO REVIEW QUESTIONS

A1. *Every minute.*

A2. *Infrared beam.*

A3. *Tactical Environmental Support System (TESS).*

A4. *12,000 feet.*

A5. *Via satellite.*

A6. *The display group and the sensor group.*

A7. *Time, air temperature, dew-point temperature, maximum and minimum temperature, wind direction and speed, pressure, and rainfall.*

A8. *Through a data transmission line.*

A9. *To store backup observing equipment.*

A10. *Keep the shelter free of dirt and oil the door hinges.*

A11. *Alcohol and mercury.*

A12. *NAVMETOCCOMINST 3141.2 and NAVMETOCCOMINST 3144.1.*

A13. *To the nearest 1/10 of a degree at the top of the meniscus.*

A14. *Weekly.*

A15. *NAVSEAINST 5100.3.*

A16. *Electric, sling, and rotor.*

A17. *Point the air intake into the wind and turn the fan motor on. After 60 seconds, read the wet-bulb temperature at 10-second intervals until the temperature reads the same for two consecutive readings.*

A18. *Changing the wet-bulb wick, keeping the instrument clean and dry, changing the batteries and light bulbs, and lubricating the motor.*

A19. *Twice every year aboard ship and every year at shore stations.*

A20. *Tap the face of the instrument and align the indicator arm over the reflection in the mirror. Read the scale to the nearest 0.005 inch or 0.1 hPa and apply any corrections.*

A21. *If the altimeter settings computed from the DASI are off by more than 0.02 inches from the altimeter settings computed from the aneroid barometer.*

A22. *An adjustable, grease-filled, dampening cylinder cancels out shipboard vibrations and the pitch and roll of the ship.*

A23. *Every 4 days at 1200 UTC, and on the first day of the month. Time should be set to UTC.*

A24. *Seven to eight pulls.*

A25. *Time adjustments are required when the time is in error by 15 minutes or more, and pressure adjustments are required when the pressure is off by 1.5 hPa or more.*

A26. *The AN/UMQ-5 wind-measuring set, the Type B-3 wind-measuring system, and the AN/PMQ-3 hand-held anemometer.*

A27. *At 0000 UTC on the first day of each month and at intermediate times to prevent loss of data.*

A28. *At the beginning and ending of each chart roll, near the time of each synoptic hour, when notified of an aircraft mishap, for each disruption in the trace, and at the first observation of the day when the station does not operate 24 hours a day.*

A29. *Relative wind.*

A30. *Observe the sea height and direction.*

A31. *The true wind will be less than the apparent wind.*

A32. *Obtain the ship's course and speed, and then proceed to an unobstructed area of the ship near the windward side. Align the sights with the true-wind reference or the bow of the ship parallel to the centerline.*

A33. *Only when the wind speed is lower than 15 knots.*

A34. *1 inch.*

A35. *The ML-217 rain gauge.*

A36. *It measures the transmissivity of light through air to determine visibility.*

A37. *The term "sector visibility" with respect to the AN/GMQ-32 indicates how far an unlighted object can be seen during the day. Runway visual range is how far runway lights can be seen by a pilot.*

A38. *NAVMETOCCOMINST 3141.2.*

A39. *Transmissivity values are presented on the AN/GMQ-32 in terms of percentages and converted to reportable values (i.e., statute miles) using tables in NAVMETOCCOMINST 3141.2.*

A40. *Replacing the recorder paper, rewinding the recorder, adjusting the recorder zero adjustment, and refilling/cleaning the inkwell and pens.*

A41. *Rotating beam projector, cloud height detector, and the ceiling height recorder.*

A42. *Cloud height is determined by the angle of the projected light beam reflected from the base of the clouds directly overhead the receiver.*

A43. *A correction factor.*

A44. *When applying the correction factor would yield an elevation angle less than 45°, the correction is ignored and the angle is read as is.*

A45. *Clinometer.*

A46. *NAVMETOCCOMINST 3144.1.*

A47. *After sighting the light spot at the cloud base, determine cloud height by multiplying the baseline distance by the tangent of the elevation angle.*

A48. *When clouds are thin, or on hazy days.*

A49. *High winds may carry lighter 10-gram balloons out of sight horizontally before they reach the cloud layer.*

A50. *Cloud height is determined by the elapsed time, in minutes and seconds, that the balloon takes to enter into the cloud base. The standard ascension rate of the balloon used is compared with height values listed in NAVMETOCCOMINST 3141.2 or NAVMETOCCOMINST 3144.1.*

A51. *To compute sea-level pressure, altimeter setting, and pressure altitude from station pressure.*

A52. *Dew-point temperature, relative humidity, and conversion values for the Fahrenheit and Celsius temperature scales.*

A53. *The CP-718/UM Density Altitude Computer.*

A54. *It may also be used to compute the ship's course and speed necessary for optimum relative wind direction and speed when conducting flight operations.*

A55. *The date and time the outage began or ended, the reason for the outage (if known), and the name of the person making the entry.*

A56. *METEM-qualified technicians.*

A57. *To ensure that routine maintenance is completed as required and that maximum equipment operational readiness is achieved.*

CHAPTER 3

SURFACE OBSERVATION CODES

INTRODUCTION

The World Meteorological Organization (WMO) is an international organization located in Geneva, Switzerland. Operating as part of the United Nations, its purpose is to provide international exchange of meteorological, oceanographic, and geophysical data and to conduct research in these areas. Most members of the United Nations are also members of the WMO, and have agreed to an international exchange of data in code forms specified by the WMO. These codes are used throughout the world and are known as the *WMO International Codes.* International codes have been established for reporting surface weather conditions, aviation weather conditions, upper atmospheric conditions, climatic conditions, oceanographic conditions, earthquakes, and volcanic activity. In this chapter, we will discuss weather logs for recording observations, the applicable reference sources, and the four surface observation codes. Now let's take a closer look at the WMO regions and code forms.

WMO REGIONS AND CODE FORMS

LEARNING OBJECTIVES: Recognize the seven WMO regions. Distinguish between the various regional and national codes. Identify the four different code forms used by weather observers. Identify the primary references for weather observations used by Navy and Marine Corps personnel.

The WMO has divided the world into the following seven regions:

- Region I - Africa
- Region II - Asia
- Region III - South America
- Region IV - North and Central America
- Region V - South-west Pacific
- Region VI - Europe
- Region VII - Antarctic

Within each region, certain codes are used that are not used in any other region. These codes are called "Regional codes." When information is included in these codes that does not conform to international code formats, the format difference is called a "Regional coding practice."

Many countries are contained in each WMO region. When a particular country elects to report additional information in an International code that does not conform to either the Regional coding practice or to the International code format, it is known as a "National coding practice." Similarly, when a particular country chooses not to use an International code but reports conditions by using their own code, the code is known as a "National code form."

The WMO International codes are explained in detail in WMO Publication 306, Manual on Codes, Volume I, *International Codes.* This publication has been republished by Commander, Naval Meteorology and Oceanography Command (CNMOC) as NAVAIR 50-lP-11. (See appendix IV, WMO Code Tables.) A more complete listing of regional and national coding practices is contained in WMO Publication 306, Manual on Codes, Volume II, *Regional Codes and National Coding Practices.* Both publications have been distributed to all Navy and Marine Corps observation sites.

Weather observers throughout the world record and report surface weather observations in four different international code forms. The four code forms are

- METAR Code;
- SPECI Code;
- Land Synoptic Code; and
- Ship Synoptic Code.

A modified version of the METAR and SPECI code is used by federal agencies in the United States. Surface *Weather Observations and Reports,* Federal Meteorological Handbook No. 1 (FMH-1) is a publication developed by the National Oceanic and Atmospheric Administration (NOAA) for use by the National Weather Service. It contains detailed instructions for the METAR and SPECI codes as used in the United States. Other federal agencies may develop their own observing handbooks. However, they must comply with the basic standards set forth in FMH-1.

A slightly more modified version of the METAR/SPECI codes has been developed for use by Navy and Marine Corps activities. The primary reference manuals outlining these procedures are NAVMETOCCOMINST 3141.2, *Surface METAR Observations User's Manual,* used by shore activities, and NAVMETOCCOMINST 3144.1, *United States Navy Manual for Ship's Surface Weather Observations,* used by shipboard observers. These provide detailed instructions on recording and encoding observed surface aviation weather observations using the METAR/SPECI format.

Before we discuss the different codes, let's take a closer look at the forms used to record observations and how these observation records are handled and archived.

REVIEW QUESTIONS

Q1. What does the acronym WMO mean?

Q2. The United States is located in what WMO region?

Q3. What are the four code forms used by Navy weather observers?

Q4. What instruction governs weather observation procedures at Navy and Marine Corps shore stations?

Q5. What instruction governs weather observation procedures for U.S. Navy ships?

OBSERVATION RECORDS

LEARNING OBJECTIVES: Describe the forms used to record weather observations. State the primary purpose of collecting accurate weather data. Explain three methods used to correct observation forms. Describe how observation records are maintained and archived.

Throughout the Navy and Marine Corps, observations are recorded on weather observation forms, also known as weather logs. These observation forms are permanent official records. At shore stations, observations are recorded on CNMOC Form 3140/12. See figure 3-1. This is a modified version of the Federal Meteorological Form 1-10 used by the National Weather Service. Shipboard observations are recorded on CNMOC Form 314113 (fig. 3-2). The data columns on the forms are generally identified by a column number in parenthesis, and the columns are generally arranged in numerical sequence. There are also sections for recording synoptic data, hourly and daily summaries, and runway condition summaries. A complete discussion of the actual recording of observation elements and the precision required will not be discussed here. Refer to either NAVMETOCCOMINST 3141.2 or NAVMETOCCOMINST 3144.1

FORM ENTRIES

METAR/SPECI aviation weather observations are recorded on the observation forms previously discussed. Entries should be neat and legible, and only a pencil with black, grade 2, medium lead or a 0.5 mm mechanical pencil with black lead should be used. Missing data is indicated by an "M" in the appropriate column. In the block labeled "STATION," enter the type (NAVLANTMETOC DET, NAVPACMETOCCEN, etc.), the official station name, and the state or country abbreviation. Also enter the latitude, longitude, station elevation, time conversion (LST to UTC), magnetic to true conversion, the date, and the ICAO call sign.

Only observers certified in accordance with NAVMETOCCOMINST *1500.2, Naval Meteorology and Oceanography Command Training and Certification Program, can* make entries on the form. Noncertified observers may make entries on the form under the immediate supervision of a certified observer. The certified observer will then assume responsibility for the validity of the entries by initialing in Column 15. Noncertified observers may initial the observation, but the certified observer must initial first.

CORRECTIONS TO OBSERVATION FORMS

The primary purpose of collecting and transmitting weather data is the safety of aircraft, ships, and personnel. Therefore, you must make every effort to ensure that data is entered correctly. At times, even the best observer will make an error, but you should make every effort to detect errors before the data is distributed. When an error is discovered, a corrected observation should be transmitted as soon as possible.

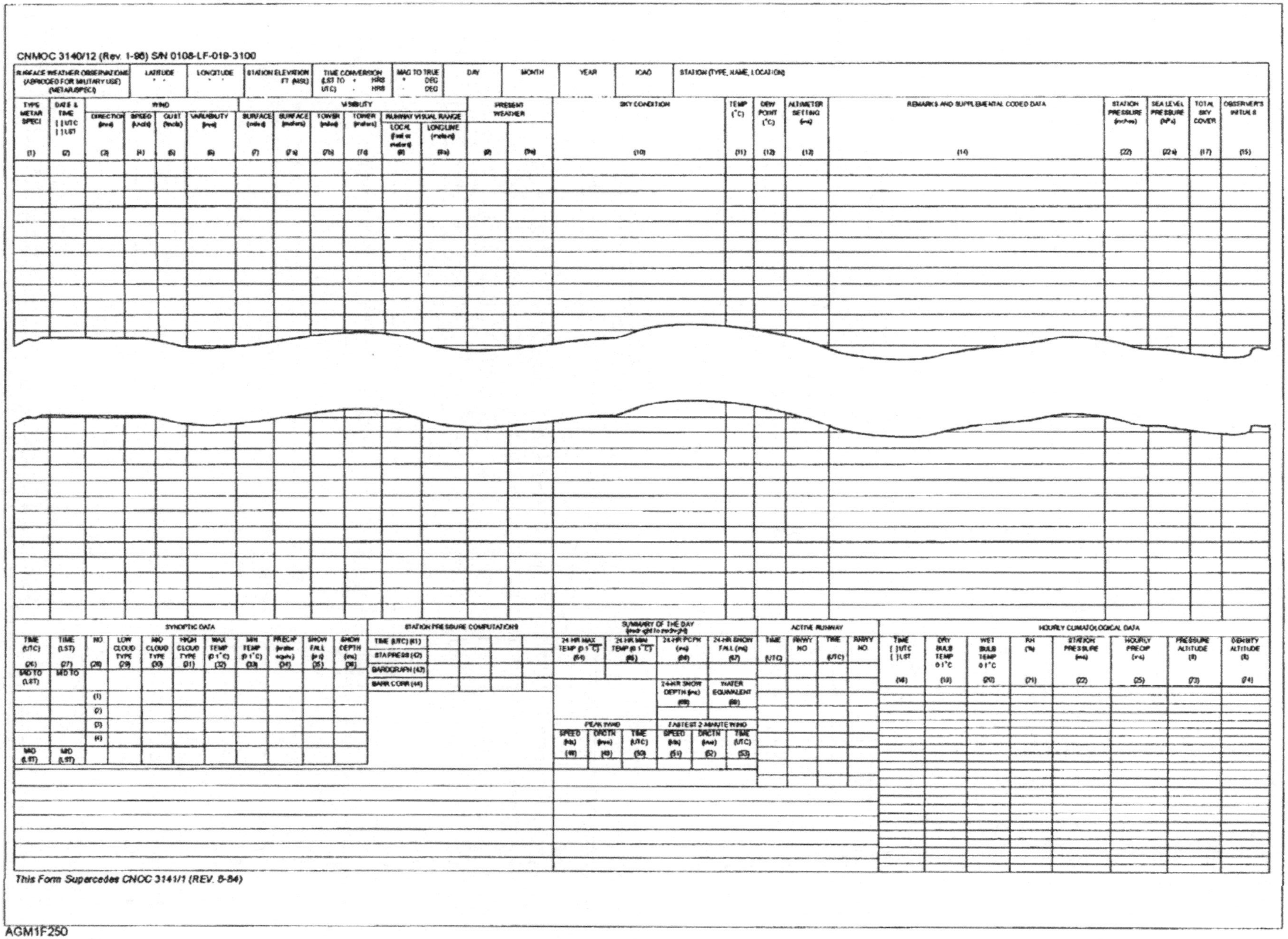

Figure 3-1.—CNMOC Form 3140/12, surface weather observations (shore station).

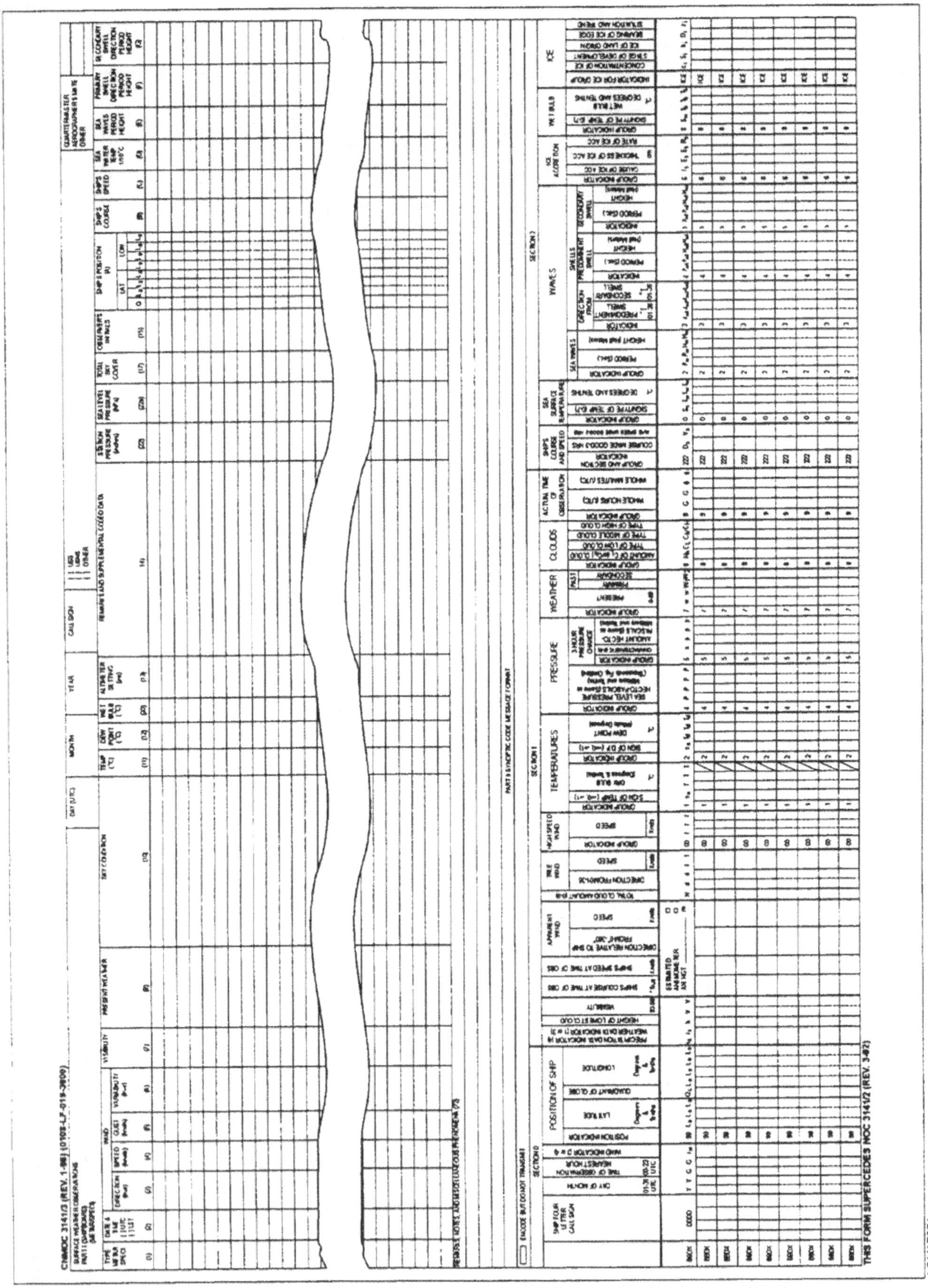

AGM1F251

Figure 3-2.—CNMOC Form 3141/3, surface weather observations (shipboard).

Since all observations are entered into the climatic data base, errors on the observation form should be corrected even if discovered too late to provide timely and accurate information to aircraft and ships. You should correct errors in the following three ways:

- If an error has not been transmitted, erase and reenter data correctly.
- If an error is discovered after transmission, line out the error, enter the correction in red, transmit the correct data, and then enter the *correction time,* which is the time of transmission of the corrected data.
- If an error has been transmitted but superseded with a later observation, simply line out and enter the correction in red.

See NAVMETOCCOMINST 3141.2 or NAVMETOCCOMINST 3144.1 for additional details on entering corrected data. To help reduce errors, all data should be reviewed by another qualified observer. Most sites have an established quality control person who routinely reviews the observation forms. Other than the observer, the quality control person is the only individual who may be authorized to make corrections to an observation record.

MAILING OF RECORDS

At the end of each month, the original observation forms are forwarded to the Fleet Numerical Meteorology and Oceanography Detachment, Asheville, North Carolina. Refer to NAVMETOCCOMINST 3141.2 or NAVMETOCCOMINST 3144.1 for detailed information on mailing observation forms. Additional information is also contained in NAVMETOCCOMINST 3140.1, *U.S. Navy Meteorological and Oceanographic Support System Manual,* and FNMODASHEVILLENOTE 3140, *Procedures for Submitting Meteorological Records.* The observation forms are microfiched at Asheville, and the information is included in the National Climatic Records data base. Duplicate copies of the observation forms may be retained on board until no longer needed; however, most sites retain the duplicate copies for at least 1 year.

After recording the observation, it must be encoded for local and/or longline (regional or national) dissemination. In the next section, we will discuss the codes used for dissemination purposes.

REVIEW QUESTIONS

Q6. *How should missing data be indicated on observation forms?*

Q7. *What is the primary purpose of collecting and transmitting weather data?*

Q8. *How is an observation error corrected on an observation form after the observation has been transmitted?*

Q9. *Where are observation forms mailed at the end of each month?*

SURFACE AVIATION WEATHER – METAR/SPECI CODE

LEARNING OBJECTIVES: Identify the applications of the format of the METAR and SPECI codes. Describe the elements of the METAR and SPECI codes and define the meaning of each element.

The METAR code is an International Observation code used to record and disseminate routine surface aviation weather observations. The SPECI code is a related International code used to record and disseminate selected surface aviation weather observations marking significant changes in the weather conditions. It is used to supplement the hourly observations in the METAR code.

The METAR and SPECI codes discussed in this section are used by all Navy and Marine Corps activities worldwide and are a modified version of the WMO METAR and SPECI codes used by most of the other countries in the world.

As previously mentioned, the primary reference manuals for Navy and Marine Corps activities are NAVMETOCCOMINST 3141.2 and NAVMETOCCOMINST 3144.1. You, as an observer, must be thoroughly familiar with these instructions. We will now discuss the encoding of the individual elements of a METAR observation.

TYPES OF OBSERVATIONS

In METAR, there are only two types of observations:

- Routine observations (METAR)—routine observations, taken each hour on the hour. The observation time is noted when the last element was observed and must be within 5 minutes before the scheduled observation hour.

- Special observations (SPECI)—observations taken to note any significant change in the weather as defined by NAVMETOCCOMINST 3141.2 and NAVMETOCCOMINST 3144.1. The observation time is the time that the significant weather element is observed. SPECI observations are also used in the event of an aircraft mishap, volcanic eruptions, and any other phenomena designated by local authority.

While METAR observations contain complete observation data, SPECI observations usually contain data pertinent only to the significant event or local requirement. Table II-1-1 of NAVMETOCCOMINST 3141.2 and table 2-1 of NAVMETOCCOMINST 3144.1 list which elements are reported for each type of observation.

BASIC CODE FORM

The METAR code contains both a basic report section and a supplemental or additive data section, as shown in table 3-1. In addition, remarks may be coded or in plain language. By regional decision, coded additive data groups may be added to the reports. Any missing elements are simply left out of the report. Keep in mind there are minor code differences between reporting stations in the United States and activities located outside of the continental United States (OCONUS).

BASIC METAR CODED REPORT

The basic coded METAR report contains identification data, wind, visibility, weather, sky coverage, temperature, dew-point temperature, and altimeter setting.

Identification Data

Identification data in a METAR or a SPECI observation consists of the observation type (METAR or SPECI), the station identifier, and the observation time.

ICAO STATION IDENTIFIER.—CCCC. The first letter indicates the country, while the remaining three letters identify the actual station. These identifiers are designated by the International Civil Aviation Organization (ICAO) and are used to identify each aviation terminal, including military stations. A list of approved identifiers can be found in FAA Order 7350.6, *Location Identifiers.*

UTC TIME.—YYGGggZ. METAR observations are identified with the date and actual time of the observation in UTC time. SPECI observations are identified with the time that the SPECI observation criteria was met. *YY* is the day, *GG* is the hours, and *gg* is the minutes (00 in a METAR observation, the actual time in minutes for a SPECI observation). *Z* is the indicator for UTC time. The date is not recorded on either observation form but is for dissemination purposes only.

Report Modifiers

A report modifier may or may not appear in the report. The report modifier AUTO is used by automated stations. The report modifier COR indicates the report was retransmitted with corrected data. Use the same time as the original report.

Table 3-1.—METAR and SPECI Code Format

SYMBOLIC FORMAT OF METAR/SPECI CODES: | Basic Reporting Code
METAR or **SPECI** CCCC YYGGgg**Z** AUTO/COR dddff(f)G$f_mf_m(f_m)$KT
$d_nd_nd_nVd_xd_xd_x$ VVVVVSM $RD_RD_R/V_RV_RV_RV_R$FT or
$RD_RD_R/V_NV_NV_NV_NVV_XV_XV_XV_X$FT w'w' $N_SN_SN_Sh_Sh_Sh_S$ or $VVh_Sh_Sh_S$ or
SKC $T'T'/T'_dT'_d$ $AP_HP_HP_HP_H$ RMK SLPppp
Supplemental Reporting Code
Note: When a specific phenomenon does not occur, the group or group suffix is not reported.

Wind

Wind is reported by the groups $dddff(f)Gf_mf_m(f_m)KT$ and $d_nd_nd_nVd_Xd_Xd_X$. The first wind data group includes reported true wind direction, wind speed, and wind character; the second group is used to report variable wind direction. While the wind direction and speed must be included in every METAR report even when the wind is calm, the wind character and wind direction variability are only reported when significant.

WIND SPEED, DIRECTION, AND CHARACTER.—$dddffGf_mf_mKT$. The true wind direction, *ddd,* is reported to the nearest 10 degrees, and may be encoded as *VRB* when wind speed is less than 6 knots. For example: Wind from 90° is reported "090." The *ff* is the 2-minute average wind speed in knots. If no gusts are reported, the identifier KT follows without a space. Calm winds are encoded "00000KT." Wind speeds exceeding 99 knots are reported in three figures. Gusts are only reported if winds exceed the average wind speed by 10 knots or greater. The *G* is the indicator for wind gusts, and f_mf_m is the maximum gust speed observed during the last 10-minute period. Gusts exceeding 99 knots are also reported in three figures; for example, a wind from 270 at 25 knots with gusts to 40 knots would be encoded as 27025G40KT.

While *KT* is the indicator for wind speed (in knots) as used in the United States, wind speeds may be reported by other countries in kilometers per hour or in meters per second. *KMH* indicates wind speed in kilometers per hour, and *MPS* indicates wind speed in meters per second.

VARIABLE WIND DIRECTION GROUP.—dndndnVdxdxdx. The variable wind direction group is only reported if the winds vary by 60° or more and the winds are >6 knots. The dndndn is the "left" direction limit (in true [T] azimuth degrees), while the *V* is an indicator for "variable," and dxdxdx is the "right" limit. For example, if the winds are variable between 123°T and 191°T, the group would be encoded 120V190. Wind direction may also be reported as variable if the wind speed is less than 6 knots, for example, VRBO4KT.

Visibility Groups

The visibility groups, VVVVVSM, $RD_RD_R/V_RV_RV_RV_RFT$, or $RD_RD_R/V_NV_NV_NV_NVV_XV_XV_XV_XFT$ are used to report *horizontal surface visibility* and *runway visual range,* respectively.

HORIZONTAL SURFACE VISIBILITY.—VVVVVSM group is the minimum significant surface horizontal visibility and is reported by *VVVVV* in statute miles (SM) with reportable increments, as described in chapter 1. This element is reported in *meters by* OCONUS stations.

RUNWAY VISUAL RANGE.—$RD_RD_R/V_RV_RV_RV_RFT$ or $RD_RD_R/V_NV_NV_NV_NVV_XV_XV_XV_XFT$. Runway visual range (RVR) is only reported when the RVR on any active runway is 6,000 feet or less or the visibility is less than 1 mile. Groups may be repeated as required for each runway. The *R* is the indicator for RVR, while D_RD_R is the runway identifier, plus designator *L* (left), *C* (center), or *R* (right) as appropriate. The $V_RV_RV_RV_R$ is the average touchdown RVR during the 10-minute observation period, in hundreds of feet.

If the Runway visual range is variable, the second RVR group, $RD_RD_R/V_NV_NV_NV_NVV_XV_XV_XV_XFT$, is used in place of the first. To be considered variable, 1-minute average readings must differ by more than 50 meters or, more than 20% of the mean value during the 10-minute observation period. The lowest reportable 1-minute mean minimum and the highest reportable 1-minute mean maximum visual ranges, respectively, are reported during variable RVR conditions.

When an observed RVR is less than the RVR sensor's established accurate minimum reading, the minimum accurate reading is reported and preceded by the letter *M.* Example: The sensor reports a 40-meter visibility but the sensor's minimum limit is 50 meters; $V_RV_RV_RV_R$ is encoded M0050FT. Likewise, if the observed RVR is greater than the established accurate RVR-sensor maximum, then the sensor's accurate maximum reading is reported and preceded by the letter *P.* Again, OCONUS stations will report RVR using meters.

Present Weather Groups

This group is used to report present weather—w'w'. This group is only reported when significant weather is occurring at the station or in the vicinity (within 10 miles of the station) at the time of the observation.

The aviation present weather group may be used three times to include all significant weather. Each usage consists of from two to nine characters to describe present weather. Each group may contain, in order, an intensity symbol "+" for heavy or "-" for light (no intensity symbol means the precipitation is moderate) or the indicator *VC* for vicinity, a two-letter qualifier

(*MI, PR, DR, BL, BC, SH, TS,* or *FZ*), and a two-letter weather descriptor. When more than one type of precipitation is falling at the same time, up to 3 two-letter precipitation descriptors may be combined in the same group. Obscurations are normally only reported when the prevailing visibility is less than 7 miles. Note that no intensity qualifier may be coded with *VC*. Table 3-2 provides a list of METAR/SPECI code weather phenomena type entries.

REVIEW QUESTIONS

Q10. How many types of observations are in the METAR code?

Q11. What elements make up the identification section of a METAR observation?

Q12. How should a wind direction of 183° a wind speed of 105 knots, and gusts at 120 knots be encoded?

Q13. How should a variable wind of 090° to 150° at 15 knots be encoded?

Q14. What does the symbol VVVVVSM indicate?

Q15. How should the RVR information "left runway 02, visibility 1,000 feet, varying 3,000 feet" be encoded?

Q16. What does the symbol BR indicate?

Q17. How should heavy showers in the vicinity be encoded?

Q18. What does the symbol PO indicate?

Sky Coverage/Height Group

Either the group $N_SN_Sh_Sh_Sh_S$ is used up to three times to report cloud layer coverage and height or the group VVhshshs is used to report conditions when the sky is obscured. The abbreviation "SKC" is used when the sky is clear.

CLOUD COVERAGE / HEIGHT.— $N_SN_SN_Sh_Sh_Sh_S$. This group is used to report coverage of cloud layers or cloud masses, not specifically for individual cloud types. The summation principle is used to evaluate sky coverage in ascending order, as discussed in chapter 1. The same sky coverage abbreviations for $N_SN_SN_S$ are also used on the form; that is, *FEW* for 1/8 to 2/8 coverage, *SCT* for scattered (3/8 to 4/8), *BKN* for broken (5/8 to 7/8), and *OVC* for overcast (8/8). The cloud base height, $h_Sh_Sh_S$, is encoded in hundreds of feet, as discussed in chapter 1. At mountain stations, if the layer is below station level, the height of the layer will be coded as ///.

This group may be used six times for masses of clouds at six different levels. When CB or TCU clouds are observed, the cloud abbreviation is added to the end of the group, as in "BKN035CB." Stations overseas are

Table 3-2.—METAR/SPECI Code Weather Phenomena Type Entries

QUALIFIER		WEATHER PHENOMENA		
Intensity or Proximity 1	**Descriptor** 2	**Precipitation** 3	**Obscuration** 4	**Other** 5
- Light Moderate + Heavy VC Vicinity	MI Shallow PR Partial BC Patches DR Low Drifting BL Blowing SH Shower(s) TS Thunderstorm FZ Freezing	DZ Drizzle RA Rain SN Snow SG Snow Grains IC Ice Crystals PE Ice Pellets GR Hail GS Small Hail and/or Ice Pellets	BR Mist FG Fog FU Smoke VA Volcanic Ash DU Widespread Dust SA Sand HZ Haze PY Spray	PO Well-dvlpd Dust/Sand Whirls SQ Squalls FC Funnel Cloud Tornado Waterspout SS Sandstorm DS Duststorm

NOTE: The weather groups are constructed by considering columns 1 to 5 in the table above in sequence, i.e., intensity followed by description, followed by weather phenomena. Example, heavy rainshower(s) is coded as **+SHRA.**

restricted to three reportable layers except when CB or TCU are present, in which case a fourth layer may be used.

SURFACE-BASED OBSCURATIONS.—Surface-based obscurations will also use $N_SN_SN_Sh_Sh_Sh_S$ group. However, $h_Sh_Sh_S$ will be encoded as 000 for heights of less than 50 feet. Stations overseas will encode 000 as ///.

When the sky is totally obscured, the VVhshshs group is used in place of the NsNsNshshshs group, with Wthe indicator for an indefinite ceiling, and $h_Sh_Sh_S$ the vertical visibility into the ceiling.

CAVOK STATEMENT.—In many countries, a statement for Ceiling And Visibility O.K. (CAVOK) may be substituted for the visibility, weather, and sky coverage when (1) the visibility is 10 km or greater, (2) there are no clouds below 5,000 feet (1,500 m) and no CB clouds, and (3) there is no significant weather occurring. This statement is NOT acceptable for use by Navy and Marine Corps activities.

Temperature

The temperature and dew-point temperature group $T'T'/T'_dT'_d$ is always included in a METAR observation report. T'T' is the temperature rounded up to the nearest whole degree Celsius, while $T'_dT'_d$ is the dew-point temperature rounded up to the nearest whole degree Celsius. Prefix single-digit temperatures with a zero. Negative temperatures are preceded by an *M.* For example, -9°C is encoded M09. If the temperature is not available, omit the entire group. If only the dew point is not available, use a single solidus after the temperature (10/).

Altimeter Setting

The altimeter setting group $AP_HP_HP_HP_H$ is always included in a METAR observation report. The altimeter setting following the *A* indicator is reported in inches of mercury rounded to the nearest hundredth of an inch. Do not use a decimal point. For example, an altimeter setting of 29.242 inches is encoded A2924.

REMARKS AND ADDITIVE DATA

In addition to the regularly reported data identified in table 3-1, the METAR code may also contain many remarks. Some of the remarks are required at different scheduled observation times, while others are required to provide amplification for certain significant meteorological events. Some remarks may be in the format of coded data groups (known as *additive data groups* and *supplemental data groups*), or as abbreviated plain language. NAVMETOCCOMINST 3141.2 and NAVMETOCCOMINST 3144.1 provide general guidance for plain language remarks and the proper abbreviations to use. FAA Order 7340.1, *Contractions,* contains a word-to-contraction encoding and contraction-to-word decoding listing for all allowable contractions. See table 3-3 for a listing of reportable remarks.

Table 3-3.—METAR and SPECI Code Remarks

REMARKS (RMK)		
Element	METAR	SPECI
Volcanic Eruptions	X	X
Tornadic Activity	X	X
Type of Station	X	X
Peak Wind	X	
Wind Shift, FROPA	X	X
Tower Visibility	X	X
Variable Prevailing Visibility	X	X
Sector Visibility	X	X
Lightning	X	X
Beginning/Ending of Thunderstorms and/or Precipitation	X	X
Thunderstorm Location	X	X
Hailstone Size	X	X
Virga	X	X
Variable Ceiling	X	X
Obscuration(s)	X	X
Variable Sky Condition	X	X
Significant Cloud Types	X	X
Pressure Rising/Falling Rapidly	X	X
Sea Level Pressure	X	X
Aircraft Mishap	X	X
No SPECI Taken	X	
Snow Increasing Rapidly	X	
Runway Condition	X	X
Breaks or Thin Spots in Overcast	X	X
First and Last Remark	X	X
X indicates element included at all stations		

All remarks and additive data groups used in an METAR coded observation must be indicated by the abbreviation "RMK" and should be in the order listed in table 3-3. Some general comments on certain remarks appear below.

SEA LEVEL PRESSURE (SLPppp). This remark is MANDATORY. Sea level pressure is encoded as SLPppp where SLP is the remark identifier and ppp is the sea-level pressure coded by using the tens, units, and tenths digits in hectopascals. For example, a sea level pressure of 998.2 hectopascals would be encoded as "SLP982." If sea level pressure is not available, it is coded as "SLPNO."

AIRCRAFT MISHAP (ACFT MSHP). If a report is taken to document weather conditions when notified of an aircraft mishap, the remark ACFT MSHP will be included on the observation form but will not be transmitted.

Table 3-4.—Additive Data Groups

Additive Data (RMK)		
Element	**METAR**	**SPECI**
Hourly Precipitation Amount	X	
3- and 6-hourly Precipitation Amount	X	
24-hour Precipitation Amount	X	
Snow Depth on the Ground, 4/sss	X	
Water Equivalent of Snow on the ground, 933RRR	X	
Cloud Types 8/$C_LC_MC_H$	X	
Hourly Temperature and Dew Point	X	
6-Hourly Maximum Temperature	X	
6-Hourly Minimum Temperature	X	
24-Hour Maximum and Minimum Temperature	X	
3-Hour Pressure Tendency 5appp	X	
X indicates element included at all stations		

FIRST and LAST (FIRST and LAST). At part-time stations, the first and last reports transmitted will be identified by including the word "FIRST" in the report of the day after a break in observing coverage and/or the word "LAST" in the last report of the day before a break in observation coverage.

Additive Data Groups

Additive data consists of supplementary precipitation, temperature, cloud, and pressure information reported hourly, every 3 hours or every 6 hours. At 0300Z, 0900Z, 1500Z, and 2100Z, the 3-hourly additive data is included in observation reports. At 0000Z, 0600Z, 1200Z, and 1800Z, the 6-hourly additive data is included in the observation reports.

The 3-hourly additive data consists of the 3-hour pressure tendency, cloud type information, and precipitation data reported in the symbolic format shown in table 3-4. The 6-hourly additive data shown in this table is identical except it contains 6-hour precipitation data.

Other supplementary information consists of snow depth, maximum or minimum temperature, and 24-hour precipitation. The snow depth group is reported (in inches) only if there is snow or frozen precipitation on the ground. It is normally reported in the OOOOZ and 1200Z observations, and at subsequent 6-hourly observations (18002, OOOOZ, 06002) if measurable precipitation has occurred during those periods. The maximum and minimum temperatures during the past 6 hours are reported at 0000Z, 0600Z, 1200Z, and 1800Z, while maximum and minimum temperatures during the past 24 hours are reported at 0600Z. The 24-hour precipitation (liquid equivalent) is reported only at 1200Z. If a station is closed down for a weekend or holiday, the 24-hour precipitation should also be used to report total precipitation since the last reported 24-hour precipitation In this case, the 24-hour precipitation may actually be used to report a 72-hour total precipitation.

Runway Conditions

Runway surface condition (RSC) and average runway condition reading (RCR) are included in a METAR coded report when runway conditions produce less than normal braking for landing aircraft. Different codes may be combined to describe conditions. Each condition should be followed by a decelerometer reading (RCR) from 02, poor braking action, to 25, excellent braking action. Codes used for RSC include "WR" for wet runway, "SLR" for slush on runway,

"LSR" for loose snow on runway, "PSR" for packed snow on runway, and "IR" for ice on runway. The code "RCRNR" is used when braking action is impeded but accurate decelerometer readings are not available. For example, a runway with packed snow and a decelerometer reading of 15 would be reported as PSR15.

NAVMETOCCOMINST 3141.2 contains many additional examples of runway condition reports, including descriptive terms used when a decelerometer is not available.

Regional Differences

Specific data that may be included in the supplemental data section within each WMO region are normally described in WMO Publication 306, Manual on Codes, Volume II, *Regional Codes and Coding Practices.* Within WMO Region IV, North America, at the time of this writing, no decisions have been made on regional additions to the METAR code.

SPECI CODE OBSERVATION REPORTS

The SPECI code is used to report any significant changes in the weather at any time other than the scheduled hourly observation. Both NAVMETOCCOMINST 3141.2 and NAVMETOCCOMINST 3144.1 outline criteria and requirements for special observations that are reported using the SPECI code. The SPECI code uses the same groups in the same order as reported in the METAR code except that for a special observation, SPECI replaces METAR. The time reported in a SPECI report is the time that the change was observed to have occurred, not the time that the report is transmitted.

REVIEW QUESTIONS

Q19. How should scattered cumulonimbus at 2,500 feet be encoded?

Q20. How should a surface-based partial obscuration (3/8) of fog be encoded?

Q21. In what situation is the group $VVh_sh_sh_s$ used?

Q22. How should an air temperature of -4.4°C be encoded?

Q23. What remark in the METAR code is mandatory for all observations?

Q24. What does the additive data group 4/006 indicate?

Q25. The total 24-hour precipitation is reported at what time?

Q26. What does the additive data remark SLR12 indicate?

Q27. What publications contain criteria and requirements for SPECI observation?

SHIPBOARD SURFACE AVIATION WEATHER—SHIP AVIATION CODE

LEARNING OBJECTIVES: Identify the manuals that provide instructions for recording weather elements on the U.S. Ship Aviation Code observation form. Identify the applications of the U.S. Ship Aviation Code observation form. List the weather elements required on the U.S. Ship Aviation Code form and describe how the elements are entered and encoded.

Shipboard weather observers use CNMOC Form 3141/3 to record weather observation data in the METAR/SPECI code. Unlike shore sites, the METAR code aboard ship is used only to record data, not to encode data for transmission. For transmission to data collection centers, these observations are encoded into WMO Code FM 13-XI, the ship synoptic code, normally transmitted every 6 hours. The bottom portion of CNMOC Form 3141/3 is used to record the encoded synoptic observations. We will discuss the Ship Synoptic code later in this chapter. In this section, we briefly cover how the data is recorded on the form.

The shipboard observation form (fig. 3-2) is very similar to the shore station METAR code form. While some columns for recording weather elements at sea are the same as those used ashore, the elements that are unique to observations at sea are designated with letters. Additionally, all columns are identified with the type of data to enter, and, in many cases, with the system of measurement and the degree of precision required.

NAVMETOCCOMINST 3144.1, *Manual for Ship's Surface Weather Observations,* provides detailed instructions for recording observed weather elements on the form. Remarks that apply to aircraft operations are entered in column 14. Remarks are generally entered in the same order that the basic coded information is entered.

The additive data, supplemental data, remarks about RVR, and runway conditions as used in the METAR/SPECI code at shore stations are NOT used by shipboard observers.

REVIEW QUESTIONS

Q28. What column on CNMOC Form 3141/3 is used to record remarks?

Q29. What column is used to record sea surface temperature?

Q30. What column is used to recordsea wave height?

Q31. What publication provides detailed instructions for taking and recording ship observations aboard U.S. Navy ships?

SYNOPTIC CODES

LEARNING OBJECTIVES: Identify the manuals that provide instruction on encoding and decoding land and ship synoptic observations. Explain the meaning of each code figure in the land and ship synoptic code. Describe the code used to relay reports from moored coastal observation buoys.

Where the METAR/SPECI codes are designed to report aviation weather, the Synoptic code is specifically designed to include data for use in analyzing the current overall weather situation. It is a numerical code that consists mostly of groups of five digits, specifically designed. to permit automatic loading of computer data bases.

The Synoptic reports are transmitted by selected stations worldwide at 0000Z, 0600Z, 1200Z, and 1800Z—the "Synoptic Hours"—with the reports generally called "Main Synoptic" reports. Significant reports called "Intermediate Synoptic" reports may be transmitted at the "Intermediate Synoptic" hours: 0300Z, 0900Z, 1500Z, and 2100Z.

Although the Synoptic code transmitted by land stations and by ships report many of the same weather elements by using the same symbolic groups, there are some differences in the way the station is identified. There are also different types of data that are only reported by land stations, just as there is some data that is only reported by ships.

The primary reference manual used to encode weather elements observed during an aviation hourly weather observation at land stations into land Synoptic codes is the Federal Meteorological Handbook No. 2 (FMH-2), *Surface Synoptic Codes.* This publication describes which elements are encoded, and also covers supplemental data, which is added to the International code form as Regional data in WMO Region IV — North America, (and U.S. stations in WMO Region V — Central and Southern Pacific, including Hawaii, Guam, and the Philippines). It also includes National Data, which is added to the code in the United States.

The primary reference used when encoding weather observations into the ship Synoptic code is NAVMETOCCOMINST 3144.1, *United States Nary Manual for Ship'.s Surface Weather Observations,* although the FMH-2 also provides guidance on the ship Synoptic code. The publications that are most useful in decoding Synoptic reports received from overseas stations are the WMO Publication 306, Manual on Codes, Volume 1, *International Codes,* and Volume 2, *Regional Codes and National Coding Practices. The International Codes* provides symbolic formats for all of the International codes, an alphabetical listing of each symbolic element, with an explanation of each code element and reference to the appropriate code table, as well as a section providing all of the WMO code tables.

In the following text, we discuss the land Synoptic code, the ship Synoptic code, and the code used to relay reports from moored coastal observation buoys. Most of the applicable code tables are listed in Appendix IV. First, let's consider the land Synoptic code.

LAND SYNOPTIC CODE

In addition to reporting surface aviation hourly observations, many Navy and Marine Corps land stations, both in the United States and overseas, also report Synoptic weather observations in WMO Code FM 12-XI SYNOP.

Table 3-5 provides a breakdown of the symbolic code format of the land Synoptic code. Only those groups considered significant are included in a report. If the type of data the code group requires is not normally available at a station, the entire group is not reported. A solidi (forward-slash /) is used in place of each number that is normally reported, but is unobservable because of weather conditions or equipment failure.

Table 3-5.—Symbolic Format of Land Station Synoptic Weather Observation Report (WMO Code FM 12 XI SYNOP)

AAXX $YYGGi_w$ IIiii i_Ri_Xh Nddff (00fff) ls_nTTT $2s_nT_dT_dT_d$ $3P_OP_OP_OP_O$ 4PPPP (or $4a_3hhh$) 5appp $6RRRt_R$ $7wwW_1W_2$ $8N_hC_LC_MC_H$ 9GGgg 333 0. . . . $ls_nT_XT_XT_X$ $2s_nT_nT_nT_n$ 3Ejjj 4E'sss $5j_1j_2j_3j_4$ $(j_5j_6j_7j_8j_9)$ $6RRRt_R$ $7R_{24}R_{24}R_{24}R_{24}$ $8N_sCh_sh_s$ $9SPSPS_pS_P$ 444 $N'C'H'H'C_t$ 555;

Identification Data

The first few groups of the Synoptic code represent Section 0, the Identification Data section. The Identification Data section contains a data identifier, a date-time group, and a station identifier.

DATA IDENTIFIER.—The data identifier for a land synoptic report is the first group, which is always AAXX.

DATE-TIME GROUP.—$YYGGi_W$. This group provides the day of the month and the hour of the report, as well as an indicator for the wind speed. *YY* is the UTC day of the month (two digits), *GG* is the UTC hour of report (00, 03, 06, etc.), and i_W is the indicator for source and units of wind speed (WMO code table 1855).

STATION IDENTIFIER.—IIiii. The station identifier is composed of the WMO block number and the three-digit station number. *II* is the WMO block number, and *iii* is the WMO station number. Each WMO region is subdivided into data block areas. Large countries may be designated as a data block, or several smaller, adjoining countries may be grouped together to form a block. In North America the United States is block 72, with supplemental stations belonging to block 74. Alaska is block 70. Canada is block 71; Mexico is block 76; and all the countries of the Caribbean and Central America are grouped as block 78.

International Data

Section 1 of the Synoptic code contains meteorological data for international exchange, and immediately follows the identification data. This section consists of cloud height and visibility, winds, sky coverage, air temperature, dew-point temperature, station pressure and sea-level pressure, pressure change, precipitation, and weather groups. In North America (WMO Region IV), it also contains a group for cloud type. Most of the five-digit groups begin with a group identifier number that does not change.

CLOUD HEIGHT/VISIBILITY.—i_Ri_XhVV. The i_R is the precipitation data (group $6RRRt_R$) indicator (WMO code table 1819), while i_X is the indicator for station type (manned or unmanned) and for present and past weather (group $7wwW_1W_2$) (WMO code table 1860). The *h* is height (AGL) of lowest cloud (WMO code table 1600), and *VV* is horizontal surface visibility (WMO code table 4377).

SKY COVERAGE/WIND.—Nddff. The *N* is total sky cover in eighths or oktas (WMO code table 2700), while *dd* is the wind direction (hundreds and tens of degrees true) to the nearest 10 degrees, and *ff* is the sustained wind speed in the units indicated by i_W. If winds exceed 99 (knots or meters per second), *ff* is encoded 99 and the *00fff* group is included, with 00 as an indicator and *fff* as the wind speed in hundreds, tens, and units.

AIR TEMPERATURE.—$1s_nTTT$. The s_n is the temperature sign, 0 for positive (or 0) and 1 for negative. This indicator is used throughout the Synoptic code to indicate the temperature sign. The *TTT* is the temperature in tens, units, and tenths of a degree Celsius.

DEW-POINT TEMPERATURE.—$2s_nT_dT_dT_d$ The $T_dT_dT_d$ is the dew-point temperature in tens, units, and tenths of a degree Celsius.

STATION PRESSURE.—$3P_OP_OP_OP_O$. The $P_OP_OP_OP_O$ is the station pressure in hundreds, tens, units, and tenths of hPa (thousands value omitted).

SEA-LEVEL PRESSURE—4PPPP. The PPPP is the sea-level pressure in hundreds, tens, units, and tenths of hPa (thousands value omitted).

STANDARD LEVEL HEIGHT.—$4a_3hhh$. Reported by mountain stations in place of sea-level pressure, a_3 is the standard isobaric surface reported (WMO code table 0264) and *hhh* is the geopotential height in meters, omitting thousands value.

3-HOUR PRESSURE CHANGE.—5appp. The *a* is the pressure tendency (WMO code table 0200), and *ppp* is the 3-hour pressure change in tens, units, and tenths of hPa.

PRECIPITATION TOTAL.—6RRRt_R. The *RRR* is the liquid equivalent of the precipitation amount reported in hundreds, tens and units of millimeters (exception: 990 is a trace, and codes 991 to 999 are used to report tenths of a millimeter total precipitation from 0.1 to 0.9 respectively). And t_R is the duration of the reference period ending at the reporting time (WMO code table 4019). Precipitation amounts observed in inches may be converted to millimeters by multiplying 25.4 mm/inch. In WMO Regions IV and V, this group is normally used to report 6-hour precipitation at each synoptic observation with t_R encoded 1, except the 0000Z report from U.S. stations in Region V reports 24-hour precipitation with TV encoded 4.

WEATHER.—7wwW_1W_2. The *ww* is the present weather code (WMO code table 4677), and W_1 and W_2 are past weather codes (WMO code table 4561). See appendix IV.

PREDOMINANT CLOUD TYPE.—8$N_hC_LC_MC_H$. The N_h is the summation coverage of all low etage clouds present, or, if no low clouds are present, the summation of all the mid-etage cloud coverage, in oktas of the sky covered (WMO code table 2700). C_L is the predominant low-etage cloud type (WMO code table 0513). C_M is the predominant mid-etage cloud type (WMO code table 0515). And C_H is the predominant high-etage cloud type (WMO code table 0509).

OBSERVATION TIME.—9GGgg. This group is only encoded when the actual observation time (time the last element in the observation was observed) differs from the standard observation time (the synoptic or intermediate synoptic hour) by 10 minutes or more. The *GG* is the UTC hour, while the *gg* is the minutes of the actual observation time.

REVIEW QUESTIONS

Q32. What is the purpose of the Synoptic code?

Q33. What are the intermediate synoptic hours?

Q34. What is the primary reference manual used for encoding land synoptic observations in the United States?

Q35. What groups are included in the identification section of a land synoptic observation?

Q36. How would i_Ri_XhVV be encoded for a manned station with no precipitation or significant weather in the past 6 hours, a low overcast cloud deck at 3,500 feet, and visibility at 7 miles?

Q37. How would a station with 6/8 total sky coverage and a wind direction of 240° and a wind speed of 103 knots be encoded?

Q38. What do the sections $1S_nTTT$ and $2S_nT_dT_dT_d$ indicate?

Q39. What do the sections $3P_OP_OP_OP_O$ and 4PPPP indicate?

Q40. If the station pressure at 0300Z was 1020.6 hPa, then increased to a high of 1021.5 hPa at 0400Z, and is now 1019.8 hPa at 0600Z, how should the group 5appp be encoded?

Q41. If a station received 1.2 mm of precipitation between 0000Z and 0600Z, how should $6RRRt_r$ be encoded?

Q42. If a station had heavy snow showers at the time of observation and rain showers and overcast conditions in the previous 2 hours, how should the group $7wwW_1W_2$ be encoded?

Q43. If a station has no low clouds, 2/8 altocumulus castellanus, 3/8 thin altostratus, and 3/8 cirrostratus not covering the whole sky and not invading the celestial dome, how should the group $8N_hC_LC_H$ be encoded?

Regional Data Section

The regional data section, following the group 333, includes both internationally established codes and codes unique to each region.

WMO REGION IV STATE OF THE SKY IN THE TROPICS.—0$C_SD_LD_MD_H$. The C_S is the cloud state from FMH-2, table 6-4; while D_L, D_M, and D_H are the direction from which the low-, mid-, and high-etage clouds, respectively, are moving (using the 8-point compass, 1 = NE, 2 = E, and so forth; 0 is no movement, and 9 is unknown). This group is only reported by stations in the southern portion of WMO Region IV within 310 statute miles of the seacoast during the tropical season.

MAXIMUM TEMPERATURE.—1$s_nT_XT_XT_X$. The $T_XT_XT_X$ is the maximum temperature in tens, units and tenths of a degree Celsius. The time period is specified by regional agreement; see table 3-6 for WMO Region IV and V conventions.

MINIMUM TEMPERATURE.—2$s_nT_nT_nT_n$. The $T_nT_nT_n$ is the minimum temperature in tens, units, and tenths of a degree Celsius.

Table 3-6.—Synoptic Reporting Times for Maximum and Minimum Temperatures for U.S. Stations in WMO Regions IV and V

REPORT	WMO REGION	REPORTED INFORMATION
0000Z	IV only	$T_XT_XT_X$ for past 12 hours and $T_nT_nT_n$ for past 18 hours.
	V only	$T_nT_nT_n$ for past 24 hours.
0600Z	IV only	$T_XT_XT_X$ for past 24 hours and $T_nT_nT_n$ for past 24 hours.
1200Z	IV only	$T_XT_XT_X$ for previous calendar day ending 2400 LST, and $T_nT_nT_n$ for past 12 hours .
	V only	$T_XT_XT_X$ past 24 hours.
1800Z	IV only	$T_XT_XT_X$ for past 12 hours and $T_nT_nT_n$ for past 18 hours .

STATE OF THE GROUND WITHOUT SNOW OR ICE.—3Ejjj. The *E* is the state of the ground (WMO code table 0901), and *jjj* is regional information. (Group is not reported in WMO Region IV.)

SNOW OR ICE ON THE GROUND.—4E'sss. The *E'* is state of snow or ice covering the ground (WMO code table 0975) and is reported in WMO Region IV as /. The *sss* is the average snow/ice depth in hundreds, tens, and units of centimeters.

SUPPLEMENTARY INFORMATION.—$5j_1j_2j_3j_4$. The j_1 is the information designator (WMO code table 2061), while $j_2j_3j_4$ and, if necessary, the following group, $j_5j_6j_7j_8j_9$, contains the coded information.

Cloud Movement.—$56j_2j_3j_4$. In WMO Region V, U.S. stations use this group to report cloud movement pertaining to the clouds reported in the 8NhCLCMCH group for the low-, mid-, and high-etage clouds, respectively, and is not reported in Region IV.

24-Hour Pressure Tendency.—The $58p_{24}p_{24}p_{24}$ group is used to report a net increase or no net change in the 24-hour pressure tendency in tens, units and tenths of hPa. The $59p_{24}p_{24}p_{24}$ group is used to report a net decrease in the 24-hour pressure tendency. The 24-hour tendency is reported only in each of the main synoptic reports in the southern part of Region IV instead of the 5appp group in section 1. In Region V, U.S. stations report the 5appp group but additionally report the 24-hour tendency at 0000Z and 1200Z.

PRECIPITATION TOTAL.—$6RRRt_R$ If this group is not previously reported in section 1, some regions report this group in the supplemental data section.

24-HOUR PRECIPITATION.—$7R_{24}R_{24}R_{24}R_{24}$. This group is included in all synoptic reports in Region IV unless no precipitation was received. The $R_{24}R_{24}R_{24}R_{24}$ is the 24-hour total liquid equivalent precipitation in hundreds, tens, units, and tenths of millimeters. A trace (<0.005 inch or <0.05 millimeter) is encoded 9999.

CLOUD LAYER DATA.—$8N_SCh_Sh_S$. This group reports the amount, predominant type, and height of cloud in a layer, and may be used four times to report four individual layers. The N_S is the amount of cloud in oktas, the C is the predominant cloud type (WMO code table 0500), and h_Sh_S is the height of the cloud base (WMO code table 1677). This group is used by select stations in WMO Region IV.

SPECIAL INFORMATION.—$9S_PS_Ps_ps_p$. This group may be used to report special information in the International code, as selected by regional decision. The S_PS_P indicates the type of data (WMO code table 3778) while the s_ps_p is the specific data. This group is not generally used in Region IV or by U.S. stations in Region V.

Regional Mountain Station Data

Indicated by the group 444, section 4 of the report contains information generally reporting clouds with bases below the station level (mountain stations). All data other than $N'C'H'H'C_t$ is in regionally decided code. The International code $N'C'H'H'C_t$ is described by WMO code tables 2700, 0500, and 0552 for N', C', and C_t, with *H'H'* the top of the cloud mass in hundreds of meters. It is not generally in use in Region IV or by U.S. stations in Region V.

National Code Groups

All data in section 5 after the 555 indicator is in National code forms. In WMO Region IV, U.S. National Weather Service stations report certain information in this section on record temperatures, coastal tides, and coastal or lake water temperatures. The specific codes used are contained in FMH-2. Navy and Marine Corps stations do not use these codes. Canada and Mexico report different national codes, as may each country in Region V. Now let's look at the ship synoptic code.

REVIEW QUESTIONS

Q44. What information follows the 333 indicator group?

Q45. When should the 24-hour maximum and minimum temperatures be reported in WMO region IV?

Q46. When is the $58p_{24}p_{24}p_{24}$ group used?

Q47. What information follows the 555 indicator group?

SHIP SYNOPTIC CODE

The ship Synoptic code (WMO code FM 13-XI SHIP) is used aboard U.S. Naval ships to report weather as observed and recorded in the METAR/SPECI code. The ship synoptic report is encoded on the bottom portion of the ship observation form. Internationally, the code is used to disseminate meteorological data from nearly all ships that observe weather conditions.

Table 3-7 shows the symbolic format of the ship Synoptic code. Many of the data groups are the same as the land Synoptic code, especially in data section 1. The ship Synoptic code uses different environmental data in data section 2, "Maritime Data," not generally used by shore stations. While some countries may use selected groups from data section 3 as identified in the land synoptic report, U.S. Naval vessels currently do not carry the equipment required to take the appropriate measurements, and likewise, do not report total precipitation ($6RRRt_R$).

Identification Data

The identification data in the ship Synoptic code contains the message type identifier *BBXX,* the ship's International Radio Call Sign, the date-time group, and the latitude and longitude of the ship.

INTERNATIONAL RADIO CALL SIGN.—DDDD. The ship's four-letter International Radio Call Sign (IRCS) is used as identification for the station. Converting the IRCS to the ship's name and country of registration may be done by using Allied Communication Publication 100 (ACP-100), *Allied Call Sign and Address Group System - Instructions and Assignments.* The group may consist of as few as three or as many as six letters or letter-number combinations, or may use the word *SHIP* for any ship, or *RIGG* for a stationary (oil) platform.

DATE-TIME GROUP.—$YYGGi_W$. Same as for land Synoptic code.

LATITUDE/LONGITUDE.—$99L_aL_aL_a$ and $Q_CL_OL_OL_OL_O$. The 99 is an indicator for latitude, and $L_aL_aL_a$ is the latitude in degrees and tenths of a degree (the minutes divided by 60 yields tenths of a degree). The first value in the longitude group, Q_C, is the quadrant of the globe (WMO code table 3333). (In relation to the equator and the prime meridian, quadrant 1 is north and east, 3 is south and east, 5 is south and west, and 7 is north and west.) And $L_OL_OL_OL_O$ is the hundreds, tens, units, and tenths of degrees longitude.

International Data Section

All of the data in the International Data Section, Section 1 of the code, is exactly the same as the land Synoptic code. However, since most ships do not carry rain-measuring equipment, the rainfall group, $6RRRt_R$, is normally omitted from reports, and the associated indicator, i_R, is reported as 4.

Table 3-7.—Symbolic Format of Ship Synoptic Weather Observation Report (WMO Code FM 13-XI SHIP)

BBXX DDDD $YYGGi_W$ $99L_aL_aL_a$ $Q_CL_OL_OL_OL_O$ i_Ri_XhVV Nddff (00fff) $1S_nTTT$ $2S_nTdTdTd$ 4PPPP 5appp ($6RRRt_R$) $7wwW_1W_2$ $8N_hC_LC_MC_H$ 9GGgg $222D_Sv_S$ $0S_ST_WT_WT_W$ ($1P_{wa}P_{wa}H_{wa}H_{wa}$) $2P_WP_WH_WH_W$ $3d_{w1}d_{w1}d_{w2}d_{w2}$ $4P_{w1}P_{w1}H_{w1}H_{w1}$ $5P_{w2}P_{w2}H_{w2}H_{w2}$ ($70H_{wa}H_{wa)}H_{wa}$) $6I_sE_sE_sR_s$ $8S_wT_bT_bT_b$ ICE $c_iS_ib_iD_iZ_i$ (or plain language);

Maritime Data Section

The Maritime Data Section, or Section 2 of the code, follows the indicator 222, the first three digits in the 222Dsvs group. This section is normally not transmitted from land stations, since it contains course and speed of the ship, sea-surface temperature, sea and swell wave information, ice accretion information, and information on sea ice.

SHIP'S COURSE AND SPEED.—222D_Sv_S. The D_S is the direction of displacement of the ship from 3 hours before the report indicated using WMO code table 0700 (0 is no displacement, 1 is toward the northeast, 2 is toward the east, etc.). The v_S indicates the speed "made good" from 3 hours before the observation until the observation time (in 5-knot increments) using WMO code table 4451. For example, a ship may move in a large circle at 10 knots, ending up at report time at the same location it was 3 hours previous. The D_S would be encoded as 0 for "no net displacement," and v_S would be encoded as 0 for "no speed made good." A second ship may follow an erratic course, ending up at report time 2 1 nautical miles southeast of the position 3 hours previous. The D_S is encoded as 3 for southeast displacement, and the v_S is encoded as 2 for speed made good "6 to 10 knots" (21 nautical miles divided by 3 hours equals 7 knots).

SEA - SURFACE TEMPERATURE.—0$s_sT_WT_WT_W$. The code symbol s_s is the temperature sign for seawater, which is followed by the sea-surface temperature in tens, units, and tenths of degrees Celsius as $T_WT_WT_W$.

SEA WAVES.—2$P_wP_wH_wH_w$. The period of the sea waves, as determined by an observer, is reported as P_wP_W in tens and units of seconds. The height of the sea waves is reported in one-half meter units in H_wH_w. Convert observed height in feet to half-meter units by multiplying by 0.61 and rounding off. When decoding, half-meter units may be converted to feet by multiplying by 1.64. A conversion table is available in NAVMETOCCOMINST 3144.1. Ships with automatic wave sensors and buoys with sensors use group 1$P_{wa}P_{wa}H_{wa}H_{wa}$ in place of the "2" group. The $P_{wa}P_{wa}$ is the sensor-measured sea-wave period, and $H_{wa}H_{wa}$ is the sensor-measured sea-wave height in half-meter units.

SWELLWAVES.—3$d_{w1}d_{w1}d_{w2}d_{w2}$, 4$P_{w1}P_{w1}H_{w1}H_{w1}$, and 5$P_{w2}P_{w2}H_{w2}H_{w2}$, (or group 70$H_{wa}H_{wa}H_{wa}$). Two significant swell waves may be reported.

Swell-Wave Direction.—Group 3 reports the primary swell wave direction in hundreds and tens of degrees as $d_{W1}d_{W1}$ and the secondary swell wave direction as $d_{W2}d_{W2}$. When no swell waves are observed, the group is encoded as 30000, and both the 4$P_{W1}P_{W1}H_{W1}H_{W1}$ and 5$P_{W2}P_{W2}H_{W2}H_{W2}$ groups are reported as 40000 and 50000. If only one swell wave is observed, the dw_2dw_2 is encoded as 00, and the 5$P_{W2}P_{W2}H_{W2}H_{W2}$ group is reported as 50000.

Swell-Wave Period and Height.—The primary swell-wave period and primary swell-wave height are reported in group 4$P_{W1}P_{W1}H_{W1}H_{W1}$. The secondary swell-wave period and height are reported in group 5$P_{W2}P_{W2}H_{W2}H_{W2}$. The $P_{W1or2}P_{W1or2}$ is the swell-wave period in tens and units of seconds, while the $H_{W1or2}HW_{1or2}$ is the height of the swell waves in half-meter units.

Swell-Wave Height From Sensors.-Ships equipped with automatic wave-height-measuring equipment report the swell-wave height by using group 70$H_{wa}H_{wa}H_{wa}$ in place of groups 3, 4, and 5. The $H_{wa}H_{wa}H_{wa}$ is the wave height in tens, units, and tenths of meters. Swell-wave direction and period from automatic sensors are normally not reported.

ICE ACCRETION.—6$I_SE_SE_SR_S$. The ice accretion group is only included in a report when ice accretion is occurring. The I_S is the source of the ice (WMO code table 1751); the E_SE_S is the thickness of the ice, in centimeters; and the R_S is the rate of accretion (or loss) (WMO code table 3551).

WET-BULB TEMPERATURE.—8$s_wT_bT_bT_b$. The wet-bulb temperature group is indicated by code figure 8. The s_w is the sign and type of the wet-bulb temperature (WMO code table 3855) and $T_bT_bT_b$ is the wet-bulb temperature in tens, units, and tenths of a degree Celsius.

SEA ICE.—ICE $c_iS_ib_iD_iz_i$. The sea ice group is only reported when ice is observed in the sea. "ICE" is the indicator that the sea ice group follows. The c_i is the concentration or arrangement of sea ice (WMO code table 0639), S_i is the stage of development (WMO code table 3739), b_i is ice of land origin (WMO code table 0439), D_i is the true bearing of the principal ice edge (WMO code table 0739) and z_i is the ice situation and trend over the past 3 hours (WMO code table 5239). In addition to the sea ice code, any remarks considered significant by the observer may be transmitted as plain language remarks.

Regional and National Groups

Regional information may be added to ship reports following a 333 indicator group, and National coded information may be added following a 555 indicator group. The 444 indicator group is NEVER used with ship reports. Although allowable, U.S. Navy vessels do not normally include any regional or national information in transmitted ship synoptic reports.

MOORED BUOY REPORTS

Moored meteorological/oceanographic (METOC) buoys are used throughout the world, primarily in areas critical to ship navigation. In the United States coastal waters, meteorological buoys are maintained by NOAA in the near coastal waters off major sea ports and harbors on the West Coast, East Coast, and Gulf Coast. Additional meteorological buoys are maintained in a network throughout the Gulf of Mexico, and in selected locations well off the East and West Coasts.

All moored buoys use the Synoptic code to report every 3 hours. They report Synoptic Code Section 0, Identification Data, and, depending on the sensors they contain, selected groups from Sections 1, 2, 3, and 5. The primary difference is in code Section 0, Identification Data. While the reports from buoys well off the coast use the ship Synoptic code "Section 0" with three-letter/number call signs and a latitude and longitude, the coastal buoys and Gulf of Mexico buoys use the land Synoptic code "Section 0" format with the station identification given in block (block 99) and station number, or a buoy number (such as "DB273" without latitude and longitude groups. These stations may be cross-referenced to latitude and longitude, as may any station referenced by block/station number by using the *Master Weather Station Catalog,* available via the Bulletin Board System (BBS).

REVIEW QUESTIONS

Q48. What is the message type identifier for the shipboard Synoptic code?

Q49. If a ship was located at 34.10N and 020.16E, how should this be encoded on a ship synoptic observation?

Q50. What does the indicator group 222 of the ship Synoptic code signify?

Q51. How should a ship speed made good of 12 knots with a displacement of 120" be encoded for $222D_Sv_S$?

Q52. If sea waves had a period of 5 seconds and a height of 4 feet, what should be encoded for $2P_WP_WH_WH_W$?

Q53. If the primary swell-wave direction is 040° with a height of 6 feet and a period of 8 seconds, and the secondary swell-wave direction is 170° with a height of 3 feet and a period of 12 seconds, how should this be encoded?

Q54. If ice accretion from ocean *spray is occurring at a rapid rate, and 1 centimeter has already accumulated, what should be encoded for the group* $6I_SE_SE_SR_S$?

Q55. What organization is responsible for maintaining METOC buoys off the coast of the United States?

SUMMARY

The codes discussed in this section are used to disseminate surface weather observations. The U.S. METAR/SPECI code and the international METAR/SPECI codes are primarily used to support aviation operations. The international Synoptic codes are used for general meteorological applications. Information from meteorological and oceanographic (METOC) buoys are also used to augment land and ship synoptic data.

ANSWERS TO REVIEW QUESTIONS

A1. *World Meteorological Organization.*

A2. *Region IV*

A3. *METAR, SPECI Land Synoptic, Ship Synoptic.*

A4. *NAVMETOCCOMINST 3141.2.*

A5. *NAVMETOCCOMINST 3144.1.*

A6. *Annotate missing data with an "M, " and provide a brief explanation for missing data in Block 72 "Remarks, Notes, and Miscellaneous Data."*

A7. *Safety of aircraft, ships, and personnel.*

A8. *Line out the error, enter the correction in red, and annotate the time the corrected observation was transmitted.*

A9. *FNMOD, Asheville, North Carolina.*

A10. *There are two types of observations in the METAR code, METAR and SPECI.*

A11. *The observation type, the station identifier, the observation date/time, and if required, a report modifier.*

A12. *180105G120KT.*

A13. *12015KT 090V150.*

A14. *The surface visibility in statute miles.*

A15. *RO2L/1000V3000FT.*

A16. *Mist.*

A17. *VCSH.*

A18. *Well-developed dust/sand whirls.*

A19. *SCT025CB.*

A20. *SCTOOO.*

A21. *In the case of a total obscuration with an indefinite ceiling, with $h_s h_s h_s$ as the vertical visibility into the indefinite ceiling.*

A22. *M04.*

A23. *Sea-level pressure (SLPppp).*

A24. *Snow depth on ground of 6 inches.*

A25. *1200Z.*

A26. *Slush on runway with a decelerometer reading of 12.*

A27. *Both NAVMETOCCOMINST 3141.2 or NAVMETOCCOMINST 3144.1*

A28. *Column 14.*

A29. *Column D.*

A30 *.Column E.*

A31. *NAVMETOCCOMINST 3144.1.*

A32. *The Synoptic code is designed to permit automatic loading of weather information to computer data bases for use in analysis and forecasting.*

A33. *0300Z, 0900Z, 1500Z, and 2100Z.*

A34. *Federal Meteorological Handbook No. 2.*

A35. *Data identifier, date-time group, and station identifier.*

A36. *32661.*

A37. *62499 and 00103, respectively.*

A38. *Temperature and dewpoint in Celsius degrees.*

A39. *Station pressure and sea level pressure in hPa.*

A40. *58008.*

A41. *60011.*

A42. *78682.*

A43. *85088.*

A44. *Regional additive data.*

A45. *06002.*

A46. *The* $58_{p24p24p24}$ *is used to report a net increase or no net change in the 24-hour pressure tendency.*

A47. *National additive data.*

A48. *BBXX.*

A49. *99342 10203.*

A50. *Maritime data follows.*

A51. *22233.*

A52. *20502.*

A53. *30417 40804 51202.*

A54. *61012.*

A55. *NOAA.*

CHAPTER 4

PLOTTING STANDARDS

INTRODUCTION

In the past, Navy and Marine Corps weather observers spent a lot of time plotting observations on charts. With the introduction of computer processing and widespread use of computer printers and plotters, manual plotting has just about become obsolete. Many of the charts you will see are completed analyses received from the Tactical Environmental Support System (TESS), and other computer systems. TESS receives raw data, data fields, and completed analyses and prognoses via the Navy Oceanographic Data Distribution Expansion System (NODDES). However, there are still a few tasks the observer must do to prepare charts for the analyst/forecaster or for a briefing. In this chapter, we discuss the types of charts and chart projections routinely used in meteorology. Then, we discuss some specific types ofproducts used by analysts and forecasters, and the tasks routinely done by to prepare these charts for the forecaster. As we discuss each type of product, we explain the standard data plotting model or method used to display observed data.

CHARTS

LEARNING OBJECTIVES: Describe the two primary coordinate systems used to locate points on the surface of the earth. Identify the terms used to define various regions of the globe. Identify the various chart projections frequently used for meteorological and oceanographic applications. Discuss how station identifiers are used to locate weather stations and airfields.

A chart is a printed reproduction of a portion of the earth's surface, which may show water and land areas. Charts use a coordinate system to locate positions on the chart. The lines of the coordinate system and the shapes and sizes of land masses vary in appearance, depending on the projection and scale of the chart used.

COORDINATE SYSTEMS

Two coordinate systems are used by the military. At sea, naval and merchant vessels use the geographical coordinate system. Within the North Atlantic Treaty Organization (NATO) and the U.S. military, ground forces use the military grid system. Joint operations between naval and ground forces require that you be familiar with both systems of coordinates.

Geographical Coordinates

The geographical coordinate system uses a network of *parallels* (of latitude) and *meridians* (of longitude), which aid in locating the various features shown. Since earth is generally a sphere (actually it bulges slightly near the equator and is slightly flattened near the poles), the most accurate depiction of earth is a globe—a ball-shaped chart of earth. Globes are very difficult to print on flat paper or to accurately depict in a book, so they are seldom used in meteorology or oceanography. Figure 4-1 is a representation of how parallels and meridians are arranged on a globe. Notice that the meridians are farthest apart at the equator and merge at the poles, and that the parallels circle the earth parallel to the equator.

Meridians are measured in degrees of arc, with the prime meridian (0° longitude) established as a line extending north and south through the location of the Royal Greenwich Observatory in Sussex, England. Meridian degrees are referenced as east or west of the

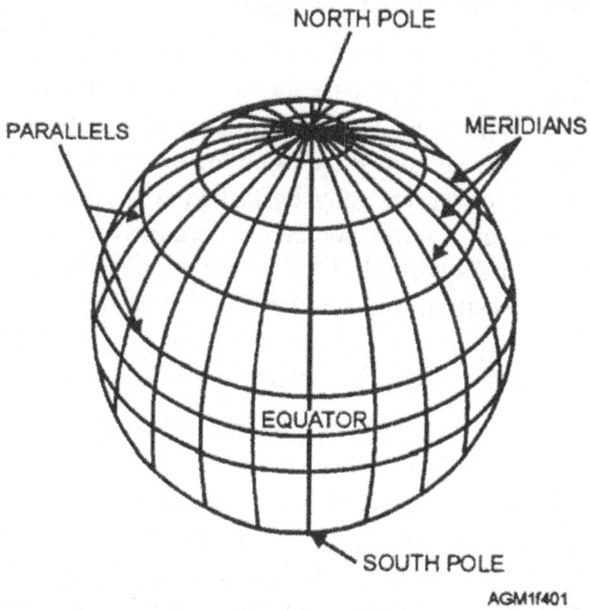

Figure 4-1.—Parallels and meridians on a globe.

prime meridian, and may be drawn on a chart at any convenient interval, such as every 5, 10, 15, or 30 degrees. The opposite of the prime meridian, either 180°E or 180°W longitude, is the international date line.

Parallels are also measured in degrees of arc, but referenced as either north or south of the equator, which is 0° latitude. The highest latitude measurements are 90° north (the North Pole), and 90° south (the South Pole). Parallels are normally drawn on charts every 5, 10, or 15 degrees.

With the system of parallels and meridians, any point on the earth's surface can be accurately located. Each degree (°) may be subdivided into 60 minutes ('), and each minute into 60 seconds ("). In standard use, a location is referenced by latitude, and then longitude, such as 43°21'13"N 073°54'03"W. A nautical mile is defined as 1 minute of arc on a great circle chart. The frequently used approximation that 1° of latitude (on any chart) is 60 nautical miles was derived from this definition. This relationship does not hold true for degrees of longitude, because the meridian lines converge toward the poles.

Locations used to plot positions on meteorological charts need only be accurate to within a few nautical miles. Most meteorological positions are converted to degrees and tenths of a degree during encoding to simplify coding. To convert to tenths of a degree, divide the minutes of both latitude and longitude by 6, and discard the remainder. For example, 43°20'13"N results in 43.3°N (20/6 = 3 remainder 2).

Military Grid Reference System

The Military Grid Reference System is used extensively by all military forces for target information and to locate positions ashore. In a warfare situation, naval guns are aimed by using the grid system, and weather observers may be tasked with providing environmental information for shore targets referenced by the grid system. In NBC warfare situations and in warfare situations requiring electro-optical or electromagnetic support, targets areas are commonly referenced in the grid system.

The Military Grid Reference System uses two separate grids to locate positions: the Universal Transverse Mercator (UTM) grid and the Universal Polar Stereographic (UPS) grid system. Nearly every military topographic chart at scales 1:50,000 and smaller already contain the military grid and geographical coordinate systems.

UNIVERSAL TRANSVERSE MERCATOR GRID.—The UTM grid system is a series of grid zone rectangles measuring 8° latitude by 6° longitude that covers earth from 80°S latitude to 84°N latitude (fig. 4-2). Columns of grid zones are numbered sequentially beginning at 180° and progressing eastward. Rows of grid zones are lettered beginning with *C* at 80°S extending northward. The letters *I* and *O* are omitted to avoid confusion with the numbers *1* and *0*. The northern-most row of grid zones, identified as row *X*, extends from 72°N to 84°N and is the only row that is not equal to 8° latitude in height. A grid zone is identified by the column number followed by the row letter, such as *34P*, which is shown shaded in the figure.

Next, each grid zone is subdivided into 100,000 meter (100 kilometer) squares, called 100,000-meter squares (fig. 4-3, view A). The 100,000-meter squares are identified by two letters. Again, the letters *I* and *0* are omitted to avoid confusion. The rest of the letters, *A* through *V*, are used to identify columns of the 100,000-meter squares, starting at 180° and extending eastward. These 20 letters are repeated every 18° longitude (every three grid zones west to east). The horizontal rows are identified from the equator northward with the letters *A* through *V*, and from the equator southward in reverse alphabetical order by the letters *V* through *A*. In the north-south orientation, there are nine 100,000-meter grid square rows in each 8° grid-zone row. In figure 4-3, view A, the shaded 100,000-meter square is identified by the grid zone (*34P*), then the column-letter (*D*), followed by the row-letter (*M*)—*34PDM*.

Next, note that each 100,000-meter grid square is divided into 10 rows and 10 columns, resulting in 10,000-meter squares, as shown in figure 4-3, view B. These squares are identified by the numbers 0 to 9 from the western-most line eastward, and then from the southern-most line northward. Point A would be referenced by the grid zone (34P), the 100,000-meter designation (DM), plus the column number (7), and row number (3)—34PDM 73.

Then, each 10,000-meter square is divided into 10 rows and 10 columns to form 1,000-meter squares, as shown in figure 4-3, view C. Again, the columns are identified from the western-most line eastward with the numbers 0 to 9, and the rows are identified from the southern-most line northward with the numbers 0 to 9. Point B is identified by the grid zone (34P), the 100,000-meter-square letters (DM), the 10,000-meter-square column number (7), followed by the 1,000-meter-square column number (6), then the 10,000-meter-square row number (3), followed by the 1,000-meter-square row number (1)—34PDM 7631.

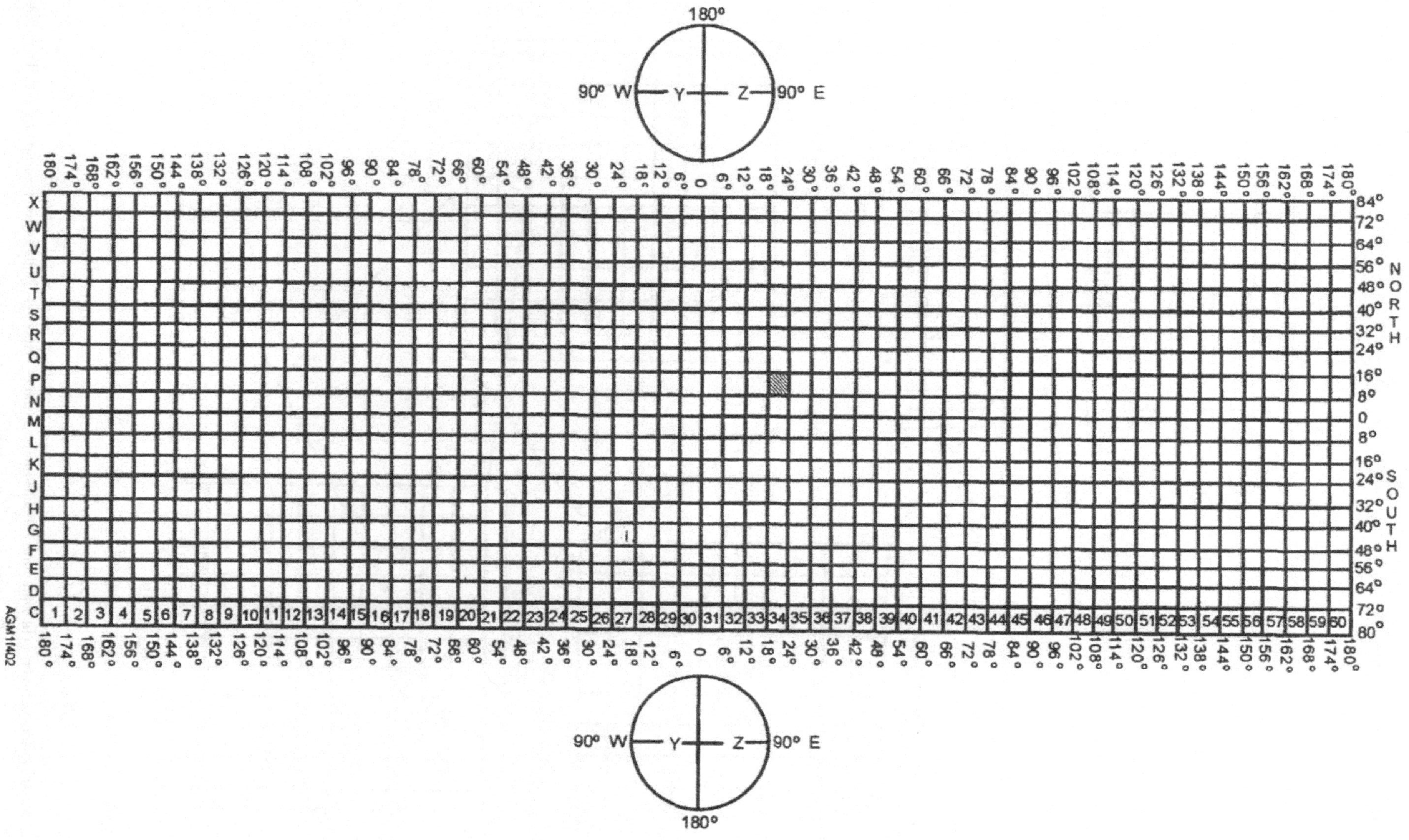

Figure 4-2.—Military Grid Reference System using Universal Transverse Mercator (UTM) grid between 84°N and 80°S latitudes and Universal Polar Stereographic (UPS) grids in the polar regions. Grid squares are identified with the column number first, and then the row letter.

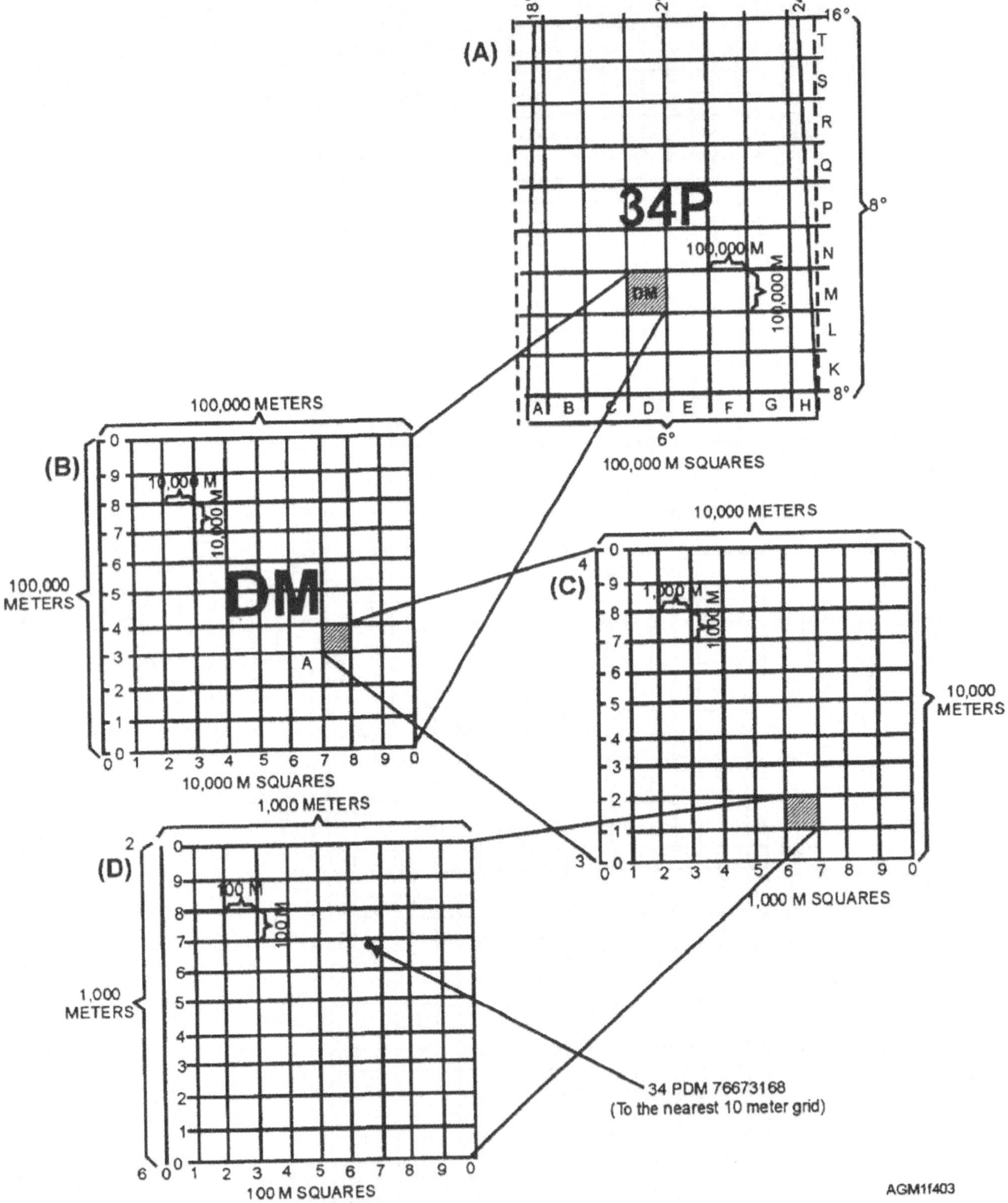

Figure 4-3.—Relationship between grid zone and 100,000-meter squares (view A), 10,000-meter squares (view B), 1,000-meter squares (view C), and 100-meter squares (view D).

This pattern of subdividing the squares and identifying the point with successively smaller columns, and then row numbers may continue indefinitely, but is normally only carried out to the nearest lo-meter square for a precision targeted position. A 10-meter position would have eight digits following the letters, such as 34PDM 76673168, as shown in figure 4-3, view D. To read this position, mentally divide the number into two groups of four digits (7667 / 3168). Think of it as moving eastward from the west side of the DM 100,000-meter grid boundary by (7) 10,000-meter-square subgrids, then (6) 1,000-meter-square grids, then (6) 100-meter-square grids, then (7) lo-meter-square grids. Then, move northward from the southern boundary of the DM grid (3) 10,000-meter grid squares, then (1) 1,000-meter grid square, then (6) 100-meter grid squares, and then (8) 10-meter grid squares.

When military units pass positions in message traffic within a region, frequently the grid zone designation is left out of the position, such as DM 40132879. This practice is common within NATO military forces in Europe.

Now let's look for a moment at figure 4-4. Since the east and west boundaries of each grid zone are specified longitude lines and the 100,000-meter squares (fig. 4-3, view A) are actual distances on earth, the 100,000-meter squares along the edges of each grid zone overlap as the longitude lines converge near the poles. Overlapping portions are truncated at the grid zone boundaries, and the charted grid zones appear to have some partial 100,000-meter squares. When identifying locations, remember that the lettering or numbering system assumes that the missing or partial portions of the 100,000-meter squares are intact.

Notice that the 100,000-meter grid lines are only parallel to the longitude lines near the center of each grid zone. This means the grid lines are oriented slightly westward of true north in the east half of a grid zone, and eastward of true north in the west half of a grid zone. The grid column lines are used as a reference called "grid north" (GN).

The UTM grid lines drawn on a chart depend on the scale of the chart. Figure 4-5 shows a portion of a coastal chart with grid lines drawn every 200 meters. On this chart, the grid location may be easily estimated to the nearest 10 meters. Based on the latitude and longitude, the grid zone is 20Q and the 100,000-meter square is PC. The numbers indicate the 10,000-meter grid coordinates. The lower-left comer of the box marked M is 20QPC 40401040.

While the grid zone designation and the letters of the 100,000-meter square may be used to approximate the general area of a UTM location in geographical coordinates, precise conversions of UTM grid coordinates to geographical coordinates must be done by referencing the UTM position on a gridded chart, and

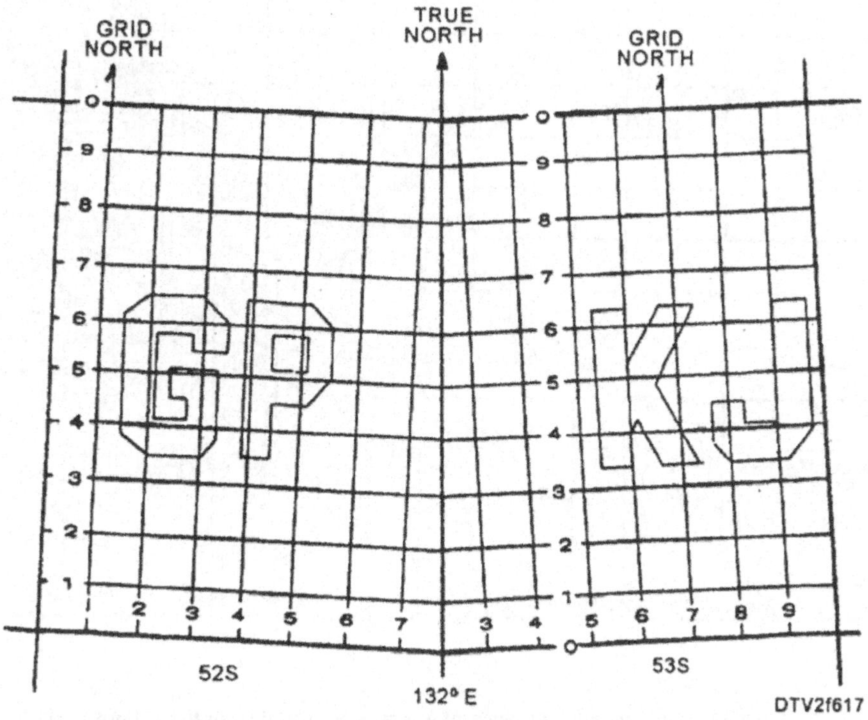

Figure 4-4.—Partial 100,000-meter squares at edges of grid zones.

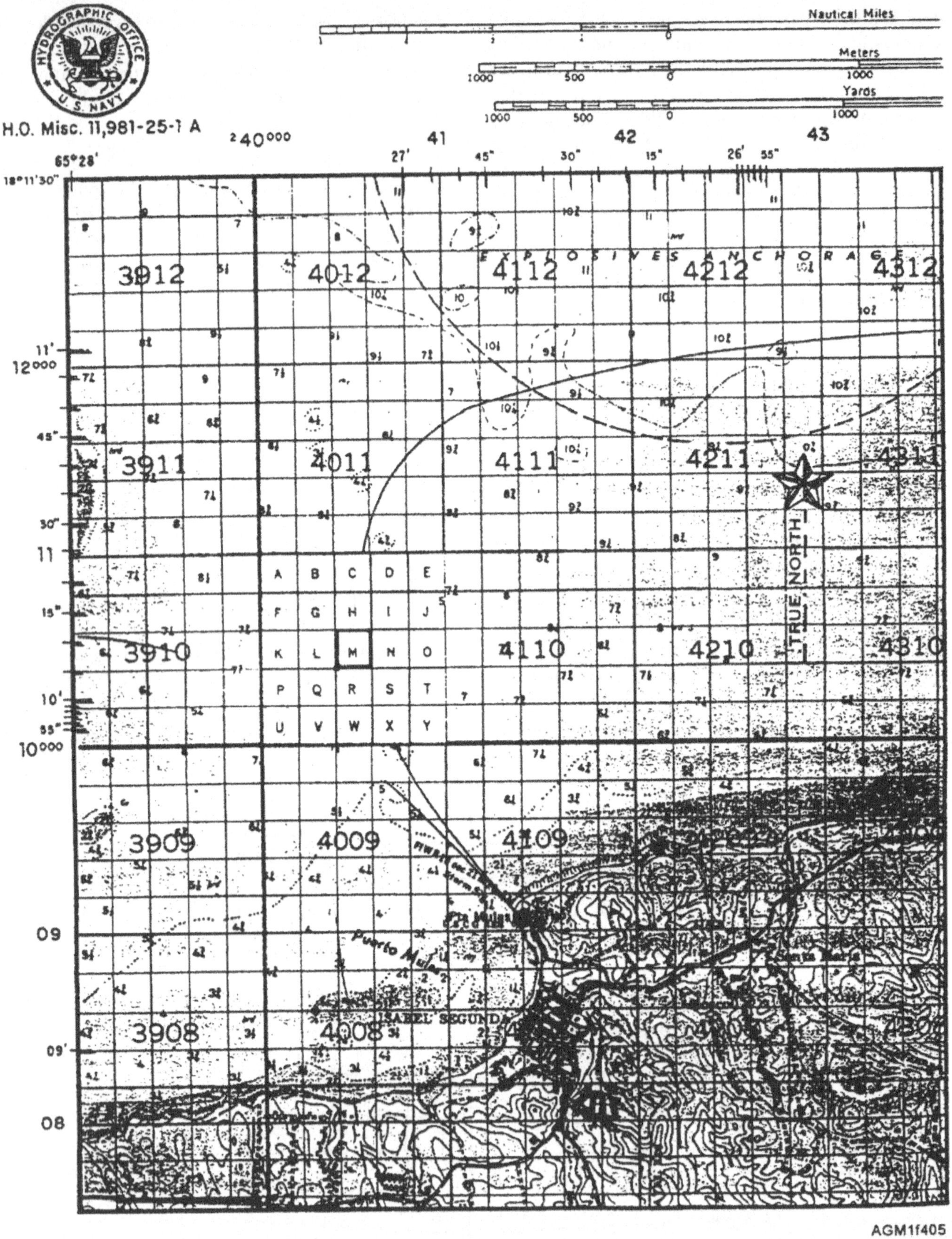

Figure 4-5.—Portion of a naval shore bombardment chart of Vieques Island, Puerto Rico, showing military grid system UTM 10,000-meter, 1,000-meter, and 200-meter grid lines.

then reading the geographical coordinates. On board ship, the Operations Specialist (OS) or Quartermaster (QM) personnel should have a complete set of charts for the ship's operating areas. These personnel should also have the necessary topographic charts for the coastal regions for which weather personnel may need to provide environmental support. Ashore, contact the National Imagery and Mapping Agency regional office or refer to the NIMA catalog for the necessary products.

UPS GRID SYSTEM.—The Universal Polar Stereographic (UPS) grid system is similar to the UTM grid system, except that only two grid zones are used in each polar region: Y and Z in the north polar region, and A and B in the south polar region. Figure 4-6 shows the grid zones and 100,000-meter squares in the north polar region (84°N to the North Pole). For the southern polar region, the grid zone designator A replaces Y, and B replaces Z, and the longitude reference lines are rotated

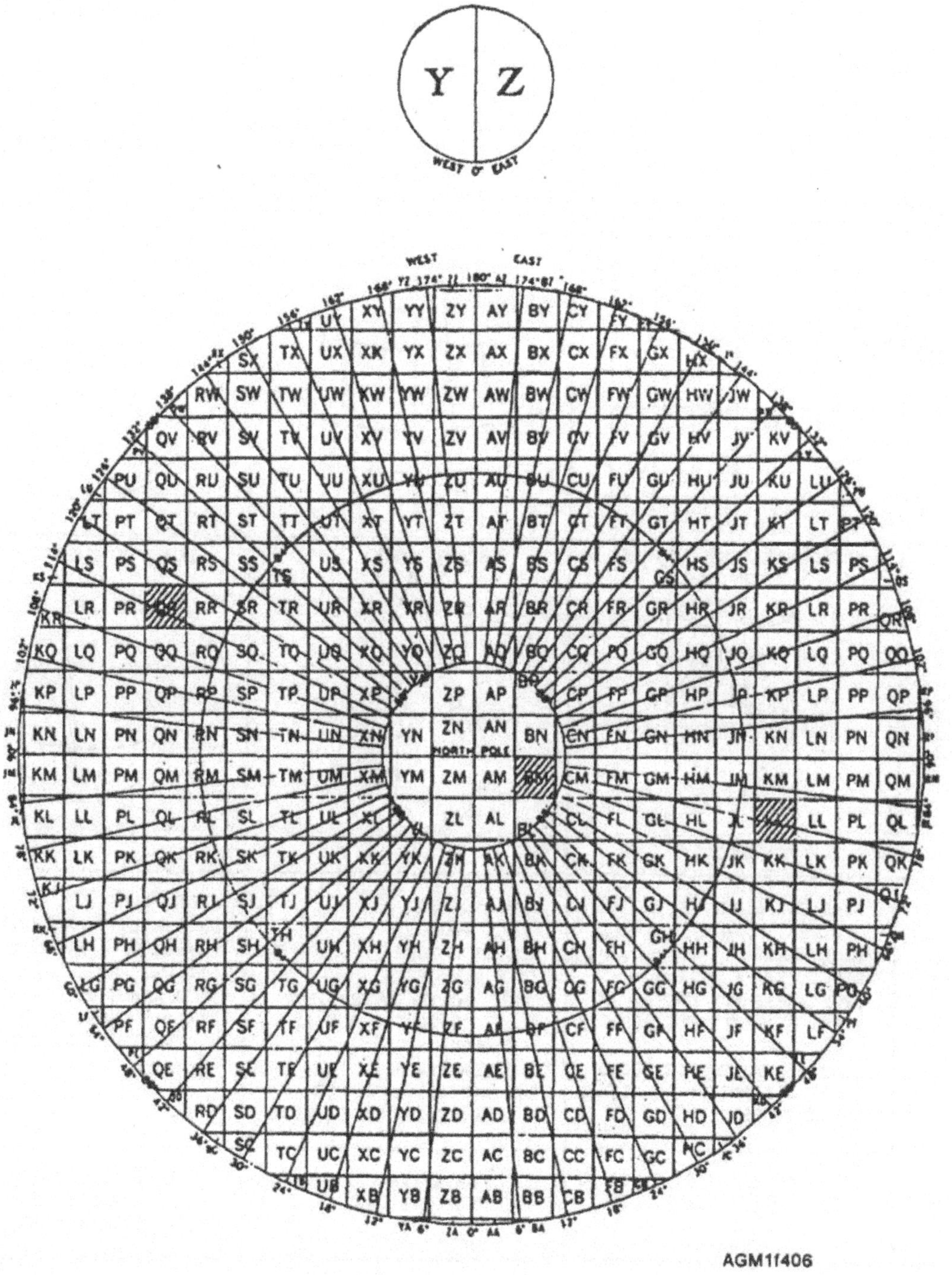

AGM1f406

Figure 4-6.—UPS grid zones and 100,000-meter squares used in the polar regions.

180°. The 100,000-meter squares from 80° to 84° are used only in the south polar region.

Because there are no column numbers in the grid zone reference, complete grid locations appear slightly different, such as ZBM 40170.

Within the full and partial 100,000-meter squares, subdivisions into 10,000-meter squares, 1,000-meter squares, and smaller squares are the same as in the UTM system.

MAP PROJECTIONS

	MERCATOR	LAMBERT CONFORMAL CONIC	POLAR STEREOGRAPHIC
PARAL-LEL	PARALLEL STRAIGHT LINES UNEQUALLY SPACED.	ARCS OF CONCENTRIC CIRCLES EQUALLY SPACED.	ARCS OF CONCENTRIC CIRCLES UNEQUALLY SPACED.
MERID-IAN	PARALLEL STRAIGHT LINES EQUALLY SPACED.	STRAIGHT LINES CONVERGING AT A POINT OUTSIDE OF MAP.	STRAIGHT LINES RADIATING FROM POLE.
APPEARANCE OF GRATICULE	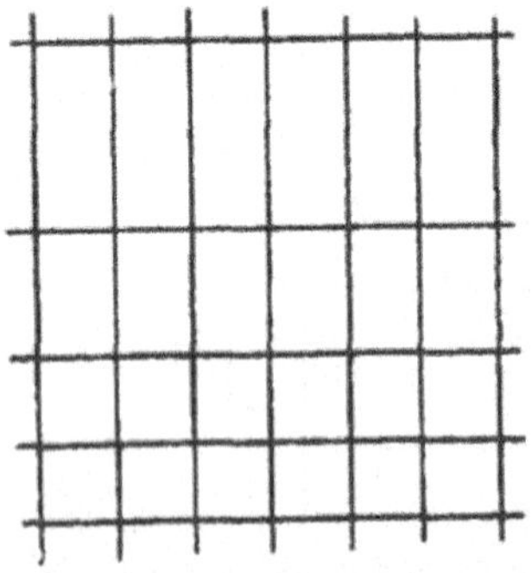	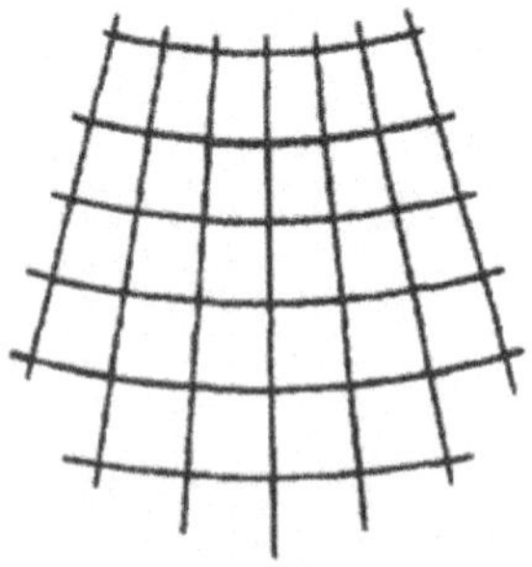	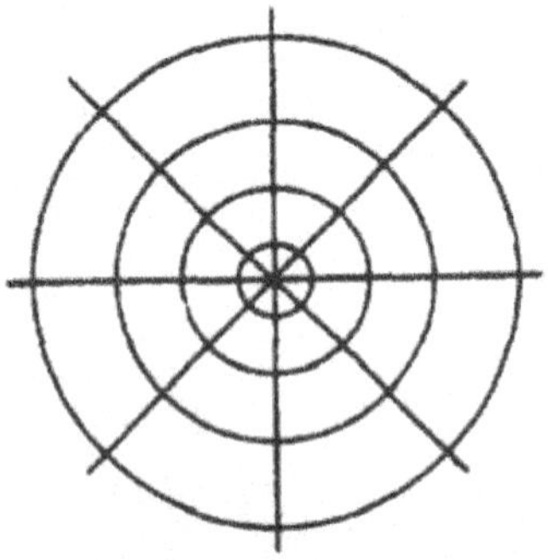
PROJECTED ON	CIRCUMSCRIBED CYLINDER 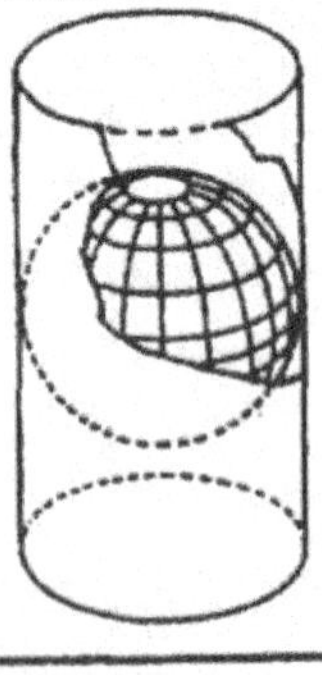	SECANT CONE 	PLANE TANGENT AT POLE 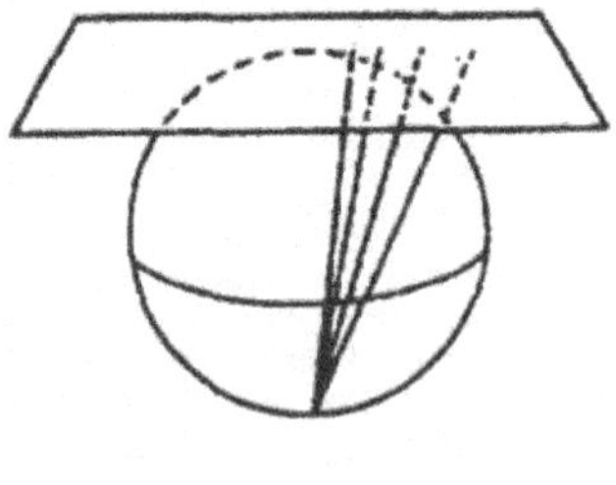
PROPERTIES	STRAIGHT LINES ARE RHUMB LINES. CONFORMAL. CONVENIENT PLOTTING. TRUE AREAS NOT SHOWN. STRAIGHT LINES NOT GREAT CIRCLES. TRUE DISTANCES NOT SHOWN. GREAT DISTORTION IN HIGH LATITUDES.	TRUE SHAPES. AREAS GOOD. DISTANCE GOOD. TRUE DIRECTIONS. SMALL NORTH-SOUTH LIMIT OF PROJECTION FOR AC-CURACY. PLOTTING FAIR. NOT SATISFACTORY FOR AREAS CLOSE TO EQUATOR.	TRUE SHAPE. ONLY AZIMUTHAL PROJECTION WITH NO ANGULAR DISTOR-TION. TRUE AREAS NOT SHOWN. SCALE INCREASES IN ALL DIRECTIONS FROM CENTER.
USES	USED FOR AREAS CENTERED IN TROPICAL LATITUDES.	USED FOR AREAS CENTERED IN MIDDLE LATITUDES.	USED FOR NORTHERN AND SOUTHERN HEMISPHERE HIGH LATITUDES.

AGM1f407

Figure 4-7.—Meteorological chart projections.

REVIEW QUESTIONS

Q1. Name the two coordinate systems used by the U.S. military?

Q2. On most charts, 60 nautical miles equals how many degrees of latitude?

Q3. Convert 23.54N and 120.38W to tenths of a degree.

Q4. When you use the Universal Transverse Mercator (UTM) grid, how is a grid zone identified?

Q5. Grid zone 27R is between what latitude and longitude?

Q6. What is normally the smallest grid zone used with the UTM grid system?

Q7. The Universal Polar Stereographic (UPS) grid system is used in what region of the globe?

AREAS OF THE GLOBE

In meteorology and oceanography, several terms are used to describe sections of the world. *Northern Hemisphere* refers to the half of earth north of the equator; *Southern Hemisphere* refers to the area south of the equator. Similarly, *Western Hemisphere* refers to the half of earth from the prime meridian westward to the international date line; *Eastern Hemisphere* refers to the half from the prime meridian eastward to the international date line. You will see frequent references to the tropical region, or the Tropics. The Tropics is the belt surrounding earth that lies between the tropic of Cancer at 23½°N and the tropic of Capricorn at 23½°S. In meteorology, however, this region may generally be considered the belt between 30°N and 30°S. Geographically, the area of earth north of the Arctic Circle at 66½°N and south of the Antarctic Circle at 66½°S are the polar regions. Between the Tropics and each polar region lies the area referred to as the middle latitudes (or the mid-latitudes). Based on annual temperature changes, the Tropics, mid-latitudes, and polar regions are also respectively called the *Torrid Zone, the Temperate Zones,* and the *Frigid Zones.* These terms are generally falling into disuse.

CHART PROJECTIONS

Several types of modifications to the shape of earth are made to allow the earth's surface to be represented on flat paper charts and displays. These modifications are called "projections." The most common projections used in meteorology and oceanography are the Mercator projections, Polar Stereographic projections, and the Lambert-Conformal Conic projections, as shown in figure 4-7. Figure 4-7 also indicates the properties of these projections and their intended uses.

You will often see references to great circle routes or tracks. A *great circle* track represents the shortest distance between two points on the surface of earth or on a globe, such as if a string were to be stretched between the two points. Great circle routes are planned on a special chart projection called a gnomonic projection. These projections are similar to polar projections, except the plain of the chart is tangent to the earth at the center of the area charted. Figure 4-8 shows a gnomonic projection with a great circle route and a rhumb line track between two ports. The *rhumb line* track is a straight line drawn between two points on a Mercator projection, and may also be called a "Mercator track." Notice that the great circle route is about 800 nautical miles shorter than the rhumb line track due to the curvature of the earth. Figure 4-9 shows the identical routes drawn on a Mercator projection.

CHART SCALE

The scale of a chart refers to a comparison of the distances shown on a chart to the actual distance on the surface of earth. A scale may be a comparative ratio

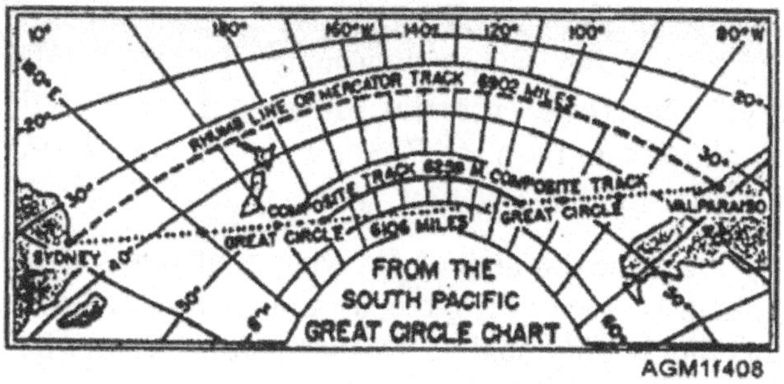

AGM1f408

Figure 4-8.—Rhumb line and great circle tracks on a great circle chart (gnomonic projection).

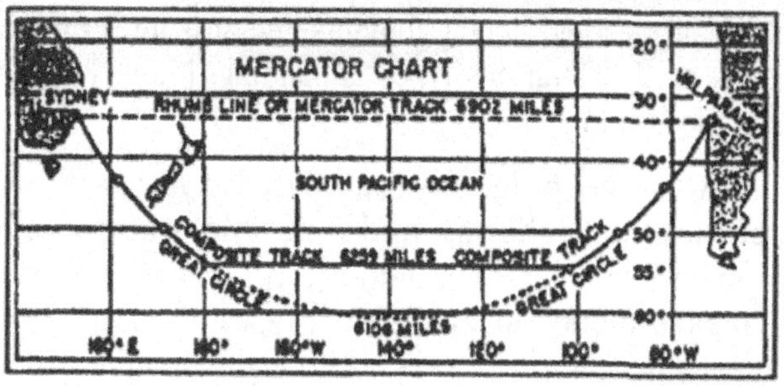

AGM1f409

Figure 4-9.—Rhumb line and great circle course on a Mercator chart.

scale, such as 1:100,000. This may be taken to mean 1 inch on the chart is equal to 100,000 actual inches on the surface of earth, or it may be interpreted as 1 meter on the chart is equal to 100,000 meters on the surface of earth. The scale may also be a distance scale, such as 1 inch = I mile or 1 centimeter = 10 kilometers, as commonly seen on road maps. The military uses the comparative ratio scale. A chart covering a relatively small area is called a small-scale chart, while a chart covering a relative large area is called a large-scale chart.

Typically, weather plotting charts range in scale from 1:1,000,000 for 3-foot by 4-foot charts of individual small countries, through 1:4,000,000 for charts showing sections of the United States. Table size charts for an entire hemisphere may be on a scale of 1:30,000,000.

TYPES OF CHARTS

There are nearly as many types of charts as there are applications for charts. The National Imagery and Mapping Agency Hydrographic/Topographic Center produces a large portion of the charts routinely used by the military. Their seven-part catalog, the *National Imagery and Mapping Agency (NIMA) Catalog of Maps, Charts, and Related Products,* provides a listing of the various charts available, as well as ordering information. The catalog Part I, *Aerospace Products,* Volume 1, contains listings for aeronautical charts-charts displaying air routes and navigational aids used for flight route planning. Volume 2 contains listings for weather plotting charts-charts showing gross topography and land/water boundaries with station circles for established weather stations. Part II of the catalog contains listings for various hydrographic charts-charts used for surface and subsurface marine navigation. These charts show sounded water depths and ocean depth contour lines as well as positions of marine navigation aids. Part III contains several volumes listing various topographic charts-charts showing detailed height contours and structures on land areas used for various types ofplanning. Many volumes are classified.

At some point in your career, you will use aeronautical charts, hydrographic charts, and topographic charts. Up until recently, every observer manually plotted many different types of observations on weather plotting charts. Although you may still on occasion, manually plot data on a weather plotting chart, and therefore should be familiar with the various charts available, nearly all charts used by analysts and forecasters today are computer plotted. Refer to Part I, Volume II, of the NIMA catalog for a listing of the various weather plotting charts available for manual plotting of data.

STATION IDENTIFIERS

Meteorological stations, when reporting weather conditions, use either a WMO block/station number or station-identifier letters. The station identifier letters are usually the International Civil Aviation Organization (ICAO) assigned station identifier, but may be an identifier assigned by the individual country. A particular station may use its WMO block/station number to report synoptic observations, but use either the ICAO identifier or the nationally assigned identifier to report weather locally.

National Imagery and Mapping Agency weather plotting charts (DOD WPCs) have predrawn station circles for nearly every known permanent weather-reporting station. Each station circle is identified with the WMO station number and/or the ICAO letter identifier, as shown in figure 4-10, view A. Mobile weather reporting stations, such as ships or temporary research sites, report location by latitude and longitude. A station circle must be added at the correct location whenever a mobile station's report is plotted.

WMO Block/Station Identification Numbers

WMO block/station numbers are used to identify locations of permanent weather reporting stations. WMO international block/station identification numbers are a series of six digits. The first two digits are the WMO block number, and the last four digits are the station number. In an area the size of a small city, the designated Synoptic Station—the single station assigned the responsibility to submit synoptic observation reports, has a four-digit station number ending in a zero. Other weather observation sites in the city that are not designated synoptic reporting sites have other numbers as the last digit. For example, figure 4-10, view B, shows the stations in the Norfolk, Virginia area. Norfolk International Airport (ORF), block/station number 723080, is the designated synoptic reporting station.

Since the last digit of the synoptic station's station number is always zero, the last digit is dropped when encoding. Additionally, weather plotting charts only indicate the first three digits of the four-digit station number, such as 308, instead of 3080, adjacent to a synoptic station circle.

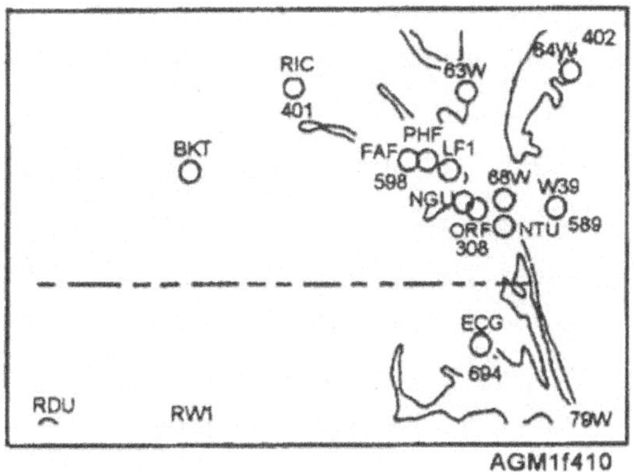

WMO STATION NUMBER	U.S. STATION ID	STATION NAME
723080	ORF	NORFOLKINTERNATIONAL
723085	NGU	NORFOLK NAS
723086	PHF	NEWPORT NEWS
723087	FAF	FORT EUSTIS/FELKER
723088	63W	MILFORD HAV EN CGS

(A) (B)

Figure 4-10.—View (A) is a section of a 1:4,000,000 scale DOD WPC (shown just smaller than actual size), and view (B), WMO international station identification numbers and U.S. national station identification letters in the Norfolk, Virginia vicinity.

WMO block numbers are indicated on the plotting charts, usually in large, pastel blue digits. And the boundaries of each block are drawn as thin pastel blue lines.

A complete numerical listing of WMO block/station numbers is contained in the *Master Weather Station Catalog* available via the Bulletin Board System (BBS) from FNMOD Tinker AFB, Oklahoma. The publication contains listings of country names cross-referenced to block number. The information provided for each station includes latitude, longitude, elevation, information on the coordinates and elevation of the upper-air observation site, and the type of data and observations available. The codes used in the listings are explained in the first section of the publication.

Other station circles may be drawn on the same chart, but identified with either a four-letter ICAO station identifier, such as EGUN, outside of the United States, or a three-letter (or letter/number) national station identifier, such as NGU or 63W within the United States.

ICAO Station Identifiers

Within most countries, except the United States, airfields and their weather observation sites are identified with a four-letter identifier assigned by the International Civil Aviation Organization (ICAO). The first letter is a regional identifier. The regional identifier is used to identify specific groups of countries in geographical areas of the world, such as M for the Caribbean and Mexico, K for the United States, or P for the Pacific. The second letter is specific to a particular country in the region, although a single large country may use several different letters as the second letter in the airfield identifier. For example, in Southeast Asia, Thailand uses ICAO four-letter identifiers beginning with the region identifier V, followed by the country identifier T. For example, Don Muong Airport, Bangkok, Thailand is VTBD. No particular relationship exists between the ICAO identifier and the city name or airfield name.

Within the Unites States, most airfields that report aviation weather use the three-letter national airfield identifiers, such as DAB for Daytona Beach Regional Airport, Daytona Beach, Florida, or BIS for Bismarck Municipal Airport, Bismarck, North Dakota, as the weather station identifier. Most of these identifiers, assigned by the Federal Aviation Administration (FAA), are similar to the city or airfield name. Naval and Marine Corps air stations are nearly all identified by a three-letter identifier beginning with an N. Many automated weather reporting stations are identified by three-letter/number combinations, such as 63W, as seen in figure 4-10. Several years ago, the FAA assigned the three-letter/number identifiers to heliports and airfields that were not manned by air-traffic controllers. Even though some of these stations do provide weather observations, many do not. (Currently, the FAA assigns unmanned airfields a combination two-letter, two-number identifier.)

In the United States, ICAO identifiers were made by simply prefixing the three-letter national identifier with the ICAO region identifier K. Airfields equipped to handle international air traffic use the ICAO identifier as both an airfield identifier and as a communications identifier for weather observations and forecasts. For example: KBIX (Keesler Air Force Base, Biloxi, Mississippi), KJFK (John F Kennedy International Airport, New York City), and KLAX (Los Angeles International Airport, Los Angeles, California) are ICAO identifiers used both to identify the station and weather products originated at the station.

Radio-navigation aids may be co-located at an airfield or be located some distance away from an airfield. These aids are identified by three-letter station identifiers. In most, but not all cases, the radio navigation aid identifier is the same as the airfield's national station identifier.

Converting Station Identifier To Location

Overseas ICAO identifiers may be cross-referenced to WMO block/station numbers and latitude/longitude by using the weather station catalog. OPARS (U.S. Navy Optimum Path Aircraft Routing System) subscribers may use the OPARS Data Base computer listing of location identifiers to cross-reference identifiers. The OPARS Data Base is the most comprehensive listing of identifiers, and is available either as a set of computer printouts, or is accessible via the OPARS computer bulletin board.

For all stations in North America and for selected Department of Defense overseas locations, FAA Order 7350.6, *Location Identifiers,* contains alphabetical name to identifier cross-references as well as alphanumeric identifiers to name cross-reference sections. The listings also include radio navigation aids. The ICAO region identifier, K, is not considered in this publication. Latitudes/longitudes are not provided.

For many countries, the DOD Flight Information Publications (En route), *IFR Supplement (Country Name),* provide alphabetical listings of airfield names andnavigation aid names. Each airfield listing provides the latitude/longitude, station elevation, and station identifier, the time zone of the airfield, communication frequencies, and navigation aid station identifiers and frequencies.

REVIEW QUESTIONS

Q8. What is the geographic definition of the tropics?

Q9. Why are great circle routes generally shorter than straight line Mercator tracks?

Q10. Which of the following scales would show greater detail; a 1:100,000 scale chart or a 1:10,000 scale chart?

Q11. What federal agency is responsible for producing maps and charts for the U.S. military?

Q12. What publication contains a listing of available weather plotting charts and hydrographic charts?

Q13. What is the significance of a WMO block/station number ending with the digit zero?

Q14. Where can a complete listing of WMO block/station numbers be found?

Q15. What is the ICAO region identifier for the United States?

Q16. What publication contains a complete listing of all station identifiers in North America?

GRAPHIC ENVIRONMENTAL PRODUCTS

LEARNING OBJECTIVES: Identify the terms used to discuss graphic meteorological and oceanographic products. Explain how legends are used to identify information presented on a graphic product. Discuss how history is displayed on a graphic product. Identify the standard representation for various meteorological parameters on graphic displays. Identify each element in a plotted meteorological report.

The majority of meteorological and oceanographic products in use are computer-produced graphic products. While nearly all of these graphic products may be reproduced on paper as a hard copy, they are designed primarily to be used in an almost paperless environment directly on a computer video display. These graphics are commonly called charts-a holdover from the time when everything was plotted on weather plotting charts or maps-referring to the geopolitical background used for location reference on graphic weather products. Many types of information may be transferred to the computer via radio, satellite link, telephone modem, telephone facsimile, LAN, or by mail or courier service on floppy or laser disk. The graphic products received by any of these methods may be in the form of digital data fields or actual graphics. Most of the graphics received are intended to be nothing more than "tools" for the forecaster to use to determine the state of the current environmental conditions or the anticipated state of conditions at some time in the future.

TERMS

A product that shows the state of the current conditions or the state of the conditions at some time in the past is called an *analysis.* A product that shows the anticipated state of conditions at some time in the future is called a *prognosis.* These terms are commonly abbreviated as *anal* and *prog,* respectively, both verbally and in text. *Outlook* and *forecast* are sometimes used interchangeably with *prognosis.* Correctly used, however, *forecast* refers to a forecaster's written or verbal interpretation of calculated graphic or alphanumeric products depicting the position and orientation of pressure centers, fronts, isobars and isoheights, etc., and *prognosis* refers to the graphic or alphanumeric product depicting those conditions. Likewise, *outlook* refers to the forecaster's interpretation of a prognosis, typically 3 to 10 days in the future, where the calculated information has a high probability of changing based on the presence of minor fluctuations in the conditions that cannot be accurately anticipated. *Outlook* implies a lower level of confidence or reliability than *forecast.* Certain graphic products depict the forecaster's interpretation of the state of the atmosphere, such as a Horizontal Weather Depiction, and is properly called a *forecast.*

The difference between graphic forecast products and prognosis products is important. You will occasionally be asked to explain certain features on graphic products. This is typically done at stations where a certified forecaster is on duty only during the day but an observer is on duty around the clock. You may explain what is actually indicated on a graphic or alphanumeric forecast product, and you may discuss what is actually depicted on analysis and prognosis products. You may not, however, provide your interpretation of how those parameters may effect weather conditions that are not depicted. Likewise, when reproducing a graphic either on paper or a video display, you may not adjust the location of any feature. In both situations your interpretation would be considered a forecast. You may not provide your own forecast until you have been certified as a analyst or forecaster in accordance with the provisions of NAVMETOCCOMINST 1500.2, *Naval Meteorology and Oceanography Command Training and Certification Program.*

Only a few of the literally hundreds of graphic products available are intended to be used, without additional processing, as briefing aids. A *briefing aid* is any product designed primarily to be used in a briefing to assist the explanation of the current and forecast conditions. Briefing aids are frequently enhanced with colors for weather systems, hazardous weather areas, land and water boundaries, and other items of interest. Briefing aids are enhanced with colors so that the person being briefed can easily see the features discussed by the briefer. The standard colors used are discussed later in this section.

GRAPHIC PRODUCT LEGENDS

All graphic meteorological and oceanographic products must have a legend block or legend line that identifies the product and the information presented on the product. A legend block, such as that shown in figure 4-11, is typically found on products received

AGM1f411

Figure 4-11.—Legend block from a NWS (National Meteorological Center) graphic product.

from the National Weather Service. Products produced by processing data fields received from FNMOC must have a legend line (or lines) added across the bottom or top of the graphic identifying the information displayed. The legend line identifies the product using standard abbreviations and the valid time. For example, an analysis of the surface pressure data field of the 1800Z 12 December 1996 observation data may be identified with a legend line SFC PRESSURE ANALYSIS VALID 08DEC96 1800Z.

The valid time is the UTC date/time that is represented by the information on the graphic. On some products, the valid time is identified with the designation VT: or the word valid; on other products, the date and time are just listed. In the case of an

TERM	SYMBOL	COLOR
Cold front at the surface		BLUE
Cold front aloft		BLUE
Cold front frontogenesis		BLUE
Cold front frontolysis		BLUE
Warm front at the surface		RED
Warm front aloft		RED
Warm front frontogenesis		RED
Warm front frontolysis		RED
Occluded front at the surface		PURPLE
Occluded front aloft		PURPLE
Stationary front at the surface		RED AND BLUE
Stationary front aloft		RED AND BLUE
Stationary front frontolysis		RED AND BLUE
Instability / Squall line		PURPLE
Trough axis		BLACK
Ridge axis		BROWN
Shear line		BLUE
Inter-tropical Convergence Zone	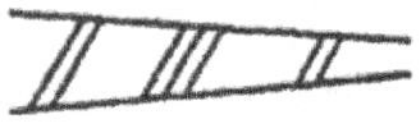	RED
Tropical Wave		BLACK
Jetstream core		RED
High Pressure/height center	H	BLUE
Low pressure/height center	L	RED
Tropical Depression center		RED
Tropical Storm center		RED
Hurricane / Typhoon center		RED

Figure 4-12.—Symbols used on surface pressure and constant-pressure level analysis and prognosis charts.

analysis, the valid time is the synoptic hour of the observations. For a prognosis, the valid time is the moment in the future when the actual conditions should be most like the conditions depicted. Certain products are called time-phased products. These products show boundaries of areas that will be effected by moving weather systems over a specified period of time. Time-phased products may either state a valid period of time, such as ACCUM PRECIP PROGNOSIS VT: 2 1 / 1200Z TO 22/1200Z DEC 96, or may indicate the time period in the title, and use a valid time for the end of the period, such as 24HR PRECIP 24 HR FCST VALID 22 DEC 96 1200Z.

When a product is used as a briefing aid, it is common practice to enlarge the valid time on the paper copy or video display so that the audience may see the time without eye strain, and to show both the UTC time and the local time. For a briefing in Norfolk, Virginia (time zone "R"), a briefing product valid at 12Z 18 Dec 96 would be identified with an enlarged valid time of

VT: | 1200Z / 0700R | 18 DEC 96

See Appendix III for a chart of time zones.

GRAPHIC HISTORY

Occasionally, you will be asked to place history on a surface chart or a constant pressure level chart prior to manual analysis. History is the past positions of pressure or height centers, fronts, troughs, or ridges. History helps the analyst or forecaster to determine past movements of major chart features. It is a valuable tool in both analysis of the current situation and the prognosis of future positions of the same features.

To place history on a chart, the previous locations of pressure centers or height centers, frontal systems and troughs, and major features, such as jet stream locations, are marked on the chart. Selected isobars, isoheights, or isotherms may also be required by the forecaster. Normally, two sets of positions are placed on a chart: either the 6- and 12-hour-old positions or the 12- and 24-hour-old positions. For instance, a chart containing surface plots of 05 December 1200Z observations may have history marking the past positions of fronts and pressure centers of 5 December at 0000Z (VT minus 12 hours) and 4 December at 1200Z (VT minus 24 hours).

When two sets of history are used, the most recent history positions are marked in orange, while the oldest positions are marked in yellow. If only one set of history is marked, the positions are marked in yellow. Standard symbols are used for all features. The symbols are discussed in the following section.

DATA DEPICTION STANDARDS

Nearly every graphic product that weather observers, analysts, and forecasters deal with is designed either to be received by electrographic methods as ready to use or to be completed by an analyst and sent to another user via electrographic methods. The methods sometimes used to disseminate charted products may be as simple as copying the chart on a photocopier and carrying it to the user. Although some of these methods allow the use of color, many graphics transmission devices operate only in black-and-white mode. Because of this, charts produced on color-capable computer displays normally use standard depictions that combine color and patterns. When charts are reproduced in black and white, the shape of the symbol alone identifies the feature depicted.

Figure 4-12 shows the standard symbols and colors used on surface and constant pressure level analysis and prognosis charts. Figure 4-13 shows the symbols and

TERM	SYMBOL	COLOR
Divergence line (asymptote)		BLUE
Convergence line (asymptote)	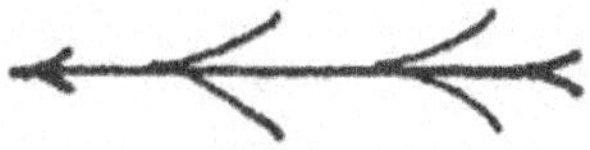	RED
Anticyclonic circulation center	A	BLUE
Cyclonic circulation center	C	RED
Neutral point	col	BLACK

Figure 4-13.—Symbols used on surface and constant-pressure level streamline analysis and prognosis charts.

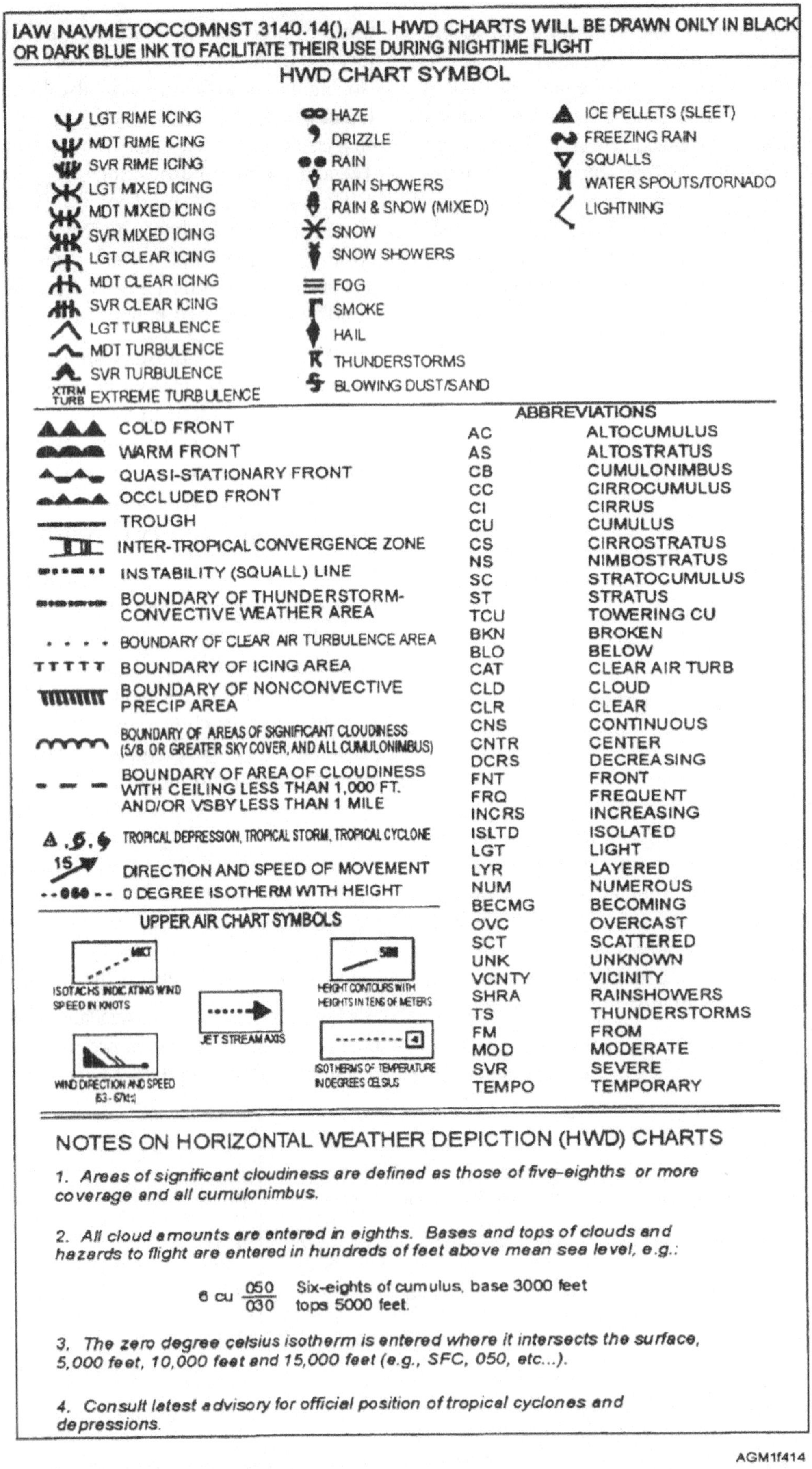

IAW NAVMETOCCOMNST 3140.14(), ALL HWD CHARTS WILL BE DRAWN ONLY IN BLACK OR DARK BLUE INK TO FACILITATE THEIR USE DURING NIGHTIME FLIGHT

HWD CHART SYMBOL

LGT RIME ICING
MDT RIME ICING
SVR RIME ICING
LGT MIXED ICING
MDT MIXED ICING
SVR MIXED ICING
LGT CLEAR ICING
MDT CLEAR ICING
SVR CLEAR ICING
LGT TURBULENCE
MDT TURBULENCE
SVR TURBULENCE
XTRM TURB EXTREME TURBULENCE

HAZE
DRIZZLE
RAIN
RAIN SHOWERS
RAIN & SNOW (MIXED)
SNOW
SNOW SHOWERS
FOG
SMOKE
HAIL
THUNDERSTORMS
BLOWING DUST/SAND

ICE PELLETS (SLEET)
FREEZING RAIN
SQUALLS
WATER SPOUTS/TORNADO
LIGHTNING

COLD FRONT
WARM FRONT
QUASI-STATIONARY FRONT
OCCLUDED FRONT
TROUGH
INTER-TROPICAL CONVERGENCE ZONE
INSTABILITY (SQUALL) LINE
BOUNDARY OF THUNDERSTORM-CONVECTIVE WEATHER AREA
BOUNDARY OF CLEAR AIR TURBULENCE AREA
TTTTT BOUNDARY OF ICING AREA
BOUNDARY OF NONCONVECTIVE PRECIP AREA
BOUNDARY OF AREAS OF SIGNIFICANT CLOUDINESS (5/8 OR GREATER SKY COVER, AND ALL CUMULONIMBUS)
BOUNDARY OF AREA OF CLOUDINESS WITH CEILING LESS THAN 1,000 FT. AND/OR VSBY LESS THAN 1 MILE
TROPICAL DEPRESSION, TROPICAL STORM, TROPICAL CYCLONE
15 DIRECTION AND SPEED OF MOVEMENT
--080-- 0 DEGREE ISOTHERM WITH HEIGHT

UPPER AIR CHART SYMBOLS

ISOTACHS INDICATING WIND SPEED IN KNOTS
JET STREAM AXIS
HEIGHT CONTOURS WITH HEIGHTS IN TENS OF METERS
WIND DIRECTION AND SPEED (63-67kts)
ISOTHERMS OF TEMPERATURE IN DEGREES CELSIUS

ABBREVIATIONS

AC	ALTOCUMULUS
AS	ALTOSTRATUS
CB	CUMULONIMBUS
CC	CIRROCUMULUS
CI	CIRRUS
CU	CUMULUS
CS	CIRROSTRATUS
NS	NIMBOSTRATUS
SC	STRATOCUMULUS
ST	STRATUS
TCU	TOWERING CU
BKN	BROKEN
BLO	BELOW
CAT	CLEAR AIR TURB
CLD	CLOUD
CLR	CLEAR
CNS	CONTINUOUS
CNTR	CENTER
DCRS	DECREASING
FNT	FRONT
FRQ	FREQUENT
INCRS	INCREASING
ISLTD	ISOLATED
LGT	LIGHT
LYR	LAYERED
NUM	NUMEROUS
BECMG	BECOMING
OVC	OVERCAST
SCT	SCATTERED
UNK	UNKNOWN
VCNTY	VICINITY
SHRA	RAINSHOWERS
TS	THUNDERSTORMS
FM	FROM
MOD	MODERATE
SVR	SEVERE
TEMPO	TEMPORARY

NOTES ON HORIZONTAL WEATHER DEPICTION (HWD) CHARTS

1. *Areas of significant cloudiness are defined as those of five-eighths or more coverage and all cumulonimbus.*

2. *All cloud amounts are entered in eighths. Bases and tops of clouds and hazards to flight are entered in hundreds of feet above mean sea level, e.g.:*

6 cu $\frac{050}{030}$ Six-eights of cumulus, base 3000 feet tops 5000 feet.

3. *The zero degree celsius isotherm is entered where it intersects the surface, 5,000 feet, 10,000 feet and 15,000 feet (e.g., SFC, 050, etc...).*

4. *Consult latest advisory for official position of tropical cyclones and depressions.*

AGM1f414

Figure 4-14.—Symbols used on Horizontal Weather Depiction (HWD) analysis and prognosis charts.

colors used on surface and upper level streamline analysis charts. Figure 4-14 shows the symbols and colors generally used on low- and high-level Horizontal Weather Depiction (HWD) analysis and prognosis charts, although there are minor variations. Many of these HWD chart symbols may also be used on surface analysis charts, especially for a small area Local Area Weather Chart (LAWC).

Contour Lines

In addition to standard symbols, most analysis and prognosis charts have one, two, or sometimes several sets of data contour lines displayed. Contour lines normally connect areas with equal data values, so may generally be called "isolines" or "isopleths"-lines of the same value. Normally, the primary set of contour lines is displayed as a solid line, while the secondary contour is displayed as a &shed line. For example, a standard 850-hPa, 24-hour prognosis chart has height contours in solid lines, temperature contours in dashed lines, and wind speed and direction as plotted data. Colors may also be used with contour lines. Normally contour lines are identified with the value at each end, or at the top of closed circular systems, as shown in figure 4-15.

The change in numerical value between each contour line is called the "contour interval." Table 4-1 lists several of the most frequently seen contours, the basic contour value, and the standard contour interval. Sometimes the distance between contours is so wide that intermediate contours are drawn. For example, one intermediate pressure contour using a 2-hPa interval may be drawn between each 4-h.Pa standard interval contour. Intermediate contours are normally dashed lines that are only drawn where they are needed to show a pattern and are normally labeled at each end.

Other Contour Lines

Other contour names not used as often as those in table 4-1 are as follows:

- Isochrone—lines of equal time occurrence of a phenomenon, such as the start or end of rainfall
- Isodrosotherm—lines of equal dew-point temperature
- Isopach—lines of equal thickness
- Isohyet—lines of equal precipitation amount
- Isohume—lines of equal humidity (relative, specific, or mixing ratio)
- Isopycnic—lines of equal or constant density
- Isogon—lines of equal vector quantity, such as wind direction
- Isovel—lines of equal vertical velocity
- Isovort—lines of equal constant absolute vorticity

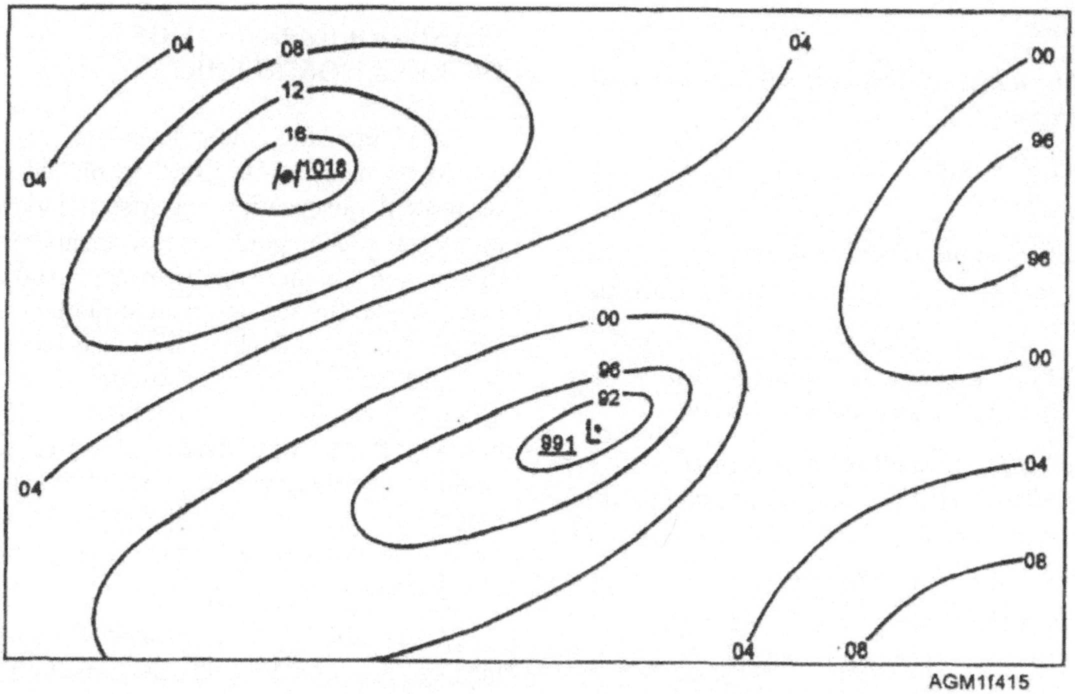

Figure 4-15.—Labeling of contour line values.

Table 4-1.—Common Contour Names, Standard Values, and Intervals

ISOLINE NAME	DRAWN FOR VALUES OF EQUAL...	BASIC CONTOUR VALUE	STANDARD CONTOUR INTERVAL	EXAMPLE OF CHART USED ON
Isobar	Pressure	04 (1004 hPa)	4 hPa	SFC PRES
Isoheight	Altitude	147 (1470 m) 306 (3060 m) 558 (5580 m) 720 (7200 m) 912 (9120 m) 176 (11760 m)	30 m 30 m 60 m 60 m 120 m 120 m	850-hPa HT 700-hPa HT 500-hPa HT 400-hPa HT 300-hPa HT 200-hPa HT
	Wave height	3 A	3 ft	Wave HT
Isotherm	Temperature	0°C	5°C	850-hPa to 200-hPa TEMP SST
		0°C	2°C	300-hPa & 200-hPa
Isotach	Wind speed	10 kts	20 kts, 50 kts	

- Isallobar—lines of equal pressure change
- Isallohypse—lines of equal height change
- Isallotherm—lines of equal temperature change

REVIEW QUESTIONS

Q17. What is the difference between an outlook and a forecast?

Q18. Define a briefing aid as it pertains to weather forecasting?

Q19. What color is used to mark only one set of history on a chart?

Q20. What colors are used to indicate (a) an occluded front and (b) a shear line, on a surface pressure chart?

Q21. What color is used to indicate a divergent asymptote on a streamline analysis?

Q22. What color is used to indicate (a) clear air turbulence (CAT) and (b) nonconvective continuous frozen precipitation on a horizontal weather depiction analysis?

Q23. What is the standard contour interval used for isobars?

Q24. What is the standard contour interval used for isoheights at 300-hPa?

Q25. What is the standard contour interval used for wave heights?

Q26. What term is used to indicate lines of equal dew-point temperature?

Q27. Lines of equal pressure change are defined what term?

STANDARD DEPICTION OF OBSERVATION REPORTS

In addition to contour lines and weather system depictions, many charts also show plotted wind speeds or plotted observation reports. Standard plotting models are used to guide the placement of information. Plotting models vary slightly by application. There are two synoptic surface observation plotting models: the surface synoptic land observation model and the surface synoptic ship observation model. There is also a synoptic upper-air observation plotting model used to plot reports on all constant pressure level charts. Additionally, there is a METAR observation plotting model, which is a slight variation from the synoptic models and is occasionally used to plot hourly aviation observations.

Observation reports plotted by computers or manually by people use the same plotting models but with one important difference. When a chart is plotted manually, all data in the observation are routinely transferred to the chart, and the single plotted chart is used for many different applications. This is done because manually plotting separate charts is very time-consuming. But when observation data is computer plotted, such as done by the TESS system, only data actually needed for a single type of analysis is routinely displayed. Separate charts may be produced very quickly for each individual analysis, especially since the computer performs the initial analysis. For instance, sea-level pressure, wind speed and direction may be the only parameters plotted for a simple pressure analysis. Only the pressure tendency and change group need be plotted for an isallobaric analysis.

The operator specifies which data to display on the video screen and how the data is to be displayed. The standard plotting models presented in this section should be used as a guide so that the forecaster, when reviewing a "plotted" chart recalled from a saved data file, or a hard-copy of the chart, will be able to interpret the information. These models are also useful in interpreting plotted data received via facsimile on either NWS charts in the United States or from foreign meteorological services.

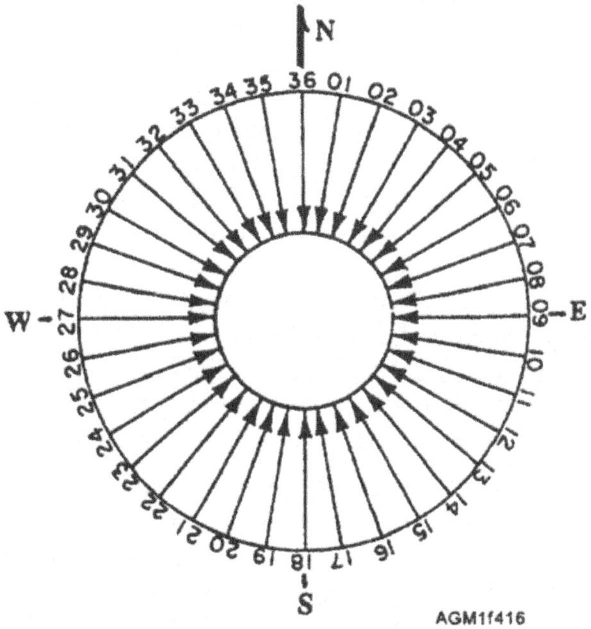

AGM1f416

Figure 4-16.—Wind direction in true degrees indicated by the orientation of the wind shaft from the station circle or station location.

Wind Plots

Many prognostic charts show plotted wind directions and speeds. Winds are always plotted with a system of winds shafts and barbs. The direction is indicated by the orientation of the shaft extending from the station circle (fig. 4-16). The wind speed is indicated by the barbs on the shaft and gusts are written at the end of the wind shaft (fig. 4-17).

TESS 3.0 only . . . A square is used to indicate each 100 knot increment	
Flagged barb indicates 50 knot increments of wind	
Single barb indicates 10 knot increments	
Half barb indicates 5 knots	
No barbs on shaft indicates less than 3 knots	
An *X* on the end of the shaft indicates a missing wind speed	
An *X* on the center of the shaft indicates a missing wind direction	
A circle drawn around the station circle indicates calm winds	
Wind barbs extending clockwise from the wind shaft, as shown above, are used for plots in the Northern Hemisphere. To plot winds in the Southern Hemisphere, plot wind barbs so that they extend counterclockwise from the wind shaft.	

AGM1f417

Figure 4-17.—Wind speed indicated by barbs. Plotted winds are rounded-off to the nearest 5 knots.

Synoptic Surface Plotting Models

The plotting models for both land stations and ship-reported synoptic surface and upper-air observations are dictated by World Meteorological Organization publication WMO No. 305, Guide on the Global Data-processing System, Volume *II, Preparation of Synoptic Weather Charts and Diagrams.* Plotted observations use symbols, numbers, and letters to depict various reported data. Figure 4-18 shows the symbolic Land Synoptic code, the symbolic plotting model, an actual report, and the plotted report. Figure 4-19 shows the symbolic Ship Synoptic code, the symbolic plotting model, an actual report, and the plotted report.

Normally, only observations from the same synoptic hour are plotted on a chart. If observations from a past synoptic hour say, 0300Z or 0000Z data plotted on a 0600Z chart, the observation hour is plotted below the station model, such as 032 or 00Z. Late reports and off-time reports may be manually plotted on a computer plotted chart to aid reanalysis in data sparse areas. When this is done, late reports corresponding to the synoptic hour of the chart are normally hand plotted in black ink, while older synoptic intermediate reports are plotted in different colored ink, such as blue, red, or green. (Note: red ink or pencil does not show up on a chart under shipboard red lighting and normally does not copy well on a photocopier.)

Quadrant-of-the-globe, latitude, and longitude, or the block and station number in the case of land observations, are not plotted, but are used to locate the station circle.

Several reported values are plotted with a graphic symbol vice the reported numeric value. Total sky cover, N, is indicated by shading portions of the station circle. Present weather, past weather, cloud types, pressure tendency, and the ships course-made-good are also indicated by symbols. These are shown in Appendix IV.

Reported temperatures are rounded off to the nearest whole degree Celsius for land station reports. For ship reports, temperatures are not rounded off, but plotted as they are reported to the nearest tenth degree Celsius. Only the sea-surface temperature contains a decimal point before the tenths value by WMO guidelines.

SYMBOLIC FORM (Data in brackets not routinely reported)

AAXX YYGGi_w IIiii $i_R i_x$hVV Nddff (00fff) 1s_nTTT 2$s_n T_d T_d T_d$ 3$P_o P_o P_o P_o$ 4PPPP (or 4$a_3$hhh) 5appp (6RRRt_R) 7ww$W_1 W_2$ 8$N_h C_L C_M C_H$ 9GGgg 333 0.... 1$s_n T_x T_x T_x$ or 2$s_n T_n T_n T_n$ (3Ejjj 4E'sss 5$j_1 j_2 j_3 j_4$ $j_5 j_6 j_7 j_8 j_9$) 6RRRt_R 7$R_{24} R_{24} R_{24} R_{24}$ 8N_sC$h_s h_s$ 9$S_p S_p s_p s_p$ 444 N'C'H'H'C_t 555;

ACTUAL REPORT

AAXX 07124

72409 21320 80327 10157 20150 39922 49928 57028 76264 87923 91156 333 10183 20091 60175 70031 87923 91038 555 90712;

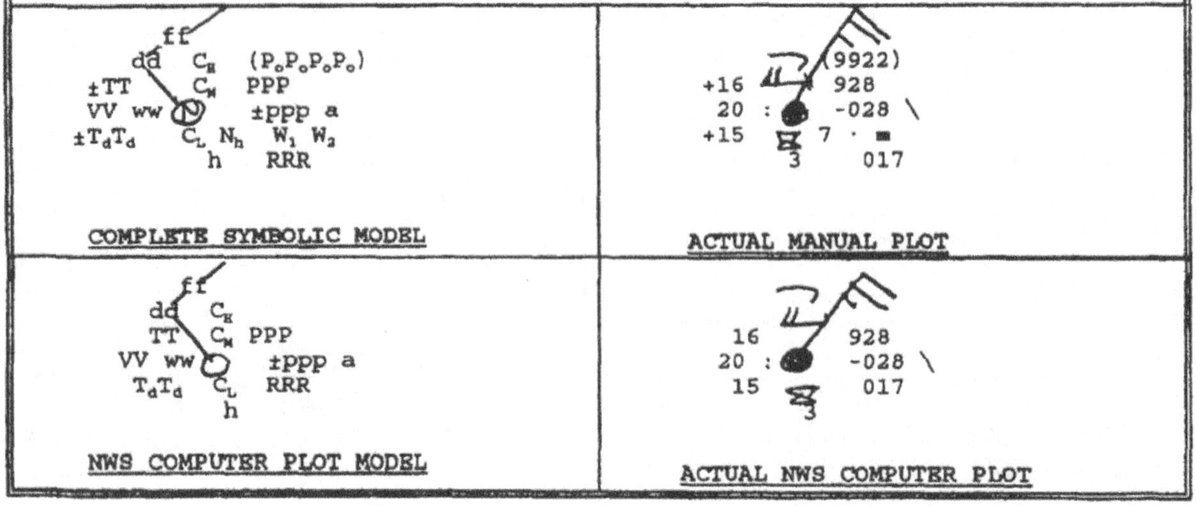

AGM1f418

Figure 4-18.—Synoptic land observation plotting models.

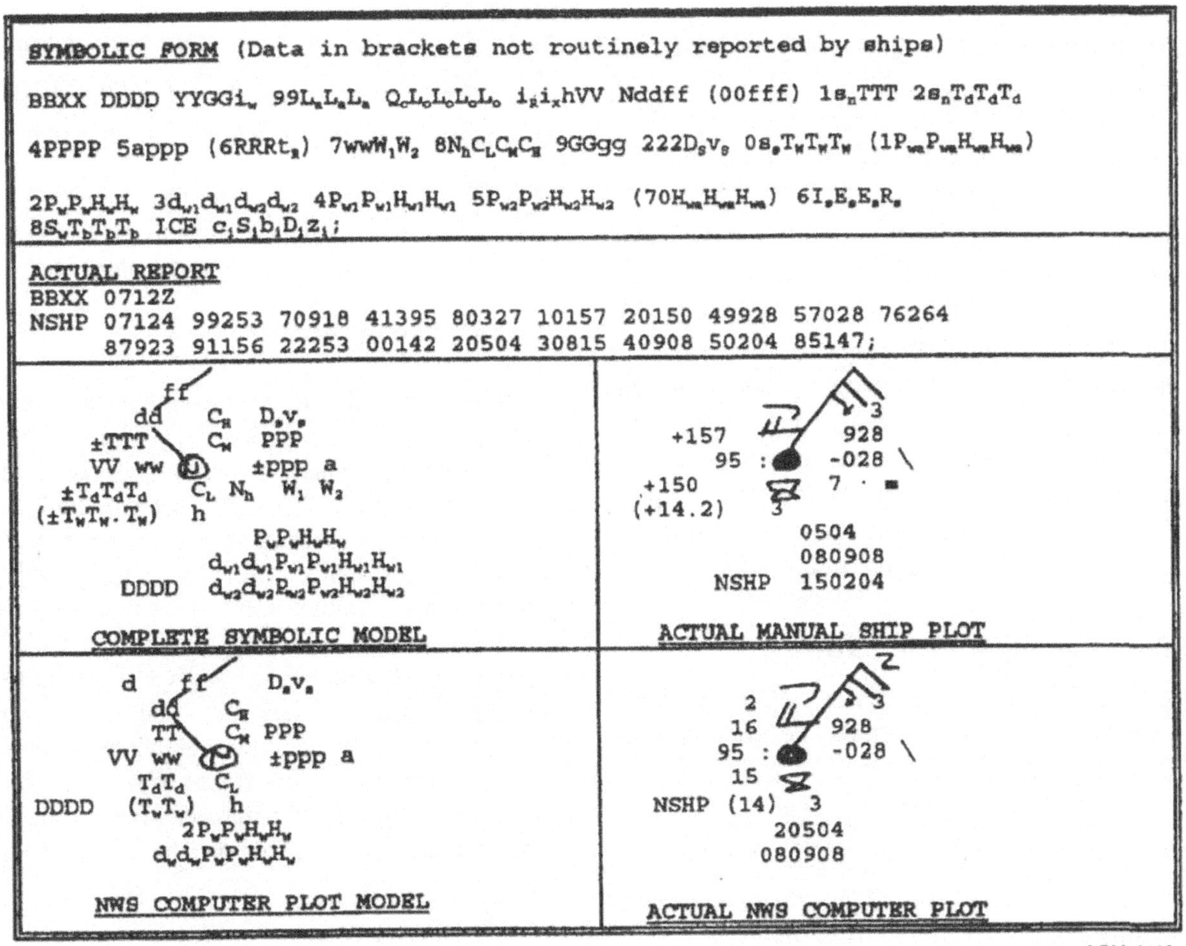

Figure 4-19.—Synoptic ship observation plotting models.

Synoptic Upper-air Plotting Model

A complete upper-air observation report may be plotted on a single chart—the Skew T, Log P diagram, discussed later in this chapter—or sections of the report pertaining to each standard level may be plotted on the appropriate constant pressure level chart. For an 850-hPa pressure level chart, only the data reported for the 850-hPa level is extracted from each individual observation for plotting. Figure 4-20 shows the

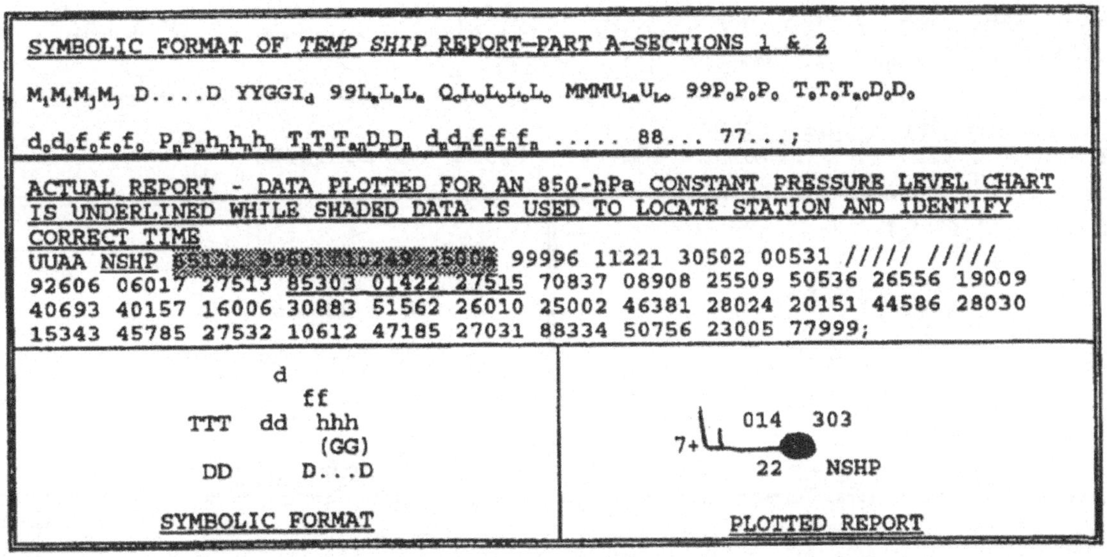

Figure 4-20.—Standard plotting model for constant-pressure level data received in upper-air observation reports.

standard plotting model for upper-air observed data. As an example, we have used a ship observation. A land station report is plotted in the same manner, using the block-station number to locate the station. Block-station number is not plotted.

Notice in the constant pressure level plotting model that the tens digit of the wind direction is normally plotted near the base of the wind barb. A plus sign (+) after the tens of wind direction is used by WMO guidelines to indicate wind directions reported at odd five-degree increments. The hour of the observation GG is plotted in parenthesis to the right of the report only when the report is more then 1 hour off the synoptic chart time. The station circle is filled in when the dew-point depression is less than or equal to 5°C.

REVIEW QUESTIONS

Q28. Besides speed and accuracy, what else is a significant advantage of the TESS computer plotting system?

Q29. What does an 5" on the center of the shafi of a wind plot indicate?

Q30. Where is air temperature information placed on a surface observation plotting model?

Q31. Where is pressure change information placed on a plotting model?

Q32. Where would the observation time be plotted if a particular observation was from a different synoptic period than the overall chart?

Q33. How is total shy cover information indicated on a METAR surface observation plotting model?

Q34. How is ship movement plotted on a ship synoptic model?

Q35. What does a filled in station circle indicate on a constant pressure standard plot?

METAR Plotting Models

The synoptic surface and synoptic constant pressure level plotting models are used every 6 hours to plot data for the synoptic hours. In the United States, the NWS also provides Synoptic Intermediate Surface Analysis charts, so surface analyses are available every 3 hours. Occasionally, a forecaster may need hourly plots of data for small areas. This is usually done to allow the analyst to preform a Local Area Weather Chart (LAWC) analysis. LAWCs are used to track weather features whenever a detailed picture of developing weather conditions is required for a small area, usually only a few states in size. LAWCs are usually not done for ocean areas due to the lack of observations. Figure 4-21 shows the METAR plotting model.

In METAR code plotting, the weather is either plotted in the letter abbreviations as reported, or converted to standard symbology at the forecaster's discretion. See Appendix IV for the standard weather symbology. The first reported cloud layer is plotted above the station circle with each successive layer plotted below the first. If a layer is identified as CB or

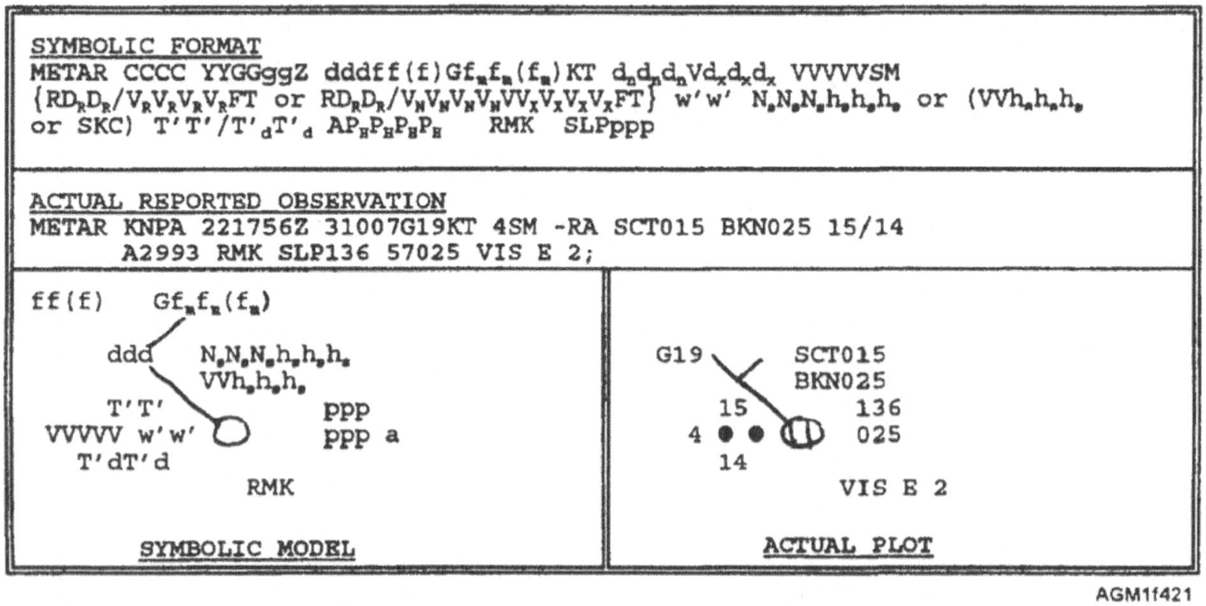

Figure 4-21.—METAR code plottiog model.

TCU, the designation is plotted to the right of the layer height. Total cloud layer coverage is plotted as the symbols 0 for clear (no cloud layers reported), 0 for few or scattered, ◑ for broken, ⊕ for overcast, or ⊗ for obscured. The sea level pressure in hectopascals is plotted in the upper right of the station circle.

Visibility is plotted in statute miles. Weather may be either plotted in the letter code as reported or converted to standard symbology at the forecaster's discretion (see Appendix IV). If an overseas location reports CAVOK instead of visibility and cloud layers, plot 9999 for visibility and OK to the right of the station circle. Runway visual range and remarks are plotted below and to the right of the station circle.

SEA-SURFACE TEMPERATURE CHART

The sea-surface temperature (SST) chart may be manually plotted or computer-plotted. Reports of satellite-sensed sea-surface temperature or synoptic-observation sea-surface temperatures are used. If plotted by a computer, the chart is routinely supplemented with any and all data available from other sources. An SST chart that uses various sources of information is called a "composite" SST chart. If based on satellite observations, the chart is supplemented with ship synoptic SST reports as well as ship and aircraft bathythermograph observation SSTs. Normally, up to 3 days of collected data is used for a manually plotted SST chart. Only in the most data-sparse areas is it considered acceptable to use reportsup to 10 days old.

The analyst normally draws isotherm contours on the plotted chart at 2-degree Celsius intervals for an SST analysis.

Since both new and older data are used and several types of data are combined, each type and age of data must be identified to allow the analyst to weight each report for accuracy. Data are routinely coded with symbols to indicate both the source and the age of the report. Recent calibrated infrared satellite observations of the sea surface in cloud-free and low-humidity areas are considered the most accurate and reliable. Recent bathythermograph observations are next in accuracy, while recent ship-observed SST reports are least reliable. Satellite observations are usually computer-plotted on the base chart, with the temperature (in degrees Celsius) plotted next to a small square. When supplemented with ship SST observations, the ship-reported SST is plotted following a Day Code symbol, as shown in figure 4-22. For example, ship-reported SSTs from the current day may be plotted as -12.3 (12.3°C) or --0.3 (-0.3°C), while a report of an -0.5°C ship-reported SST 3 days old is plotted as +-0.5. Aircraft or ship bathythermograph SST reports are also plotted by following the date codes shown in the table, but the bathythermograph observations are always circled. For example, Δ14.6 indicates a bathythermograph reported 14.6°C SST 1 day old.

DAYS OLD	SYMBOL
Current Day	–
1 day old	Δ
2 days old	∇
3 days old	+
4 days old	×
5 days old	• –
6 days old	• Δ
7 days old	• ∇
8 days old	• +
9 days old	• ×

Figure 4-22.—SST data age plotting symbols.

If the validity of an SST report is in question, the report is underlined, such as ∇17.7 . If a report is corrected before plotting, the report is double-underlined, such as x22.2.

Enter the symbols and age of the data in a legend block in the top or bottom margin of the chart.

As with the SST chart, there are a few special practices followed with the sea height chart.

SEA HEIGHT CHART

The sea height chart is usually a computer plot of ship and coastal station locations for a given synoptic time, showing only the wind speed and direction, the sea-wave height group ($P_wP_wH_wH_w$ from the $2P_wP_wH_wH_w$ group), the primary swell-wave direction/period/height group ($d_{w1}d_{w1}P_{w1}P_{w1}H_{w1}H_{w1}$), and the secondary swell-wave direction/period/height group ($d_{w2}d_{w2}P_{w2}P_{w2}H_{w2}H_{w2}$). The computer plotted chart may also have a preliminary pressure analysis printed to aid the analyst. Ship reports received late and classified ship reports may require manual plotting on the chart following the format of the computer plot.

Unless the computer is instructed to convert reported heights in half-meter units to feet, the observer must convert the heights. Heights are converted to feet by multiplying the height in half-meters by 3, then dividing by 2. Converted heights are usually plotted by writing the sea height over the swell height as in a fraction immediately to the right of the plotted report. Figure 4-23 shows a computer-plotted report of the sea and swell waves. The observer has converted the height in half-meter units to height in feet, and has written the converted height to the right of the report.

Sometimes the analyst may ask the observer to draw arrows for the wave directions. When this is required, the observer draws a "straight" arrow pointing away from the station circle directly opposite the wind direction as shown. Since seas are caused by the wind, the direction the seas are moving toward is opposite the wind direction (direction winds are blowing "from"). The swell-wave directions report the direction the swells are coming from. Swell-wave directions are indicated by drawing a "wavy" arrow from the station circle pointing opposite the reported direction. The primary and secondary swell-wave directions (080° and 150°) are drawn above. A shorter wavy arrow is sometimes used to indicate the secondary swell-wave direction.

Two charts are routinely derived from sea- and swell-wave reports: the significant wave height analysis and the combined sea-height analysis. The computer systems at FNMOC routinely produce significant wave height analyses. The combined sea-height analysis is normally analyzed from computer plotted observation data. Aboard ship, sea and swell heights are normally plotted by the TESS system, but are occasionally manually plotted. The analyst may analyze both sea and swell wave directions and heights independently to produce a sea and swell wave analysis, or analyze just the higher of the sea or swell wave heights for a significant wave height analysis. In all types of wave height analyses, isoheight contour lines are drawn at 3-foot intervals, normally as solid black lines. Arrows may be added by the analyst to indicate prevailing wave directions.

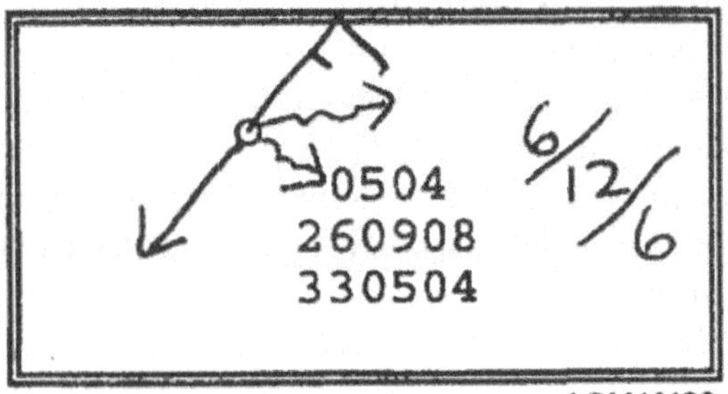

Figure 4-23.—Computer-plotted wave report with observer-entered height conversions and direction arrows.

The Combined sea-height, often identified as the "C-height," is defined as the square root of the sea-wave height squared plus the swell-wave height squared. The combined sea-height represents an approximation of the height of the highest crests of water formed when the crests of sea waves interact with the crests of swell waves. For example, a reported sea height of 3 feet with a reported swell height of 4 feet would result in a C-height equal to the square root of 3^2 plus 4^2, or 5 feet. If the C-height is not computer-calculated and plotted, the analyst or the observer calculates the values and plots them to the right of each observation plot. This is normally done by using a table that is produced locally.

REVIEW QUESTIONS

Q36. Where are remarks plotted on a METAR code plotting model?

Q37. What does the symbol [64] indicate?

Q38. What does the symbol [13] indicate?

Q39. What does the symbol [22] indicate?

Q40. What is the normal contour interval for an SST analysis?

Q41. How is 4-day old SST information annotated on *an SST analysis?*

Q42. How are sea and swell wave directions depicted on a sea height chart?

SKEW T, LOG P DIAGRAM

LEARNING OBJECTIVES: Identify the scales used to plot reported information on the Skew T, Log P diagram. Discuss how the Pressure Altitude scale is constructed. Describe how reported information is plotted on the diagram and how the temperature, dew-point temperature, and pressure altitude curves are completed by using plotted information.

One of the most useful products for analysis of the state of the atmosphere over a single location is the Skew T, Log P diagram. The standard Skew T, Log P diagram is one of the few products that provides a complete profile of the atmosphere in the vertical, from the surface to as high as 25 hectopascals.

Environmental computer systems, such as TESS, have built in programs that will plot a complete upper-air observation report on a background representative of the Skew T, Log P and calculate many values routinely found by using the diagram. However, the computer plot does not provide the overall detail provided by a paper copy of the Skew T chart. An actual DOD-WPC 9-16 Skew T, Log P Diagram should be available at your command.

Detailed information on plotting, analysis, and forecasting techniques using the Skew T, Log P diagram is contained in NAVAIR 50-lP-5, *The Use of the Skew T, Log P Diagram in Analysis and Forecasting.*

DIAGRAM FAMILIARIZATION

The Skew T, Log P diagram is a set of curved and straight lines that graphically represent the physical processes that occur in the atmosphere as moist air rises and descends. In the upper-right comer of each diagram is a detailed explanation of each set of lines on the diagram. The Skew T, Log P diagram is printed in black, pastel green, and light brown. DOD WPC (Department of Defense, Weather Plotting Chart) 9-16 is the basic chart. Several different variations of the chart are available and listed in the NIMA Catalog. All versions have the basic plotting scales discussed below, superimposed over each other (fig. 4-24). These scales are used as guides for plotting observed information, and then for deriving other values when analyzing the plotted information. The use of the Skew T, Log P diagram allows analysts to find important values and indicators without having to preform complex thermodynamic calculations.

Isobar Scale

Isobars are lines of equal pressure. The scale is a series of horizontal, solid brown lines spaced logarithmically at 10-hectopascal intervals from 1050

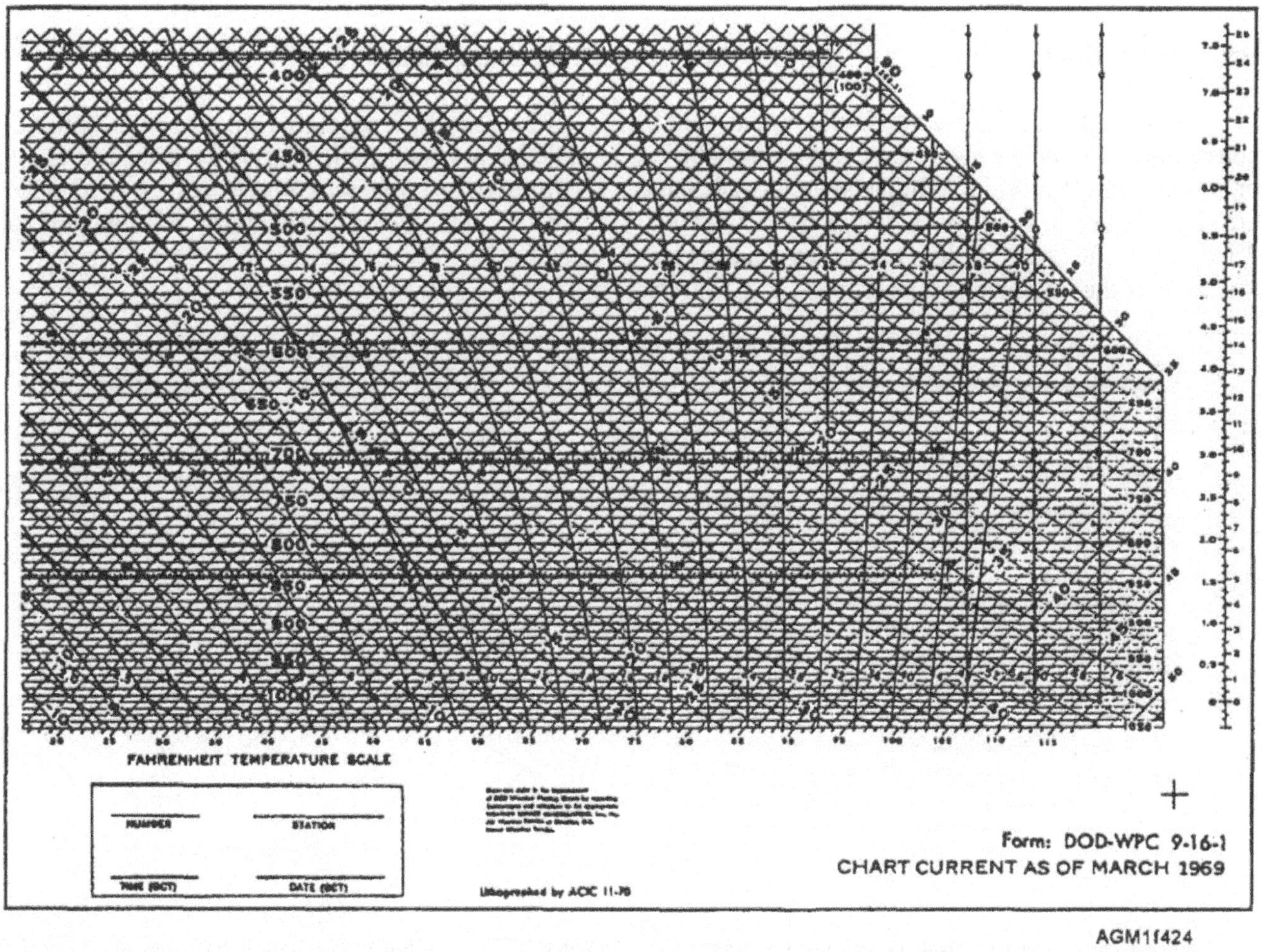

Figure 4-24.—A portion of a Skew T, Log P diagram (version 9-16-1) showing the superimposed plotting scales.

hPa at the bottom of the chart to 100 hPa at the top (fig. 4-25). Pressure values are printed in the center of the diagram as well as at both the left and right sides. The upper portion of the isobar scale from 400 to 100 hectopascals is also used to plot pressure levels 100 to 25 hectopascals. Most of the mandatory reporting levels are indicated by slightly thicker brown lines. When the station pressure is lower than 1000 hPa, the height of the 1000 hPa surface may be obtained from the nomogram in the upper lefthand comer of the diagram.

Isotherm Scale

Isotherms are lines of equal temperature. The scale is depicted as a series of straight, solid brown, right-diagonal lines (fig. 4-26). Isotherms are evenly spaced at l-degree Celsius intervals ranging from warm temperatures (around 50°C) at the lower right to cold temperatures (near minus 120°C) at the upper left. The temperature range varies on the different versions of the chart. Isotherms are labeled every 5 degrees, and a Fahrenheit temperature conversion scale is printed across the bottom edge of the isotherm scale.

Dry Adiabat Scale

The dry-adiabatic lapse rate is the rate at which nonsaturated air cools when it rises in the atmosphere, or warms when it subsides.' On the scale, dry adiabats are slightly curved, solid brown, left-diagonal lines at a 2°C interval (fig. 4-27). Values for the dry adiabats are the same as the intersecting isotherm at the 1,000-hPa level, and are printed across the top edge and right side of the diagram.

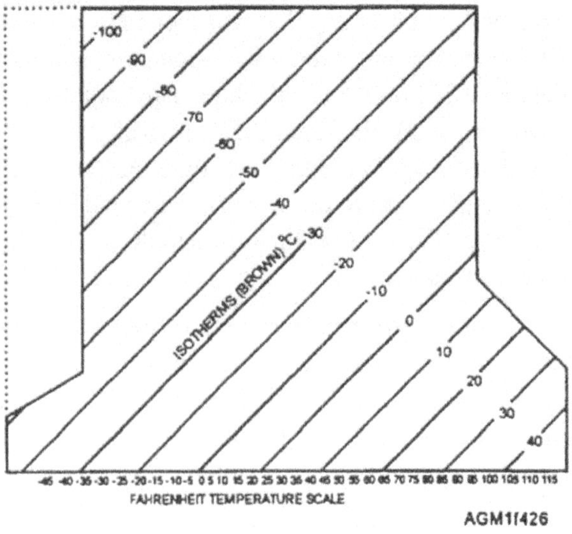

Figure 4-26.—Example of isotherms on a Skew T, Log P diagram.

Saturation Adiabat Scale

The saturation-adiabatic lapse rate (sometimes called a moist-adiabatic lapse rate), is the rate at which saturated air cools when it rises, or warms when it subsides. On the scale, saturation adiabats are solid,

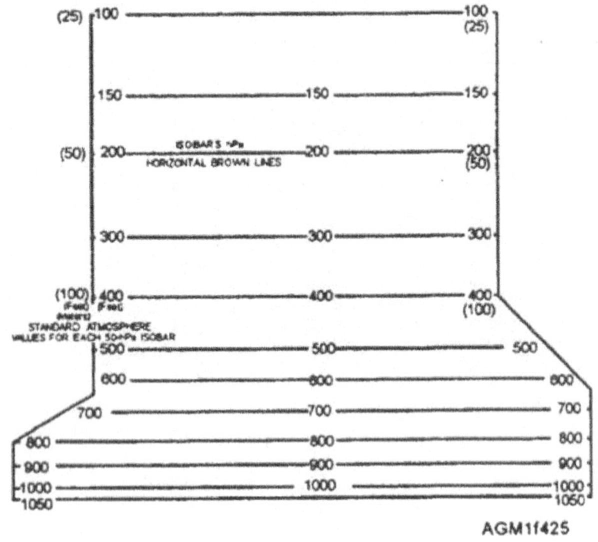

Figure 4-25.—Example of isobars on a Skew T, Log P diagram.

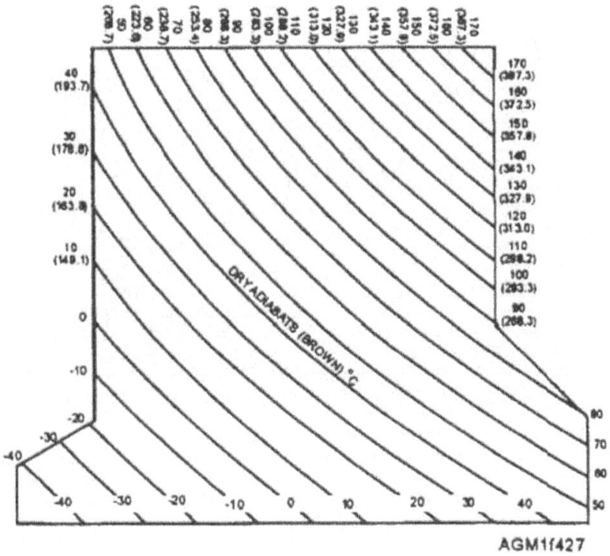

Figure 4-27.—Example of the dry adiabat scale on a Skew T, Log P diagram.

curved green lines (fig. 4-28) at a 2°C interval. Each saturation adiabat is labeled with the Celsius temperature value of its point of intersection with the 1,000-hPa isobar, and is also indicated by the green numbers at the 200-h.Pa level where the scale ends.

Mixing Ratio Lines

The saturation mixing ratio is the greatest amount of water vapor that may be contained in a parcel of air, expressed in units of grams of water vapor kilogram of dry air (g/kg). Dashed, right-diagonal green lines are used to represent various mixing ratio values (fig. 4-29). These values are printed in green between the 1,000-hPa and 950-hPa isobars.

U.S. Standard Atmosphere

The U.S. standard atmosphere is a representation of the annual average temperatures at various levels in the atmosphere for the continental United States. It is shown on the diagram as a heavy brown line (fig. 4-30). The height scale on the right side of the diagram, in both meters and feet, is calibrated to the U.S. Standard Atmosphere. The heights provided on the left side of the diagram under the pressure level labels are based on the U.S. Standard Atmosphere. The values in parenthesis are provided in geopotential feet, while the values in the square brackets [ ] are given in geopotential meters.

Wind Plotting Scale

A wind plotting scale is also provided on the right side of the diagram (fig. 4-30). Three identical wind scales are provided because up to three successive soundings may be plotted on the same diagram. This is

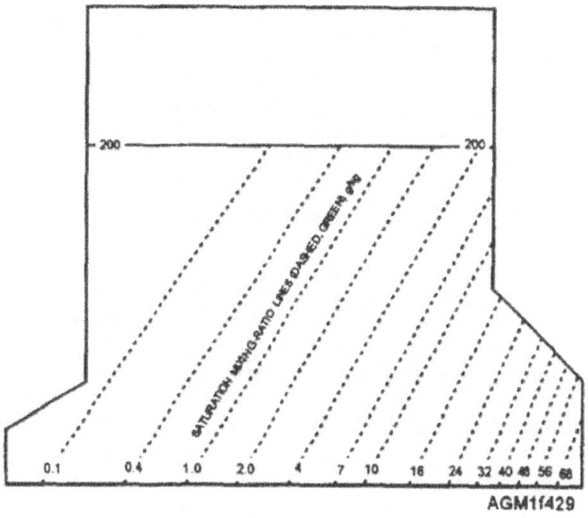

Figure 4-29.—Example of the saturation mixing ratio scale on a Skew T, Log P diagram.

a procedure used by forecasters to study the changes in the atmosphere with time for a single location.

Later, as you progress as an Aerographer's Mate, you will learn more about the other scales on the diagram and their uses.

PRESSURE-ALTITUDE SCALE

A modified procedure to accurately determine heights of features analyzed on a Skew T diagram was introduced in 1987. This procedure requires you to construct a height scale on the diagram before plotting information. You will use the scale to plot reported altitudes of mandatory pressure levels. These levels are then used to construct a pressure-altitude (PA) curve.

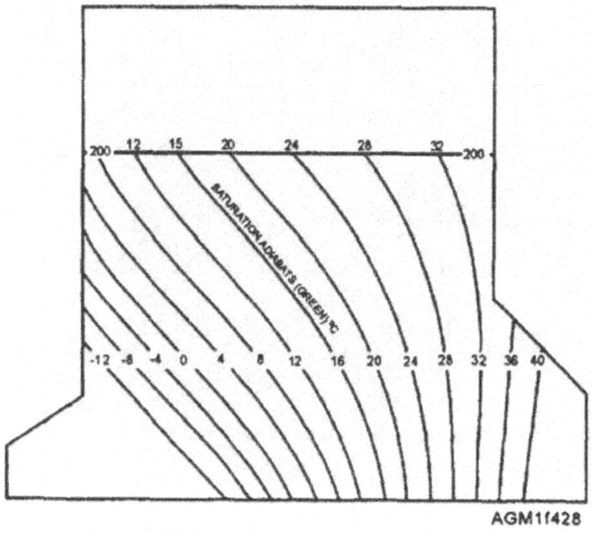

Figure 4-28.—Example of the saturation adiabat scale on a Skew T, Log P diagram.

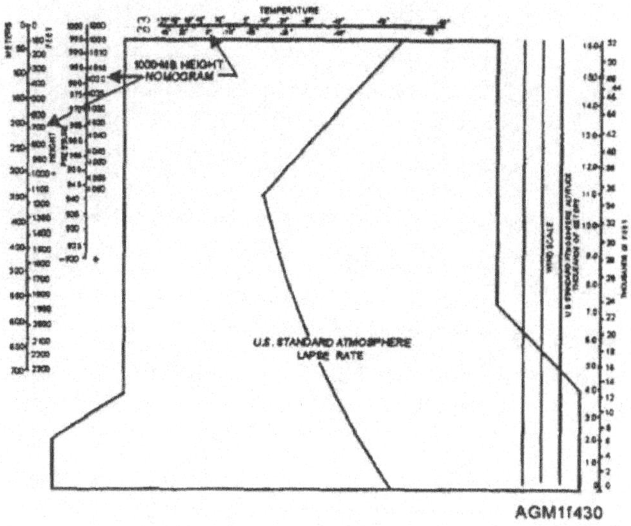

Figure 4-30.—Example of the U.S. standard atmosphere and related height scale, as well as the wind plotting scale.

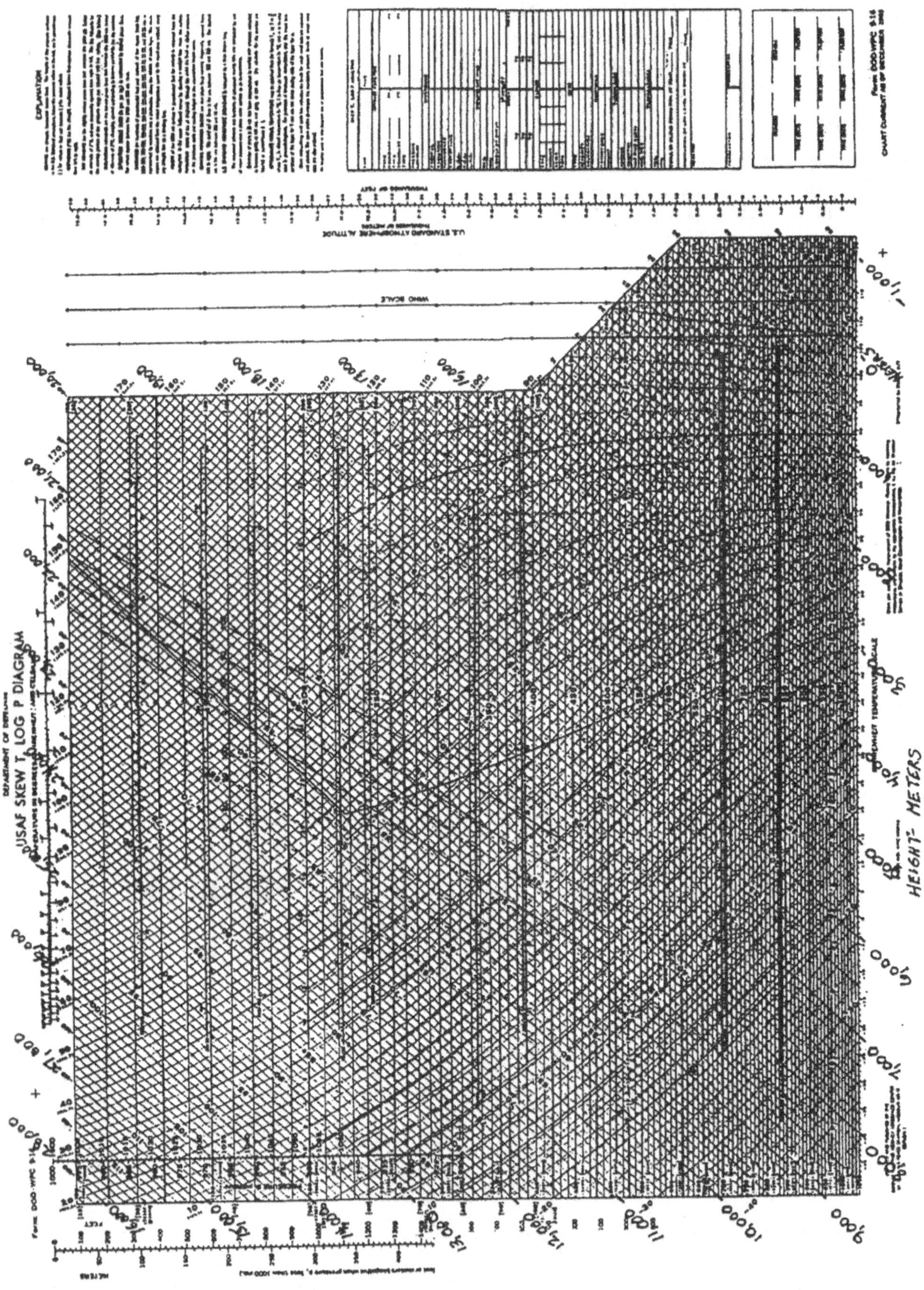

AGM1f431

Figure 4-31.—Relabeling the Skew T, Log P diagram for the pressure-altitude scale.

TEMP CODE PART A (The indicators for the mandatory pressure levels are bolded)

TTAA 64121 72306 **99**030 05050 09015 **00**211 06060 09005 **92**815 02227 08021

85490 00646 07016 **70**010 06900 08527 **50**560 22764 09047 40718 33372 09045

30916 459/ / 09071 **25**090 543/ / 09096 **20**225 581/ / 09099 **15**475 595/ / 09615

10745 575/ / 09100 88145 589/ / 09098 77132 09628 40508;

SYMBOLIC FORMAT

$M_iM_iM_jM_j$ YYGGI_d IIiii	(identification)
$99P_0P_0P_0$ $T_0T_0T_{a0}D_0D_0$ $d_0d_0f_0f_0f_0$	(surface)
$P_nP_nh_nh_nh_n$ $T_nT_nT_{an}D_nD_n$ $d_nd_nf_nf_nf_n$	(all other levels)
$88P_tP_tP_t$ $T_tT_tT_{at}D_tD_t$ $d_td_tf_tf_tf_t$ or 88999	(tropopause)
$77P_mP_mP_m$ $d_md_mf_mf_mf_m$ $(4v_bv_bv_av_a)$ or 77999;	(maximum wind)

Figure 4-32.—An example of TEMP coded mandatory level information.

Altitudes reported in upper-air observations are reported in geopotential meters. A geopotential meter is an approximation based on measured temperature, humidity, and pressure values. It is approximately equal to a standard meter.

Drawing the altitude scale requires a slight modification of the diagram. Start by relabeling the 40°C isotherm 0 meters, on the bottom margin of the diagram. Continue labeling the isotherms every 10°C across the bottom margin of the diagram, and then up the left edge of the diagram, in 1,000-meter increments, as shown in figure 4-31. The isotherm scale may now be used also as an isoheight scale, with the -120°C isotherm representing a 16,000-meter isoheight.

To plot information above the 100-hPa level, the altitude scale must be continued. The continuation scale is constructed by labeling the 0°C isotherm on the right margin of the diagram as 16,000 meters, and labeling the isotherms every 10°C in 1,000-meter increments up the right margin, and across the top of the diagram. On this continuation scale, the -120°C isotherm now also represents a 28,000-meter isoheight.

PLOTTING REPORTED INFORMATION

As an observer, you will plot information directly from Parts A, B, C, and D of the TEMP, TEMP MOBIL, and TEMP SHIP coded reports, and from Parts B and D of the PLOT, PILOT MOBIL, and PILOT SHIP coded reports on this diagram.

The usual procedure for plotting information on the diagram is to first plot mandatory pressure-level information (at a pressure level- the altitude, temperature, dew-point depression, wind direction and wind speed), as contained in Parts A and C of the appropriate TEMP, TEMP MOBIL, or TEMP SHIP report. Then, plot the significant level information (at a pressure level- the temperature and dew-point depression) reported in Parts B and D of the appropriate TEMP, TEMP MOBIL, or TEMP SHIP report. Next, the significant level winds (at a pressure level- wind direction and wind speed) from Parts B and D of the TEMP, TEMP MOBIL, or TEMP SHIP code when used outside of WMO Region IV. The fixed regional level winds from Parts B and D of the PILOT, PILOT MOBIL, or PILOT SHIP coded reports, as used within WMO Region IV, are plotted instead of significant level winds.

Use of Plotting Colors

Colors are used only to indicate the age of the trace. Black ink is normally used for history-a tracing of an older sounding onto a new diagram that is to be used to plot new information. The first plotted sounding on a diagram is normally done in blue ink or pencil. A second sounding plotted on the diagram is done in red ink or pencil. All data plotted for each sounding is plotted in the same color. If only one set of data is plotted on a diagram, blue ink or pencil is used to plot altitudes, temperature, dew-point depressions, and winds. The same color is used to enter information in the Legend block and, by the analyst, in the Analysis block. The analyst may use different colors, as desired, during analysis of information on the diagram.

Plotting Mandatory Levels

From Part A of the upper air coded message, locate the information for the surface and each mandatory pressure level to 100 hectopascals (fig. 4-32).

The reported altitude of the surface and each mandatory level is written on the right side of the colored portion of the diagram immediately above the printed isobar value (fig. 4-33). Notice that the 925-hPa mandatory level is not printed on the diagram. Mark the 925-hPa level with a tic mark on the edge of the diagram and enter the reported altitude just above this level, as shown in the figure. The three digits encoded for the altitude are converted to actual altitude values by the observer, following the guidance given in table 4-2.

When the reported surface pressure is less than 1,000 hPa, such as 976 hPa, an extrapolated 1,000-hPa height below station level is reported following the 00 indicator for the 1,000-hPa level. (In this case the 1,000-hPa level temperature/dew-point depression group and the winds group would be encoded as slashes.) For example, a reported level 00531 ///// ///// indicates a 1,000-hPa height as -03 1 meters (the 5 indicates a negative value), while a reported height of 00631 would indicate -13 1 meters. The height of the surface level is assumed to be zero.

The height is then plotted as a small dot surrounded by a *2/8*- to 3/8-inch square, at the intersection of the isobar of the mandatory pressure level and the isoheight line of the constructed height scale (fig. 4-34). Each printed isotherm is 100 meters. Estimate placement of the dot to the nearest 10 meters.

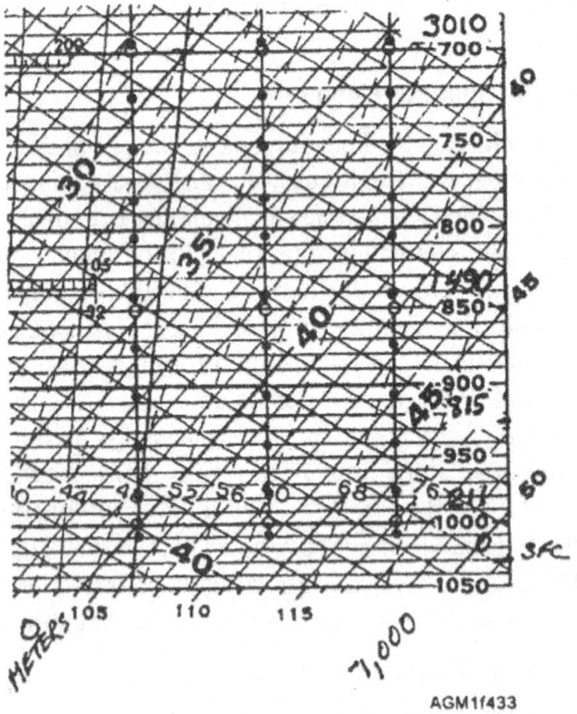

AGM1f433

Figure 4-33.—Plotted altitude values.

On the same horizontal isobar, the temperature is plotted as a small point surrounded by a 2/8- to 3/8-inch-diameter circle where the isotherm crosses the isobar. Temperatures are plotted to the nearest tenth degree Celsius (fig. 4-35). Estimate the position between the printed isotherms. Remember, if the reported tenths value is even, the temperature is above zero, and if the tenths value is odd, the temperature is below zero degrees Celsius.

To plot the dew-point depression, move along the isobar toward the left the number of degrees and tenths of degrees indicated by the encoded dew-point depression. Place a small point at the indicated value and surround the point with a 2/8- to 3/8-inch-high triangle (fig. 4-36). Remember, dew-point depression is the difference between the dew-point temperature and the temperature, and the dew-point temperature is either the same as or lower than the temperature. Also, remember that dew-point depression values are in degrees and tenths of a degree Celsius for code figures 00 to 50, while reported values 56 to 99 are in whole degrees when 50 is subtracted from the code figure. For example, the code figure 43 is a dew-point depression of 4.3°C, while 63 is (63 - 50 = 13) a dew-point depression of 13.0°C.

The next encoded value, wind direction, is plotted on the right wind scale for the first sounding plotted on a diagram, and succeeding reports are plotted on the

Table 4-2.—Rules for Converting Encoded Altitude to Actual Altitude

LEVEL (hPa)	RULE
1000 925	no change
850	prefix with a *1*
700	prefix with a *2* if ≥500, or a *3* if <500
500 400 300	Suffix with a *0*
250 200 150 100 70	prefix with a *1* and suffix with a *0*
50 30 20	prefix with a *2* and suffix with a *0*

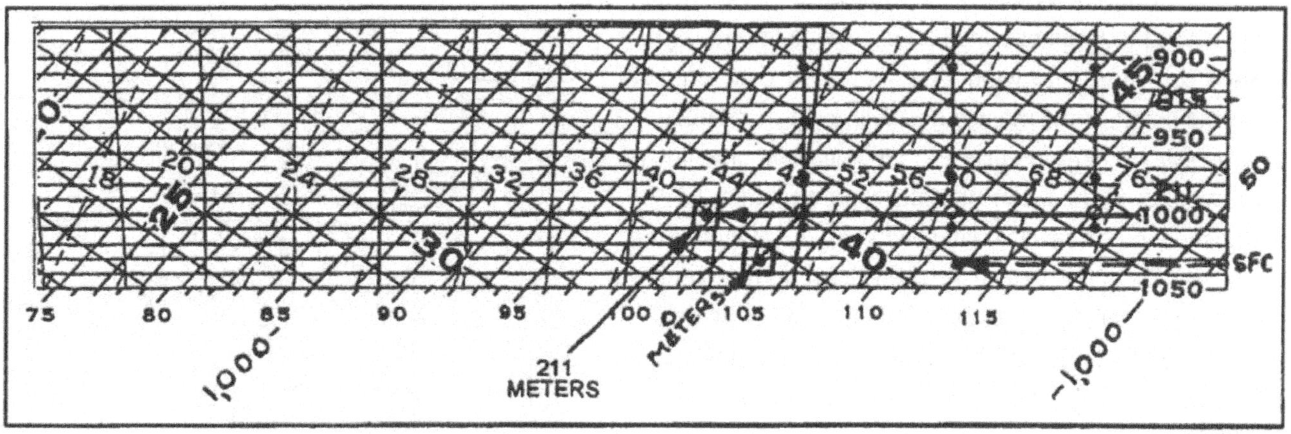

AGM1f534

Figure 4-34.—Plotting reported altitude of the surface and mandatory-pressure level.

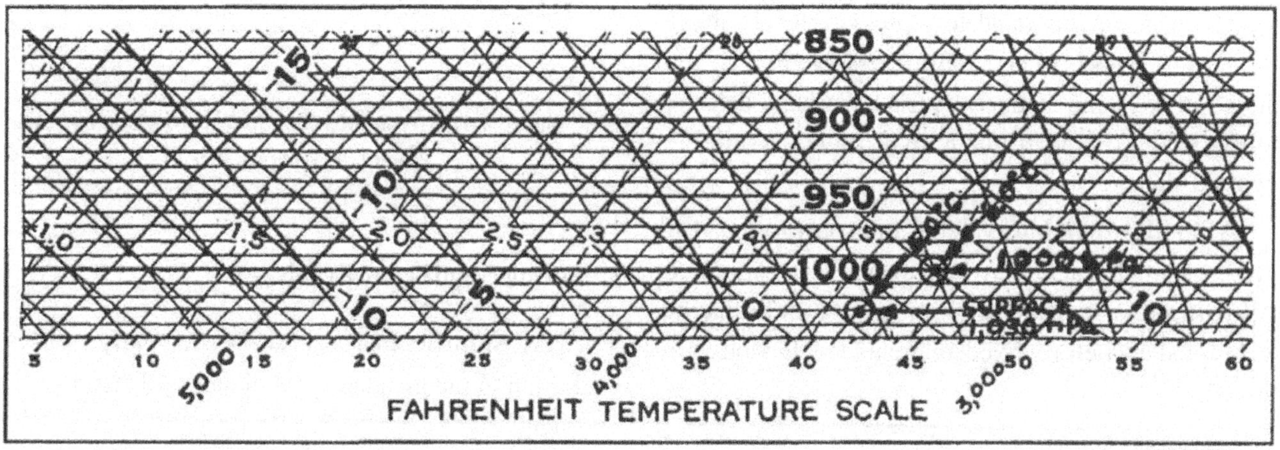

AGM1f435

Figure 4-35.—Temperature plotted on the mandatory pressure level isobar at the intersection of the isotherm value.

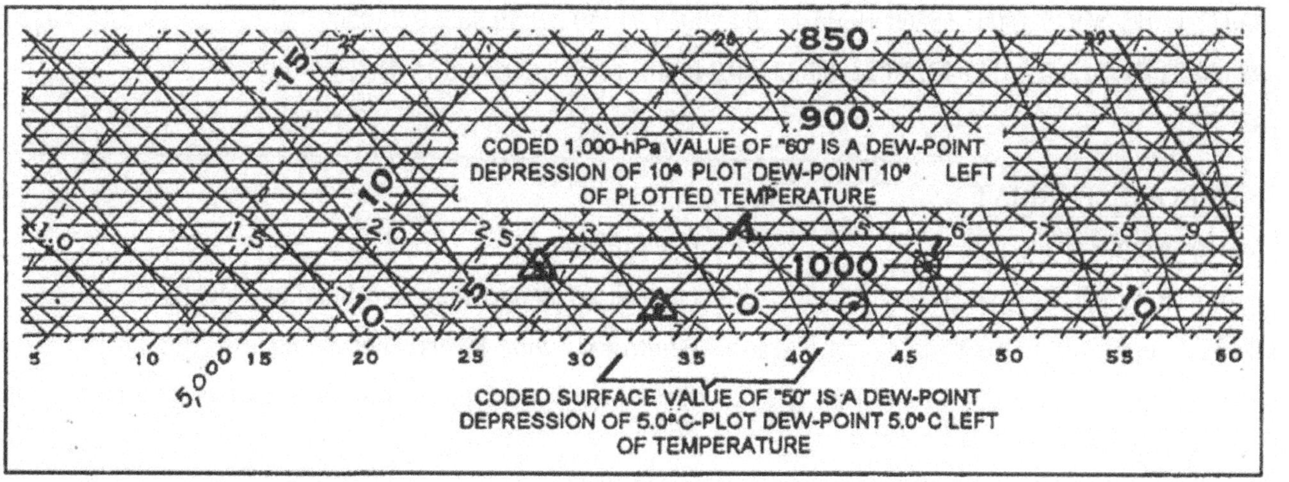

AGM1f436

Figure 4-36.—Dew-point depression plotted to left of temperature at each mandatory pressure level.

center and/or left wind scales. The right-most wind scale is used to transfer wind history from a previously plotted diagram. The wind shaft is drawn extending from the small, nondarkened circle at each mandatory level isobar in the same manner that a constant-pressure level wind is plotted. For plotting wind direction, true north is the top of the chart, and due south, the bottom of the chart. The shaft extends toward the reported wind direction, the tens value of the wind direction is written near the end of the shaft, and the odd five-degree increments of reported direction are indicated by a small plus sign (+) after the tens value of the direction.

Wind speed is indicated on the wind shaft in the same manner as surface and constant-pressure level plots (fig. 4-37).

The data for the mandatory pressure levels above 100 hectopascals is obtained from message Part C, and plotted in the same manner. Levels of missing data are noted on the diagram just above the last available reported data.

After the mandatory level data is plotted, the pressure altitude curve may be drawn. This is discussed following this section on plotting data, but must be accomplished before fixed regional level winds are plotted.

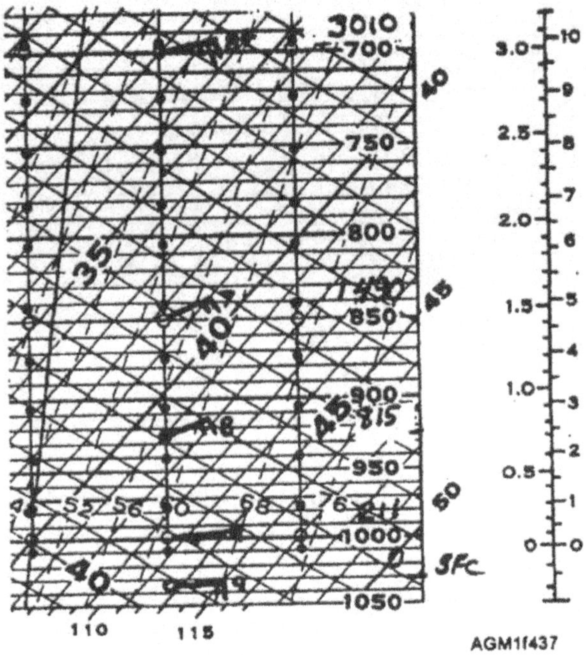

Figure 4-37.—Wind direction and speed plotted at each mandatory pressure level.

Plotting Significant Levels

All significant level data is normally plotted in addition to the mandatory level data. All stations report

TEMP CODE PART B (NON-REGION IV)—Significant level indicators for both temperature/humidity and winds are bolded.

TTBB 64120 47580 **00**030 05050 **11**930 06040 **22**770 02920 **33**650 10100 **44**600
14740 **55**435 29769 **66**358 38170 21212 **00**030 13015 **11**990 17022 **22**985 17035
33972 17015 **44**925 18005 **55**860 19015 **66**700 20025 **77**550 22040 **88**320 23050
99300 23070 **11**260 23112 **22**220 23090 **33**100 24060 31313 46105 81135 41414
43322;

SYMBOLIC FORMAT—

$M_iM_iM_jM_j$ $YYGGa_4$ IIiii	(identification)
$n_nn_nP_nP_nP_n$ $T_nT_nT_{an}D_nD_n$	(significant temperature and humidity levels)
21212 $n_nn_nP_nP_nP_n$ $d_nd_nf_nf_nf_n$	(significant wind levels - not included in Region IV)
31313 $s_rr_ar_as_as_a$ 8GGgg $(9s_sT_WT_WT_W)$	(sounding system and tracking method, observation time and SST)
41414 $N_hC_LhC_MC_H$;	(cloud data)

Figure 4-38.—Significant temperature/humidity and wind levels as reported in TEMP code.

significant temperature and humidity levels in Part B and D of the TEMP, TEMP MOBIL, or TEMP SHIP coded message.

Ships and land stations outside of Region IV report significant level winds in Parts B and D of the TEMP message following the 21212 indicator group (fig. 4-38). Within WMO Region IV, significant level winds are not reported. However, fixed regional level winds are reported in Parts B and D of the PILOT, PILOT MOBIL, or PILOT SHIP code (fig. 4-39).

The first significant level is the surface, repeating information previously reported in Part A. All remaining significant levels should be plotted. Plot the temperature, dew-point depression, and winds at the isobar level reported. You may have to estimate between isobars and between isotherms drawn on the diagram to locate the correct placement of the reported information. Plot temperatures, dew-point depressions, and significant level winds the same way as for the mandatory levels.

Plotting Fixed Regional Level Winds

The fixed regional level winds (FRLW) are plotted on the wind scale on the blackened circles. The altitude scale (kilometers and thousands of feet) located just to the right of the wind scales is used to identify the preplotted FRLW circles on the wind scale. Remember, although the indicated altitude increments in the coded fixed regional level winds may be thought of as thousands of feet, the value indicated is actually the number of 300-meter increments above the surface. For example, 90012 indicates three FRLWs follow, the first at a "00" 300-meter increment altitude (surface), the second at a "01" 300-meter increment altitude (0.3 km), and the third at a "02" 300-meter increment altitude (600 meters or 0.6 km). Alternatively, you may think of the wind reports as the surface, 1,000-foot, and 2,000-foot levels. Plot the reported wind at the blackened

TEMP CODE PART B (WMO REGION IV)

TTBB64120 72306 **00**030 05050 **11**930 06040 **22**770 02920 **33**650 10100 **44**600
14740 **55**435 29769 **66**358 38170 31313 46105 81135 41414 43322 51515 10153;

SYMBOLIC FORMAT

$M_iM_iM_jM_j$ YYGGa_4 IIiii	(identification)
$n_nn_nP_nP_nP_n$ $T_nT_nT_{an}D_nD_n$	(significant temperature and humidity levels)
31313 $s_rr_ar_as_as_a$ 8GGgg $(9S_ST_wT_{ww})$	(sounding system and tracking method, observation time and SST)
41414 $N_hC_LhC_MC_H$	(cloud data,
51515 101. . . . . . . ;	(additional data groups)

PILOT CODE PART B (FIXED REGIONAL LEVELS-REGION IV)

PPBB 64121 72306 90012 10013 09523 10020 90346 11510 12005 12505 90789
14005 15508 16510 91246 17015 18019 18523 9205/ 20025 21530;

SYMBOLIC FORMAT

$M_iM_iM_jM_j$ YYGGa_4 IIiii	(identification)
$9t_nu_1u_2u_3$ dddff dddff dddff . . . . . ;	(significant wind levels)

Figure 4-39.—Significant temperature/humidity levels in TEMP code, with fixed regional level winds reported in PILOT code.

circle horizontally even with the indicated value on the altitude scale (fig. 4-40).

When there is a significant difference between your station's elevation and sea level, or if the surface pressure is greatly different than 1,013 hPa (standard pressure), then the preplotted dots for the fixed regional level winds will not be accurately spaced with respect to the mandatory level wind plots. In these cases, the forecaster may desire that you plot the fixed-regional level winds by establishing the altitude using the pressure-altitude curve, as shown in figure 4-40.

Plotting Tropopause Height

The tropopause information is encoded in Part A or C of the TEMP coded messages following the mandatory level information. The three groups of information start with the indicator 88. The format of the data is shown in figure 4-32. The reported tropopause information (pressure level, temperature, dew-point depression, and wind direction and speed) is plotted just as a mandatory level. Additionally, in the uncolored portion of the diagram between the left wind scale and the diagram, a horizontal line is extended from the tropopause pressure level, and the block letters *TROP* are written above the line. When group 88999 is reported, the block letters *NO TROP* are written just to the right of the colored portion of the diagram at the 100-hPa level. See figure 4-41 for examples of both plotted tropopause and maximum wind information.

Plotting Maximum Wind

The maximum wind pressure level, wind direction and speed follows the 77 indicator or the 66 indicator in Part A or C of the message. At the pressure level reported, a dashed horizontal line is drawn extending from the right side of the colored portion of the diagram and the word *MAX* is written on the line. The maximum wind is plotted at that same level on the wind scale. Wind shear values, if reported, are usually not plotted.

Plotting Supplemental Information

Much of the additional data following the 51515 group in Parts B and D of the TEMP code can be plotted on the diagram. If the group refers to a single level, a straight line is usually drawn at the pressure level, and the remark entered on the diagram. If the code refers to a layer of the atmosphere, the top and bottom of the layer are indicated by drawing horizontal lines at the pressure level, and the comment is entered on the chart. These coded remarks may refer to the reason for termination, levels of doubtful or missing data, or

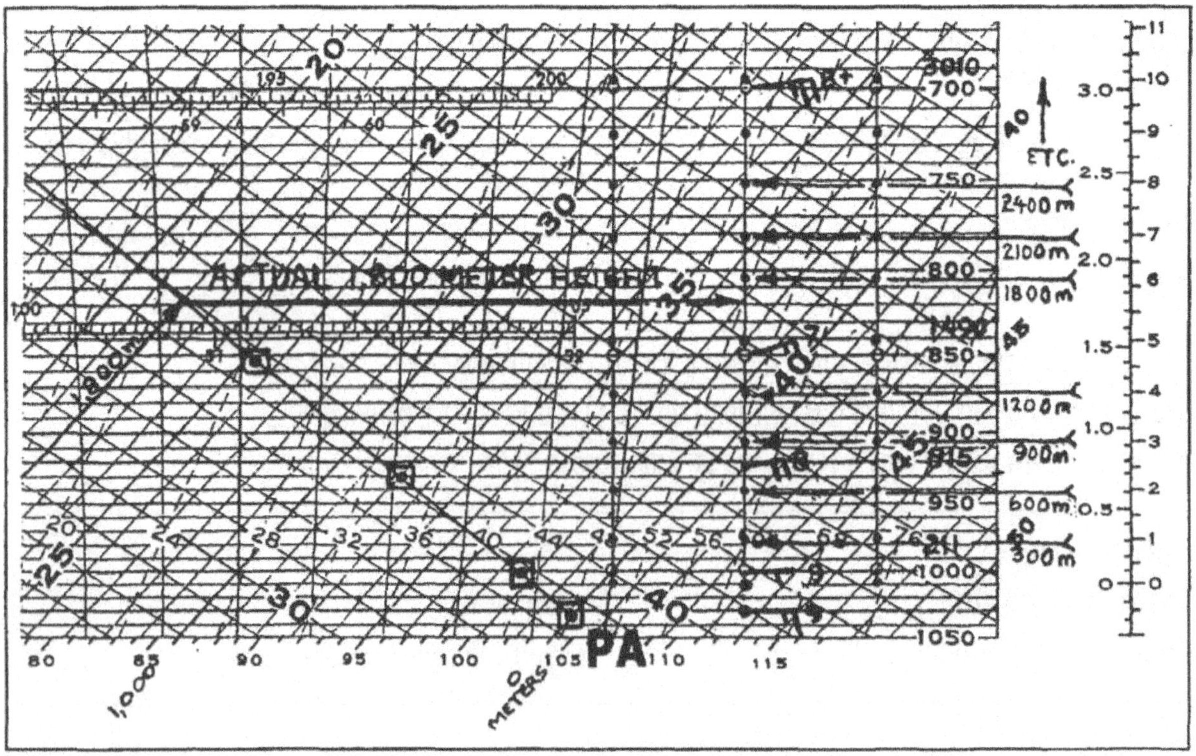

AGM1f440

Figure 4-40.—Fixed regional level winds may be plotted on the solid dots on the wind scale, or positioned using the PA curve.

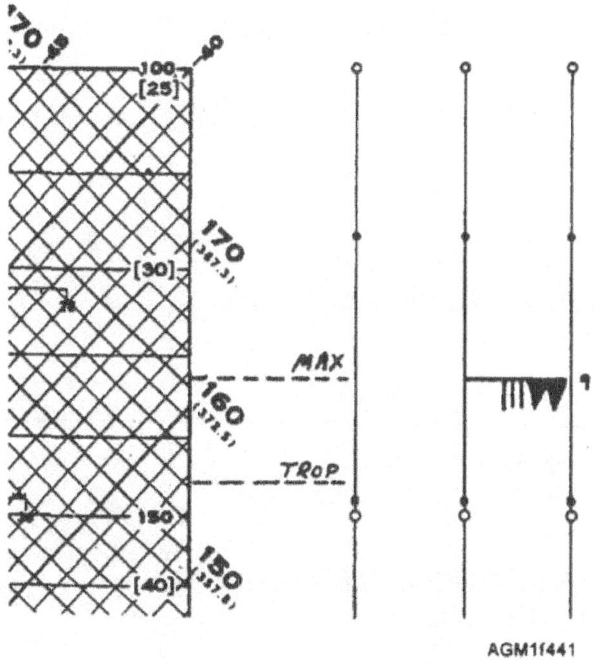

Figure 4-41.—Plotted tropopause and maximum wind information.

freezing levels. The 10190 group reports extrapolated altitude of the next higher standard pressure level after sounding termination. The altitude may be entered and plotted on the diagram, and *EXTRAP* should be written after the altitude to indicate it is extrapolated data. Sea-surface temperature, as reported following the 31313 group, may be plotted directly under the temperature trace in the bottom margin of the diagram. An SST of 15.7°C would be plotted as SST = 15.7°C.

TEMPERATURE TRACE

The temperature trace is constructed by connecting each of the plotted mandatory- and significant-level temperature plots with a thin solid line. Use a straight-edge as a guide to draw the lines between each plotted temperature, in order, from the bottom of the diagram to the top. Label the temperature trace with a block letter *T* at both the top and the bottom.

DEW-POINT TRACE

The plotted position of the reported dew-point depressions represents dew-point temperature. The dew-point temperature trace is constructed by connecting each of the mandatory- and significant level dew-point plots with a thin dashed line, in order, from the bottom of the diagram to the top. The trace is labeled with the letters *Td* at both the top and bottom.

PRESSURE-ALTITUDE (PA) CURVE

The pressure-altitude (PA) curve is constructed by connecting each of the plotted height points with a continuous solid line. When levels above the 100-hPa level are plotted on the diagram, a separate continuation PA curve is constructed connecting the consecutive altitude plots for the levels above 100 hPa. Each of the pressure altitude curves should be nearly a straight line, curving slightly upwards. A sharp bend in the curve at a particular height usually indicates either a misplotted height or a reporting error. The PA curve is labeled at both the top and bottom with the block letters *PA*.

As a Navy or Marine Corps observer, you will continue working on your requirements for advancement, and begin to analyze features on the Skew T, Log P diagram. Whether you are analyzing a manually plotted Skew T diagram or using computer assistance to analyze a Skew T displayed on a video terminal, a through understanding of the plotting methods presented in this section is essential. Information on analysis techniques on the Skew T, Log P diagram is presented in NAVAIR50-1P-5, *The Use of the Skew T, Log P Diagram in Analysis and Forecasting.*

REVIEW QUESTIONS

Q43. *What do the slightly curved, solid brown, left-diagonal lines on the Skew T diagram indicate?*

Q44. *What do the saturation mixing ratio lines indicate on a Skew T diagram?*

Q45. *How are heights of features accurately determined from a Skew T diagram?*

Q46. *What would be the height of a 1000-hPa level reported in Part A as 00526?*

Q47. *When pressure heights are being plotted on a Skew T diagram, each printed isotherm is equal to how many meters?*

Q48. *What symbol is used to indicate dew-point depression on the Skew T diagram?*

Q49. *What part(s) of an upper air TEMP coded report contain signifcant level data?*

Q50. *In what manner is the level of maximum wind annotated on a Skew T diagram?*

```
UNCLAS/ /N03123/ /
PPP MOVREP CONSTELLATION, AVWX 01/ /
ORG SP CONSTELLATION, 02/ /
ETD P SAN DIEGO CA, 191400Z5 FEB 03/ /
VIA L 32-40N9 118-11W2, 191756Z9 FEB 04/ /
VIA L 33-00N6 119-58W4, 200115Z9 FEB 05/ /
MOD LOC 32030N8 118-50W5 200516Z4 FEB OPERATING WITHIN 100NM, 06/ /
ORG SP DOWNES, 07/ /
RDV L 32-40N9 118-11W2, 191856Z0 FEB 08/ /END
```

Figure 4-42.—Typical movement report (MOVREP).

SHIP'S MOVEMENT REPORT (MOVREP) AND POINTS-OF-INTENDED MOVEMENT REPORT (PIM)

LEARNING OBJECTIVES: Discuss and define the terms used in the ship's movement report (MOVREP) messages and the points-of-intended-movement (PIM) reports. Explain how PIM are plotted on a chart or display.

Navy and Coast Guard ships usually send a Movement Report (MOVREP) message before leaving a port or an operating area. Understanding MOVREPs is not difficult if the format, terms, and standard abbreviations are known.

MOVEMENT REPORT

The MOVREP message uses a specific format and abbreviated plain language (fig. 4-42). Each line is numbered at the end of text, such as 01//, or 08//. This will enable the reader to know if any lines of information have been lost or deleted.

A MOVREP will always have END at the end of the last line to alert the reader that no additional lines follow. Check-sums, a method of verifying that no numbers in a group were deleted or changed during transmission, are used with all date-time groups and latitude or longitude positions. To obtain a check-sum, add all of the digits in a reported group together, and the units digit of the resulting sum is the check-sum.

MOVREP TERMS AND STANDARD ABBREVIATIONS

Line 1 identifies the sender of the report and contains requests for Enroute Weather Forecasts (WEAX), Aviation Enroute Weather Forecasts (AVWX), or other specialized geophysics support. Line 2, ORG, provides the ship (SP), which will change position. The ETD line provides the estimated time of departure and location (*P* for port, *L* for latitude/longitude, or *POS* for present position), and UTC departure date-time group. The date-time groups are always arranged in the format ddhhmmZk MMM, where dd is the day of the month, hh is the UTC hour, mm is the UTC minute, and MMM is the three-letter month abbreviation. The k is a check-sum. Enroute latitude and longitude positions and times follow the term *VIA*. Latitude and longitude points are provided in degrees and minutes, with check-sums. Occasionally, the type of route will be identified by GC for great circle, RB for rhumb line, DI for direct route, CO for coastal route, or STM for storm evasion. The destination follows either an ETA or MOD designation. ETA stands for estimated time of arrival, and is normally followed by a port or position and a date-time group. MOD stands for miscellaneous operational details, and is most commonly seen followed by the term *LOC,* for local operations. The last two lines in the example indicate that CONSTELLATION will rendezvous (RDV) with the DOWNS, at the position and time indicated.

POINTS OF INTENDED MOVEMENT

Aboard ship and ashore, Aerographers frequently must keep track of their own and other ships movements and locations. This is done by plotting a ship's Points of

Intended Movement (PIM) on either a paper chart or in a computer video display. These positions are used by the forecaster to determine where the ship will be at any given time, so the weather may be properly forecasted.

Your own ship's PIM may be obtained from the Quartermaster-of-the-Watch (QMOW), or the navigator at a navigation briefing. The PIM may be in the form of a series of latitude/longitude points, with UTC times, or may be stated simply as, "making course 123 true, SOA 12 knots."

PIM TERMS AND DEFINITIONS

"Making course" means the ship is actually traveling in that direction. This is different from heading or steering course, which, due to the effects of winds, seas, and currents, may be different from the direction the ship is actually moving.

The term SOA means speed of advance, and has the same meaning as speed made good—it is the speed the ship is actually moving. This is not the same as indicated speed—the speed the ship can make without considering winds, seas, and currents, based on the engine RPMs.

PLOTTING PIMs

When plotting PIMs, use a great circle chart or a polar-stereographic background to plot great circle routes. Plot all other routes on a Mercator chart or display background.

Plot PIMs either on a paper chart or the computer display by placing a dot or anx at the reported positions and writing the UTC date-time group next to that position. Indicate MOD LOC positions by drawing a circle of the diameter reported, around the latitude and longitude reported. For direct routes, connect the reported positions with straight lines; for coastal routes, connect the positions with lines running parallel to the coast. Connect reported rhumb-line route positions and great-circle route positions (on a polar stereographic chart) with straight lines. Mark the chart or display with the name of the ship or the task force/task group name, and the message date-time group of the MOVREP.

You will usually be rquired to indicate intermediate times along the ship's track, such as 12-hour increments at 0000Z and 1200Z daily. To do this, measure the length of the ship's track or a segment of the track with a set of dividers. Convert the length of the track to distance in degrees by comparing it to the latitude scale on the chart. Divide the distance, in degrees, by the length of time, in hours, indicated for the track or segment to obtain anticipated movement in degrees latitude per hour. Then, multiply by the number of hours in the desired increment to find the length of each individual segment. Use the divider to transcribe these segment lengths along the PIM track, and mark a short line across the track at each intermediate distance. Beside each mark indicate the date and time (fig. 4-43).

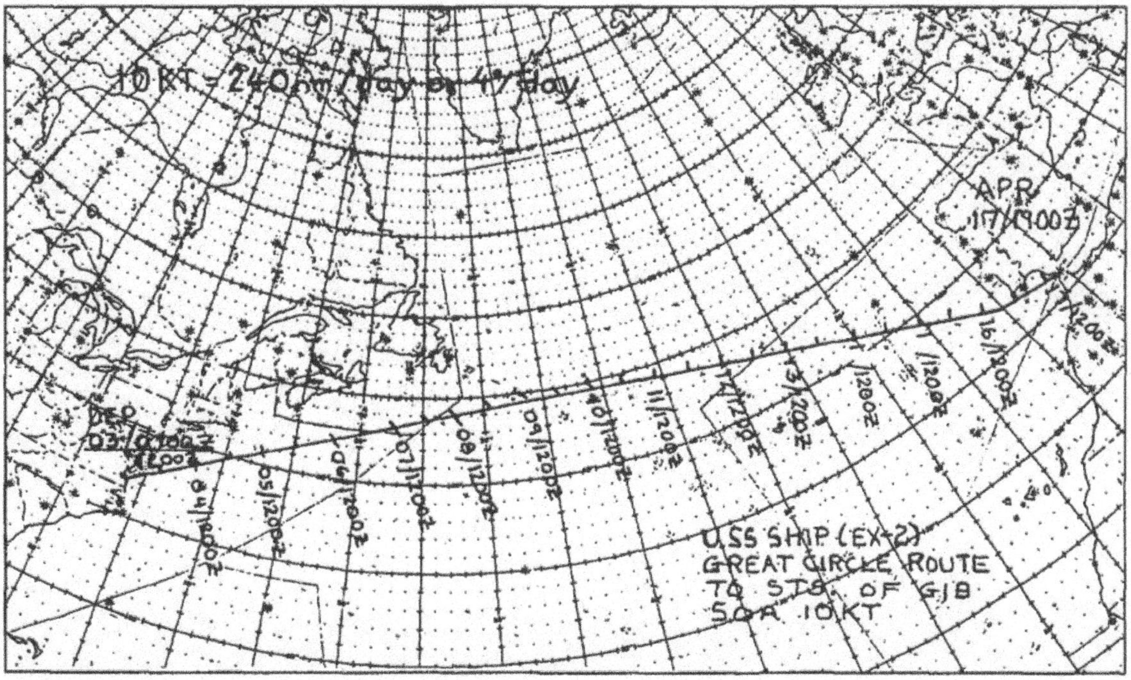

AGM1f443

Figure 4-43.—Intermediate times marked on a ship's PIM chart.

REVIEW QUESTIONS

Q.51. On what line of a movement report might you findrequestsfor an en route weather forecast?

Q.52. What does the term MOD-LOC indicate on a movement report?

Q53. What does the term SOA actually mean?

Q54. How are intermediate positions determined on a PIM chart?

SUMMARY

Many different types of graphic products are routinely produced for briefing purposes. In this chapter, we have discussed the terms used to describe various meteorological parameters on these products, the standards for the display of these parameters on graphic products, as well as the types of charts used to display graphic products. We covered in detail the plotting of both surface and upper-air observation reports, and the plotting of the Skew T, Log P diagram. In addition, we discussed the parameters of the movement report (MOVREP) and the plotting of points of intended movement (PIM) data.

ANSWERS TO REVIEW QUESTIONS

A1. *The geographical coordinate system and the grid coordinate system.*

A2.. *1°*

A3. *23.9N and 120.6W.*

A4. *The particular grid zone is identified by the column number and the row letter,*

A5. *24N and 32N latitude and 18W and 24W longitude.*

A6. *A 10 meter grid zone.*

A7. *Near the poles.*

A8. *The geographic definition of the tropics is the area of the earth lying between the Tropic of Cancer at 23 1/2°N and the Tropic of Capricorn at 23 1/2°S*

A9. *This is due to the curvature of the earth. Lines of longitude converge near the poles, and as Great Circle routes arc toward the poles, there is less distance between lines of longitude and thus a shorter distance to travel.*

A10. *1:10,000 since the smaller the scale the greater the detail.*

A11. *The National Imagery and Mapping Agency (NIMA).*

A12. *The National Imagery and Mapping Agency (NIMA) Catalog of Maps, Charts, and Related Products.*

A13. *A WMO blocck/station number ending in zero indicates a designated synoptic observation reporting station.*

A14. *The Master Weather Station Catalog.*

A15. *The letter "K".*

A16. *FAA Order 7350.6 Location Identifiers.*

A17. *A forecast normally covers a shorter period of time (less than 72 hrs) while an outlook is typically for 3 to 10 days in the future and implies a lower level of confidence.*

A18. *A briefing aid is any product designed primarily to be used in a briefing to assist the explanation of the current and forecast positions.*

A19. *Yellow.*

A20. *(a) Occluded front: purple.*

(b) Instability/squall line: blue.

A21. Blue.

A22. (a) Clear air turbulence (CAT): blue.

(b) Nonconvective continuous frozen precip: red.

A23. 4 hPa.

A24. 120 meters.

A25. 3 feet.

A26. Isodrosotherms

A27. Isallobars.

A28. TESS data can be manipulated to plot only one type of chart, such as a temperature analysis, pressure analysis, vice using a single chart to analyze all information,

A29. Wind direction is missing.

A30. Upper left of the station circle.

A31. Just to the right of the station circle.

A32. At the bottom of the station plot.

A33. By filling in the station circle with the applicable symbol.

A34. By using a small arrow to indicate direction. The speed isplotted in whole knots.

A35. The dew-point depression is less than or equal to 5°C.

A36. Below and to the right of the plot.

A37. Intermittent rain (not freezing), heavy at the time of observation.

A38. Lightning visible, no thunder heard.

A39. Snow (not falling as showers) during the past hour, but not at the time of observation.

A40. Every 2 degrees Celsius.

A41. With an X.

A42. A straight arrow is drawn for sea waves and a wavy arrow is drawn for swell waves. A smaller wavy arrow may be drawn for secondary swells. The direction of the arrow is the direction the waves are MOVING TOWARD.

A43. Dry adiabats.

A44. The greatest amount of water vapor that may be contained in a parcel of air expressed in units of grams of water vapor per kilogram of dry air.

A45. By the use of a properly constructed pressure-altitude (PA) curve.

A46. -26 meters.

A47. 100 meters.

A48. A triangle is used to indicate dew-point depression values.

A49. Parts B and D.

A50. At the pressure level reported, a dashed horizontal line is drawn extending from the right side of the colored portion of the diagram. The pressure level and the word "Mm are written on the line.

A51. Line 1.

A52. Miscellaneous operational details/local operations.

A.53. Speed-of-advance. It is the speed the ship is actually moving or the speed made good.

A54. Divide the distance (in degrees) by the length of time (in hours) indicated to obtain anticipated movement in degrees latitude per hour. Multiply by the number of hours in the desired increment to find the length of each individual segment.

APPENDIX I

GLOSSARY

A

ACCRETION—Growth or increase in size by gradual addition.

ADVECTION—Horizontal transport of an atmospheric property solely by the motion of the atmosphere.

AEROSOLS—Small droplets of liquid suspended in a gas, such as water in the air.

AFWA—Air Force weather agency.

AGL—Above ground level. Measurements suffixed by the abbreviation AGL refer to height.

ALIDADE—A device frequently used by shipboard navigation personnel to sight objects and read either relative azimuth bearing or true azimuth bearing.

AMBIENT—A representative reading or measurement for a substance under surrounding conditions.

ANALOG—Proportional and continuous. An analog recorder draws continuous lines proportional to the electronic signal input. In an analog signal, the sound pitch varies proportionately with the intensity of the signal, and the signal is continuous.

ANEMOMETER—A device used to measure wind speed and/or wind direction. From the Greek word anemo, meaning wind, and modem word meter, meaning measurement device.

ANEROID—Without fluid or without water. An aneroid barometer uses no fluid (mercury).

ANOMALOUS—Irregular or abnormal.

ANTICYCLONIC—Any rotational motion in a clockwise manner in the Northern Hemisphere, or a counter—clockwise manner in the Southern Hemisphere.

APPARENT—The way something appears or is perceived, although it may not be true.

ARQ—Automatic response to query. A method of obtaining data by using AFMEDS or CMW.

ASCENSION—Rising or increasing in elevation.

ASOS—Automated surface observing system.

AWSP—Air weather service pamphlet, now AFP or Air Force pamphlet.

AZIMUTH—The horizontal angular measurement from a fixed reference to a point. The Navy uses angular measurements in clockwise degrees from 0 to 360. When 0 is referenced to true north, the result is a true azimuth bearing. When referenced to an arbitrary direction, such as the bow of a ship, the result is a relative azimuth bearing.

B

BACKING—A change in wind direction in a counterclockwise manner in the Northern Hemisphere, or a clockwise direction in the Southern Hemisphere.

BATHYMETRY—The features and depths underwater.

BATHYTHERMOGRAPH—Any device used to measure and record temperatures through a column of water.

C

CALVE—The process of splitting ice from a glacier to form icebergs.

CCTV—Closed circuit television.

CEILOMETER—A more sophisticated and automated clinometer used to measure cloud or ceiling heights.

CIC—Combat information center aboard ship.

CLIMATIC—Any element associated with the climate of an area.

CLIMATOLOGY—The study of the statistical means, frequencies, deviations, and trends of weather elements for an area over a period of time.

CLINOMETER—A device used to obtain an angular elevation measurement of a light spot on a cloud base to determine the cloud or ceiling height. The distance between the meter and the light source must be known.

CONDENSATION—The physical process by which a vapor becomes a liquid. This process releases heat energy.

CONDUCTION—Transmission of energy through a substance by direct molecular contact.

CONVECTION—Motions in a fluid, such as the atmosphere or water, which are predominantly vertical, resulting in a vertical transport of mass and eventually in a mixing of properties and energy.

CYCLONIC—Any rotational movement in a counterclockwise manner in the Northern Hemisphere, or a clockwise manner in the Southern Hemisphere.

D

DASI—Digital altimeter setting indicator (Navy).

DDN—Defense data network.

DIURNAL—Any change that follows a daily pattern, completing one cycle on a daily basis.

DMS—Position given in degrees, minutes, and seconds of latitude and longitude.

DRIBU—Drifting buoy.

DSN—Defense switched network, an upgrade and name change to the automatic voice network (AUTOVON).

E

ETAGE—The layers of the atmosphere by which the different genera of clouds are identified.

EVAPORATION—The change of state process by which a liquid or a solid is transformed into a gaseous state. This process requires the addition of heat to the substance.

F

FAA—Federal Aviation Administration.

FAX—Short form of facsimile, referring to weather facsimile or a telefacsimile transmission.

FLOES—Pieces of ice broken loose from a sheet of ice that originally formed in the sea (frozen sea water).

FNMOC—Fleet Numerical Meteorology and Oceanography Center, Monterey, California.

FRONT—The interface or transition zone between two air masses of different density. Since temperature is the most important regulator of atmospheric density, a front almost invariably separates air masses of different temperature.

FRONTAL SURFACE—Refers specifically to the warmer side of an air mass transition zone, and slopes in the vertical toward colder air.

FRONTAL SYSTEM—Simply a system of fronts as they appear on a surface analysis or prognosis chart.

FRONTOGENESIS—The formation of a front.

FRONTOLYSIS—The dissipation or weakening of a front.

G

GEOPHYSICS—Used to mean working with the physical properties of both the air, land, and water, this term is occasionally used to describe the occupational field of Navy and Marine Corps weather personnel.

GEOPOTENTIAL HEIGHT—The height of a given point in the atmosphere calculated with respect to the energy in the column of air beneath the point, relative to sea level. In other words, an approximation of the height based on measured temperatures, pressures, and humidity content of the supporting air column, and not necessarily an exact measured height.

GHSI—General heat stress index, similar to apparent temperature.

GMT—Greenwich mean time, a term replaced by Coordinated Universal Time (UTC).

GPM—Geopotential meters, also gallons per minute.

GUST—A brief, rapid increase in wind speed.

H

HECTOPASCAL (hPa)—A unit of 100 pascals used to measure pressure, exactly equivalent to 1 millibar.

HF—High frequency.

HIGH—An "area of high pressure," referring to a higher atmospheric pressure in the horizontal plane, such as a surface isobaric chart, or an "area of high heights," referring to higher heights in the vertical plane, such as on a constant pressure chart.

HYPOTHERMIA—An abnormally low body temperature.

I

ICAO—Abbreviation for International Civil Aviation Organization.

ICEBERG—Ice found floating in the sea that originated from glaciers formed from freshwater accumulations of snow.

IFR—Abbreviation for instrument flight rules.

INFRARED (IR)—The portion of the electromagnetic spectrum with wavelengths just slightly longer than visible light energy (heat energy).

INVERSION—With respect to temperature, an increase in temperature with height. Normally, temperature decreases with height in the atmosphere.

IRCS—International radio call sign.

ISOTHERMAL—Having an equal temperature throughout.

K

kn—Alternate abbreviation for knot. In meteorology, the more frequently used abbreviation is "kt," but this should not be confused with the uppercase "KT" meaning kiloton.

L

LAN—Local area network.

LAPSE RATE—The decrease of an atmospheric variable with height; the variable being temperature unless otherwise specified.

LCD—Liquid crystal display. A gray or black display of numbers or shapes commonly used in electronics.

LDATS—Lightning Detection and Tracking System.

LEWP—A radar feature, the line echo wave pattern, normally an indicator of severe weather.

LITTORAL—The coastal zone including the beach to the coastal waters.

LLWS—Low--level wind shear.

LOW—An "area of low pressure" in the horizontal plane referring to the isobars on a surface chart, or an "area of low heights," in the vertical plane referring to the height contours on a constant pressure chart.

M

MANOP—Formatted weather message header that identifies the product type, originator, and area covered by the product.

MET—U.S. Navy mobile environmental team.

METEM TECHNICIAN—A meteorological electronic equipment maintenance technician, a term usually referring to a Navy ET who has received special training on meteorological equipment in the METEM school.

METEOROLOGY—The study of phenomenon of the atmosphere.

METVANS—USMC mobile meteorological vans. Highly transportable, completely equipped meteorological facilities constructed as complete modules in cargo containers. Modules may be used independently or connected to form complete, full--spectrum, meteorological support facilities in a forward deployed environment.

MILS—An angular measurement scale in which 800 mils equals 45 degrees of arc; a circle is 6,400 mils.

MMF—U.S. Marine Corps mobile meteorological force (weather personnel who operate USMC Metvans).

MRC—Abbreviation for maintenance requirements card used with the 3-M System.

MRS—Mini rawinsonde system.

MSL—Mean sea level, a suffix used after altitude measurements.

MVFR—Marginal visual flight rules.

N

NATO—North Atlantic Treaty Organization.

NAVMETOCCOM—Short title for Naval Meteorology and Oceanography Command headquartered at the Stennis Space Center, Mississippi.

NBC—An acronym for nuclear, biological, or chemical.

NDASI—Navy digital altimeter setting indicator, also DASI.

NDB—Nondirectional beacon; a radio aircraft navigation aid that allows aircraft to determine the direction of the transmitter.

NIMA—National imagery and mapping agency.

NOAA—National Oceanic and Atmospheric Administration, a division of the U.S. Department of Commerce.

NODDES—Naval oceanographic data distribution expansion system (used with TESS).

NODDS—Navy oceanographic data distribution system.

NOMOGRAM—Any graphic product used to find solutions to complex calculations.

NOTAM—Notice to airmen.

NUC—An abbreviation for nuclear, used within some NATO messages.

NWS—National Weather Service, a division of NOAA.

O

OA—Abbreviation for shipboard aviation operations division, the shipboard division for which most Aerographer's Mates work.

OAML—Oceanographic and atmospheric master library.

OKTAS—Eighths of the sky.

OOD—Officer of the deck.

OPARS—Optimum path aircraft routing system.

P

PHEL—Physiological heat exposure limit.

PIBAL—An acronym for pilot balloon, a small balloon tracked with a theodolite to determine wind direction and speed.

PIREP—Abbreviation for pilot report.

PMSV—Pilot meteorological service, voice radio.

PREVAILING—The most frequent or most common.

PSYCHROMETER—Any device used to measure dry--bulb temperature (air temperature) and wet--bulb temperature.

Q

QFE—A signal used to indicate the value provided; the station pressure.

QFF—A signal used to indicate the value provided; the sea--level pressure.

QNH—A signal used to indicate the value provided; the minimum altimeter setting for the period of time discussed.

R

RABAL—A method using radar to track a balloon carrying a radar—-reflector, and is used to determine upper—-level winds.

RADAT—A term used to indicate freezing level information derived from a radiosonde observation.

RADFO—An acronym for radiological fallout.

RADIOSONDE—A device carried aloft by a balloon to measure pressure, temperature, and humidity in the atmosphere.

RAOB—Acronym for radiosonde observation.

RATT—Radio teletype.

RAWINSONDE—Radio--wind sounding. A device carried aloft by a balloon that measures pressure, temperature, humidity, and the slant—-range from the release point. Calculations on the change in pressure (height) and change in slant—-range (distance) yield wind speed and direction.

RBC—Rotating beam ceilometer.

RH—Usual abbreviation for relative humidity.

ROCKETSONDE—A device carried aloft by a rocket which measures pressure, temperature, and humidity as it drifts on a parachute to the ground.

ROCOB—Acronym for rocketsonde observation.

S

SALINITY—A measurement of the amount of salts dissolved in sea water.

SAR—Search and rescue.

SLUSH—A mixture of snow and liquid water on the ground.

SMOOS—Shipboard meteorological oceanographic observing system.

SSI—Showalter stability index.

SURFCAST—(or SURFCST) Acronym for surf forecast.

SUROB—Acronym for surf observation.

SYNOPTIC—In general, pertaining to or affording an overall view. In meteorology, this term has become specialized in referring to the use of meteorological data obtained simultaneously over a wide area for presenting a comprehensive picture of the state of the atmosphere over a given period of time.

T

TACAN—Tactical air navigation, a radio aircraft navigation aid used originally by the military to provide a pilot with direction and distance to a TACAN transmitter.

TIC—Type, intensity, character code used to describe frontal systems.

TRANSMISSOMETER—Any device used to measure the transmission of light through a medium.

TROUGH—(Sometimes spelled TROF) An elongated area of low atmospheric pressure or heights extending outward from a low center; the opposite of a ridge. Also, the lowest portion of a wave cycle; the dip between the wave crests.

U

UHF—Ultra--high frequency radio transmission.

UNREP—Underway replenishment.

UPS—Universal polar stereographic grid system used by the military to locate positions in the polar regions.

UTC—Coordinated Universal Time, usually suffixed with a "Z."

UTM—Universal transverse mercator coordinates; a military coordinated system based on .a series of grids used to locate positions between 84°N and 80°s.

V

VALID—Effective, good.

VEERING—A change in the wind direction in a clockwise manner in the Northern Hemisphere, or a counterclockwise manner in the Southern Hemisphere.

VELOCIMETER—In general, a device used to measure velocity (speed). In oceanography, the sound velocimeter measures the speed of sound in water.

VERTREP—Vertical replenishment by use of helicopters.

VLF—Very--low frequency.

W

WAN—Wide area network.

WBC—Weather broadcast center.

WBGT—Wet--bulb globe temperature, a combined reading of air temperature, wet--bulb temperature, and a temperature inside a black--colored metal (high heat absorbing) ball.

WEATHER—The state of the atmosphere with respect to its effect upon life and human activities.

WEFAX—An acronym for weather facsimile, specifically the NWS service providing satellite imagery and graphic products via a geostationary satellite data broadcast.

WMO—World Meteorological Organization.

WPM—Words per minute.

X

XBT—Expendable bathythermograph, usually refers to the probe that is dropped in the water and not recovered.

XSV—Expendable sound velocimeter, usually refers to the probe that is dropped in the water and not recovered.

APPENDIX II

THE METRIC SYSTEM
AND
CONVERSION TABLES

THESE PREFIXES MAY BE APPLIED TO ALL SI UNITS

Multiples and Submultiples		Prefixes	Symbols
1 000 000 000 000	= 10^{12}	tera (tĕr′ȧ)	T
1 000 000 000	= 10^{9}	giga (ji′gȧ)	G
1 000 000	= 10^{6}	mega (mĕg′ȧ)	M*
1 000	= 10^{3}	kilo (kĭl′ō)	k*
100	= 10^{2}	hecto (hĕk′tō)	h
10	= 10	deka (dĕk′ȧ)	da
0.1	= 10^{-1}	deci (dĕs′ĭ)	d
0.01	= 10^{-2}	centi (sĕn′tĭ)	c*
0.001	= 10^{-3}	milli (mĭl′ĭ)	m*
0.000 001	= 10^{-6}	micro (mī′krō)	μ*
0.000 000 001	= 10^{-9}	nano (năn′ō)	n
0.000 000 000 001	= 10^{-12}	pico (pē′kō)	p
0.000 000 000 000 001	= 10^{-15}	femto (fĕm′tō)	f
0.000 000 000 000 000 001	= 10^{-18}	atto (ăt′tō)	a

*Most commonly used

AGM10136

COMMON EQUIVALENTS

LENGTH

1 in =	25.4 mm
1 in =	2.54 cm
1 ft =	30.48 cm
1 yd =	91.44 cm
1 mi =	5280. ft
1 mi =	1.609 km
1 mi =	0.868 nmi
1 mi =	1609.344 m
1 nmi=	6076.12 ft
1 nmi=	1.1508 mi
1 nmi=	1852 m
1 nmi=	1.853 km
1 mm =	0.039 in
1 cm =	0.394 in
1 cm =	0.033 ft
1 m =	39.37 in
1 m =	3.281 ft
1 km =	3280.84 ft
1 km =	0.6214 mi
1 km =	0.5400 nmi
1 fa =	6 ft

VOLUME

1 in^3=	16.387 cc
1 ft^3=	1728 in^3
1 ft^3=	0.02832 m^3
1 ft^3=	7.48 gal
1 yd^3=	27 ft^3
1 cm^3=	1 mℓ
1 cm^3=	1 cc
1 m^3 =	1,000 ℓ
1 m^3 =	100,000 mℓ
1 m^3 =	35.315 ft^3
1 m^3 =	264.17 gal
1 oz =	2 Tbs
1 oz =	6 tsp
1 pt =	16 oz
1 pt =	2 c
1 pt =	454 mℓ
1 qt =	2 pt
1 qt =	0.9464 ℓ
1 gal=	4 qt
1 gal=	3.7854 ℓ
1 gal=	0.8327 Igl
1 ℓ =	1000 mℓ
1 ℓ =	1 dm^3
1 ℓ =	1.057 qt

MASS (WEIGHT)

1 oz =	28.25 gm
1 lb =	453.6 gm
1 lb =	7000 gr
1 lb =	16 oz
1 lb =	0.4536 kg
1 st =	2000 lb
1 st =	907.2 kg
1 st =	0.9072 mt
1 st =	0.8929 lt
1 lt =	2240 lb
1 lt =	1.12 st
1 mt =	1000 kg
1 mt =	2204.6 lb
1 kg =	2.2046 lb
sea water 1 ft^3=	64 lb
pure water (4°C) 1 ft^3=	62.428 lb
1 mℓ =	1 gm
1 ℓ =	1 kg
1 m^3 =	1 mt

ABBREVIATIONS

ac	acre
atm	atmosphere
c	cup
cc	cubic centimeter
cm	centimeter
dm	decimeter
fa	fathom
ft	foot
gal	US gallon
gm	gram
gr	grain
hPa	hectoPascal
Igl	Imperial gallon
in	inch
kg	kilogram
km	kilometer
kph	kilometer/hour
kt	knot
ℓ	liter
lb	pound
lt	long ton
m	meter
mb	millibar
mi	statute mile
mℓ	milliliter
mm	millimeter
mph	s. mile/hour
mps	meter/sec
mt	metric ton
nmi	nautical mile
oz	US ounce
Pa	Pascal
pt	US pint
qt	US quart
st	short ton
T	tablespoon
tsp	teaspoon
yd	yard

AREA

1 in^2=	6.5 cm^2
1 ft^2=	144.0 in^2
1 ft^2=	0.09 m^2
1 yd^2=	9 ft^2
1 yd^2=	1296 in^2
1 yd^2=	0.8 m^2
1 ac =	4840 yd^2
1 ac =	43560 ft^2
1 ac =	0.4 hectare
1 mi^2=	27,878,400 ft^2
1 mi^2=	2,589,988 m^2
1 cm^2=	0.16 in^2
1 m^2 =	10,000 cm^2
1 m^2 =	10.76 ft^2

VELOCITY

1 kt =	1.1507 mph
1 kt =	0.5144 mps
1 kt =	1.852 kmh
1 mph=	0.8690 kt
1 mph=	0.4470 mps
1 mph=	1.6093 kmh
1 mps=	3.6 kph
1 mps=	2.2369 mph
1 mps=	1.9438 kt
1 kph=	0.6214 mph
1 kph=	0.5399 kt

FORCE (PRESSURE)

1 hPa =	1 mb
1 hPa =	0.750 mm_{Hg}
1 hPa =	0.0295 in_{Hg}
1 hPa =	100 Pa
1 Pa =	1 Newton/m^2
1 Pa =	100,000dyne/m^2
1 mm_{Hg}=	1.333 hPa
1 mm_{Hg}=	0.0394 in_{Hg}
1 in_{Hg}=	33.864 hPa
1 in_{Hg}=	25.4 mm_{Hg}
1 atm =	1013.250 hPa
1 atm =	14.7 lb/in^2
1 atm =	760 mm_{Hg}
1 atm =	29.921 in_{Hg}
1 atm =	33 $ft_{sea\ water}$

AGM10137

TEMPERATURE CONVERSION

FAHRENHEIT (F) TO CELSIUS (C) DEGREES

$$F = \frac{9}{5}C + 32 \quad or \quad F = 1.8\,C + 32$$

$$C = \frac{5}{9}(F - 32) \quad or \quad C = \frac{F - 32}{1.8}$$

°F.	0.0	0.1	0.2	0.3	0.4	0.5	0.6	0.7	0.8	0.9
	°C	°C	°C	°C	°C	°C	°C	°C	°C	°C
+120	+48.89	+48.94	+49.00	+49.06	+49.11	+49.17	+49.22	+49.28	+49.33	+49.39
119	48.33	48.39	48.44	48.50	48.56	48.61	48.67	48.72	48.78	48.83
118	47.78	47.83	47.89	47.94	48.00	48.06	48.11	48.17	48.22	48.28
117	47.22	47.28	47.33	47.39	47.44	47.50	47.56	47.61	47.67	47.72
116	46.67	46.72	46.78	46.83	46.89	46.94	47.00	47.06	47.11	47.17
+115	+46.11	+46.17	+46.22	+46.28	+46.33	+46.39	+46.44	+46.50	+46.56	+46.61
114	45.56	45.61	45.67	45.72	45.78	45.83	45.89	45.94	46.00	46.06
113	45.00	45.06	45.11	45.17	45.22	45.28	45.33	45.39	45.44	45.50
112	44.44	44.50	44.56	44.61	44.67	44.72	44.78	44.83	44.89	44.94
111	43.89	43.94	44.00	44.06	44.11	44.17	44.22	44.28	44.33	44.39
+110	+43.33	+43.39	+43.44	+43.50	+43.56	+43.61	+43.67	+43.72	+43.78	+43.83
109	42.78	42.83	42.89	42.94	43.00	43.06	43.11	43.17	43.22	43.28
108	42.22	42.28	42.33	42.39	42.44	42.50	42.56	42.61	42.67	42.72
107	41.67	41.72	41.78	41.83	41.89	41.94	42.00	42.06	42.11	42.17
106	41.11	41.17	41.22	41.28	41.33	41.39	41.44	41.50	41.56	41.61
+105	+40.56	+40.61	+40.67	+40.72	+40.78	+40.83	+40.89	+40.94	+41.00	+41.06
104	40.00	40.06	40.11	40.17	40.22	40.28	40.33	40.39	40.44	40.50
103	39.44	39.50	39.56	39.61	39.67	39.72	39.78	39.83	39.89	39.94
102	38.89	38.94	39.00	39.06	39.11	39.17	39.22	39.28	39.33	39.39
101	38.33	38.39	38.44	38.50	38.56	38.61	38.67	38.72	38.78	39.83
+100	+37.78	+37.83	+37.89	+37.94	+38.00	+38.06	+38.11	+38.17	+38.22	+38.28
99	37.22	37.28	37.33	37.39	37.44	37.50	37.56	37.61	37.67	37.72
98	36.67	36.72	36.78	36.83	36.89	36.94	37.00	37.06	37.11	37.17
97	36.11	36.17	36.22	36.28	36.33	36.39	36.44	36.50	36.56	36.61
96	35.56	35.61	35.67	35.72	35.78	35.83	35.89	35.94	36.00	36.06
+95	+35.00	+35.06	+35.11	+35.17	+35.22	+35.28	+35.33	+35.39	+35.44	+35.50
94	34.44	34.50	34.56	34.61	34.67	34.72	34.78	34.83	34.89	34.94
93	33.89	33.94	34.00	34.06	34.11	34.17	34.22	34.28	34.33	34.39
92	33.33	33.39	33.44	33.50	33.56	33.61	33.67	33.72	33.78	33.83
91	32.78	32.83	32.89	32.94	33.00	33.06	33.11	33.17	33.22	33.28
+90	+32.22	+32.28	+32.33	+32.39	+32.44	+32.50	+32.56	+32.61	+32.67	+32.72
89	31.67	31.72	31.78	31.83	31.89	31.94	32.00	32.06	32.11	32.17
88	31.11	31.17	31.22	31.28	31.33	31.39	31.44	31.50	31.56	31.61
87	30.56	30.61	30.67	30.72	30.78	30.83	30.89	30.94	31.00	31.06
86	30.00	30.06	30.11	30.17	30.22	30.28	30.33	30.39	30.44	30.50
+85	+29.44	+29.50	+29.56	+29.61	+29.67	+29.72	+29.78	+29.83	+29.89	+29.94
84	28.89	28.94	29.00	29.06	29.11	29.17	29.22	29.28	29.33	29.39
83	28.33	28.39	28.44	28.50	28.56	28.61	28.67	28.72	28.78	28.83
82	27.78	27.83	27.89	27.94	28.00	28.06	28.11	28.17	28.22	28.28
81	27.22	27.28	27.33	27.39	27.44	27.50	27.56	27.61	27.67	27.72
+80	+26.67	+26.72	+26.78	+26.83	+26.89	+26.94	+27.00	+27.06	+27.11	+27.17
79	26.11	26.17	26.22	26.28	26.33	26.39	26.44	26.50	26.56	26.61
78	25.56	25.61	25.67	25.72	25.78	25.83	25.89	25.94	26.00	26.06
77	25.00	25.06	25.11	25.17	25.22	25.28	25.33	25.39	25.44	25.50
76	24.44	24.50	24.56	24.61	24.67	24.72	24.78	24.83	24.89	24.94

AGM10138

Reportable Visibility Values					
NM	SM	METERS	NM	SM	METERS
0.0	0	0000	1.4	1-5/8	2600
0.05	1/16	0100	1.5	1-3/4	2800
0.1	1/8	0200	1.6	1-7/8	3000
0.15	3/16	0300	1.7	2	3200
0.2	1/4	0400	1.8	---	3400
0.25	5/16	0500	1.9	2-1/4	3600
0.3	3/8	0600	2.0	---	3700
0.4	---	0700	2.2	2-1/2	4000
0.45	1/2	0800	---	2-3/4	4400
0.5	---	0900	2.4	---	4500
0.55	5/8	1000	2.5	---	4700
0.6	---	1100	2.6	3	4800
---	3/4	1200	2.7	---	5000
0.7	---	1300	3.0	4	6000
---	7/8	1400	4.0	---	7000
0.8	---	1500	4.3	5	8000
---	1	1600	5.0	6	9000
0.9	---	1700	6.0	7	9999
1.0	1-1/8	1800	7.0	8	9999
1.1	1-1/4	2000	8.0	9	9999
1.2	1-3/8	2200	9.0	10	9999
1.3	1-1/2	2400			

1 statute mile = 5280 feet
= 0.868976 nautical mile
= 1.609344 kilometers

1 kilometer = 3281 feet
= 0.6215 statute mile
= 0.53946 nautical mile

AGM10139

HEIGHT CONVERSION
FEET TO METERS
CONTINUED

Feet	0	100	200	300	400	500	600	700	800	900
15000-	4572	4602	4633	4663	4694	4724	4755	4785	4816	4846
16000-	4877	4907	4938	4968	4999	5029	5060	5090	5121	5151
17000-	5182	5212	5243	5273	5304	5334	5364	5395	5425	5456
18000-	5486	5517	5547	5578	5608	5639	5669	5700	5730	5761
19000-	5791	5822	5852	5883	5913	5944	5974	6005	6035	6066
20000-	6096	6126	6157	6187	6218	6248	6279	6309	6340	6370
21000-	6401	6431	6462	6492	6523	6553	6584	6614	6645	6675
22000-	6706	6736	6767	6797	6828	6858	6888	6919	6949	6980
23000-	7010	7041	7071	7102	7132	7163	7193	7224	7254	7285
24000-	7315	7346	7376	7407	7437	7468	7498	7529	7559	7590
25000-	7620	7650	7681	7711	7742	7772	7803	7833	7864	7894
26000-	7925	7955	7986	8016	8047	8077	8108	8138	8169	8199
27000-	8230	8260	8291	8221	8352	8382	8412	8443	8473	8504
28000-	8534	8565	8595	8626	8656	8687	8717	8748	8778	8809
29000-	8839	8870	8900	8931	8961	8992	9022	9053	9083	9114
30000-	9144	9174	9205	9235	9266	9296	9327	9357	9388	9418
31000-	9449	9479	9510	9540	9571	9601	9632	9662	9693	9723
32000-	9754	9784	9815	9845	9876	9906	9936	9967	9997	10028
33000-	10058	10089	10119	10150	10180	10211	10241	10272	10302	10333
34000-	10363	10394	10424	10455	10485	10516	10546	10577	10607	10638
35000-	10668	10699	10729	10759	10790	10820	10851	10881	10912	10942
36000-	10973	11003	11034	11064	11095	11125	11156	11186	11217	11247
37000-	11278	11308	11339	11369	11400	11430	11461	11491	11521	11552
38000-	11582	11613	11643	11674	11704	11735	11765	11796	11826	11857
39000-	11887	11918	11948	11979	12009	12040	12070	12101	12131	12162
40000-	12192	12223	12253	12283	12314	12344	12375	12405	12436	12466

AGM10140

TEMPERATURE CONVERSION
FAHRENHEIT (F) TO CELSIUS (C) DEGREES
CONTINUED

°F	0.0	0.1	0.2	0.3	0.4	0.5	0.6	0.7	0.8	0.9
	°C	°C	°C	°C	°C	°C	°C	°C	°C	°C
-35	-37.22	-37.28	-37.33	-37.39	-37.44	-37.50	-37.56	-37.61	-37.67	-37.72
36	37.78	37.83	37.89	37.94	38.00	38.06	38.11	38.17	38.22	38.28
37	38.33	38.39	38 44	38.50	38.56	38.61	38.67	38.72	38.78	38.83
38	38.89	38.94	39.00	39.06	39.11	39.17	39.22	39.28	39.33	39.39
39	39.44	39.50	39.56	39.61	39.67	39.72	39.78	39.83	39.89	39.94

PRECIPITATION AMOUNT CONVERSION
INCHES TO CENTIMETERS

1 inch = 2.54 centimeters
1 centimeters = 0.3937 inch

Inch	Centimeter	Inch	Centimeter	Inch	Centimeter	Inch	Centimeter
0	0	0.26	0.66	0.52	1.32	0.78	1.98
0.01	0.03	0.27	0.69	0.53	1.35	0.79	2.01
0.02	0.05	0.28	0.71	0.54	1.37	0.80	2.03
0.03	0.08	0.29	0.74	0.55	1.40	0.81	2.06
0.04	0.10	0.30	0.76	0.56	1.42	0.82	2.08
0.05	0.13	0.31	0.79	0.57	1.45	0.83	2.11
0.06	0.15	0.32	0.81	0.58	1.47	0.84	2.13
0.07	0.18	0.33	0.84	0.59	1.50	0.85	2.16
0.08	0.20	0.34	0.87	0.60	1.52	0.86	2.18
0.09	0.23	0.35	0.89	0.61	1.55	0.87	2.21
0.10	0.25	0.36	0.91	0.62	1.57	0.88	2.24
0.11	0.28	0.37	0.94	0.63	1.60	0.89	2.26
0.12	0.30	0.38	0.97	0.64	1.63	0.90	2.29
0.13	0.33	0.39	0.99	0.65	1.65	0.91	2.31
0.14	0.36	0.40	1.02	0.66	1.68	0.92	2.34
0.15	0.38	0.41	1.04	0.67	1.70	0.93	2.36
0.16	0.41	0.42	1.07	0.68	1.73	0.94	2.39
0.17	0.43	0.43	1.09	0.69	1.75	0.95	2.41
0.18	0.46	0.44	1.12	0.70	1.78	0.96	2.44
0.19	0.48	0.45	1.14	0.71	1.80	0.97	2.46
0.20	0.51	0.46	1.17	0.72	1.83	0.98	2.49
0.21	0.53	0.47	1.19	0.73	1.85	0.99	2.51
0.22	0.56	0.48	1.22	0.74	1.88	1.00	2.54
0.23	0.58	0.49	1.24	0.75	1.91		
0.24	0.61	0.50	1.27	0.76	1.93		
0.25	0 64	0.51	1.30	0.77	1.96		

AGM10141

AZIMUTH MEASUREMENTS - DEGREES TO MILS							
deg	Mils	deg	Mils	deg	Mils	deg	Mils
0	0	90	1600	180	3200	270	4800
5	89	95	1689	185	3289	275	4889
10	178	100	1778	190	3378	280	4978
15	267	105	1867	195	3467	285	5067
20	356	110	1956	200	3556	290	5156
25	445	115	2045	205	3645	295	5245
30	533	120	2133	210	3733	300	5333
35	622	125	2222	215	3822	305	5422
40	711	130	2311	220	3911	310	5511
45	800	135	2400	225	4000	315	5600
50	889	140	2489	230	4089	320	5689
55	978	145	2578	235	4178	325	5778
60	1067	150	2667	240	4267	330	5867
65	1156	155	2756	245	4356	335	5956
70	1245	160	2845	250	4445	340	6045
75	1333	165	2933	255	4533	345	6133
80	1422	170	3022	260	4622	350	6222
85	1511	175	3111	265	4711	355	6311
90	1600	180	3200	270	4800	360	6400

AGM10142

TEMPERATURE CONVERSION
FAHRENHEIT (F) TO CELSIUS (C) DEGREES
CONTINUED

°F	0.0	0.1	0.2	0.3	0.4	0.5	0.6	0.7	0.8	0.9
	°C	°C	°C	°C	°C	°C	°C	°C	°C	°C
+75	+23.89	+23.94	+24.00	+24.06	+24.11	+24.17	+24.22	+24.28	+24.33	+24.39
74	23.33	23.39	23.44	23.50	23.56	23.61	23.67	23.72	23.78	23.83
73	22.78	22.83	22.89	22.94	23.00	23.06	23.11	23.17	23.22	23.28
72	22.22	22.28	22.33	22.39	22.44	22.50	22.56	22.61	22.67	22.72
71	21.67	21.72	21.78	21.83	21.89	21.94	22.00	22.06	22.11	22.17
+70	+21.11	+21.17	+21.22	+21.28	+21.33	+21.39	+21.44	+21.50	+21.56	+21.61
69	20.56	20.61	20.67	20.72	20.78	20.83	20.89	20.94	21.00	21.06
68	20.00	20.06	20.11	20.17	20.22	20.28	20.33	20.39	20.44	20.50
67	19.44	19.50	19.56	19.61	19.67	19.72	19.78	19.83	19.89	19.94
66	18.89	15.94	19.00	19.06	19.11	19.17	19.22	19.28	19.33	19.39
+65	+18.33	+18.39	+18.44	+18.50	+18.56	+18.61	+18.67	+18.72	+18.78	+18.83
64	17.78	17.83	17.89	17.94	18.00	18.06	18.11	18.17	18.22	18.28
63	17.22	17.28	17.33	17.39	17.44	17.50	17.56	17.61	17.67	17.72
62	16.67	16.72	16.78	16.83	16.89	16.94	17.00	17.06	17.11	17.17
61	16.11	16.17	16.22	16.28	16.33	16.39	16.44	16.50	16.56	16.61
+60	+15.56	+15.61	+15.67	+15.72	+15.78	+15.83	+15.89	+15.94	+16.00	+16.06
59	15.00	15.06	15.11	15.17	15.22	15.28	15.33	15.39	15.44	15.50
58	14.44	14.50	14.56	14.61	14.67	14.72	14.78	14.83	14.89	14.94
57	13.89	13.94	14.00	14.06	14.11	14.17	14.22	14.28	14.33	14.39
56	13.33	13.39	13.44	13.50	13.56	13.61	13.67	13.72	13.78	13.83
+55	+12.78	+12.83	+12.89	+12.94	+13.00	+13.06	+13.11	+13.17	+13.72	+13.28
54	12.22	12.28	12.33	12.39	12.44	12.50	12.56	12.61	12.67	12.72
53	11.67	11.72	11.78	11.83	11.89	11.94	12.00	12.06	12.11	12.17
52	11.11	11.17	11.22	11.28	11.33	11.39	11.44	11.50	11.56	11.61
51	10.56	10.61	10.67	10.72	10.78	10.83	10.89	10.94	11.00	11.06
+50	+10.00	+10.06	+10.11	+10.17	+10.22	+10.28	+10.33	+10.39	+10.44	+10.50
49	9.44	9.50	9.56	9.61	9.67	9.72	9.78	9.83	9.80	9.94
48	8.89	8.94	9.00	9.06	9.11	9.17	9.22	9.28	9.33	9.39
47	8.33	8.39	8.44	8.50	8.56	8.61	8.67	8.72	8.78	8.83
46	7.78	7.83	7.89	7.94	8.00	8.06	8.11	8.17	8.22	8.28
+45	+7.22	+7.28	+7.33	+7.39	+7.44	+7.50	+7.56	+7.61	+7.67	+7.72
44	6.67	6.72	6.78	6.83	6.89	6.94	7.00	7.06	7.11	7.17
43	6.11	6.17	6.22	6.28	6.33	6.39	6.44	6.50	6.56	6.61
42	5.56	5.61	5.67	5.72	5.78	5.83	5.89	5.94	6.00	6.06
41	5.00	5.06	5.11	5.17	5.22	5.28	5.33	5.39	5.44	5.50
+40	+4.44	+4.50	+4.56	+4.61	+4.67	+4.72	+4.78	+4.83	+4.89	+4.94
39	3.89	3.94	4.00	4.06	4.11	4.17	4.22	4.28	4.33	4.39
38	3.33	3.39	3.44	3.50	3.56	3.61	3.67	3.72	3.78	3.83
37	2.78	2.83	2.89	2.94	3.00	3.06	3.11	3.17	3.22	3.28
36	2.22	2.28	2.33	2.39	2.44	2.50	2.56	2.61	2.67	2.72
+35	+1.67	+1.72	+1.78	+1.83	+1.89	+1.94	+2.00	+2.06	+2.11	+2.17
34	+1.11	+1.17	+1.22	+1.28	+1.33	+1.39	+1.44	+1.50	+1.56	+1.61
33	+.56	+.61	+.67	+.72	+.78	+.83	+.89	+.94	+1.00	+1.06
32	.00	+.06	+.11	+.17	+.22	+.28	+.33	+.39	+.44	+.50
31	-.56	-.50	-.44	-.39	-.33	-.28	-.22	-.17	-.11	-.06
+30	-1.11	-1.06	-1.00	-.94	-.89	-.83	-.78	-.72	-.67	-.61
29	1.67	1.61	1.56	1.50	1.44	1.39	1.33	1.28	1.22	1.17
28	2.22	2.17	2.11	2.06	2.00	1.94	1.89	1.83	1.78	1.72
27	2.78	2.72	2.67	2.61	2.56	2.50	2.44	2.39	2.33	2.28
26	3.33	3.28	3.22	3.17	3.11	3.06	3.00	2.94	2.89	2.83
+25	-3.89	-3.83	-3.78	-3.72	-3.67	-3.61	-3.56	-3.50	-3.44	-3.39
24	4.44	4.39	4.33	4.28	4.22	4.17	4.11	4.06	4.00	3.94
23	5.00	4.94	4.89	4.83	4.78	4.72	4.67	4.61	4.56	4.50
22	5.56	5.50	5.44	5.39	5.33	5.28	5.22	5.17	5.11	5.06
21	6.11	6.06	6.00	5.94	5.89	5.83	5.78	5.72	5.67	5.61

AGM10143

TEMPERATURE CONVERSION
FAHRENHEIT (F) TO CELSIUS (C) DEGREES
CONTINUED

°F	0.0	0.1	0.2	0.3	0.4	0.5	0.6	0.7	0.8	0.9
	°C	°C	°C	°C	°C	°C	°C	°C	°C	°C
+20	-6.67	-6.61	-6.56	-6.50	-6.44	-6.39	-6.33	-6.28	-6.22	-6.17
19	7.22	7.17	7.11	7.06	7.00	6.94	6.89	6.83	6.78	6.72
18	7.78	7.72	7.67	7.61	7.56	7.50	7.44	7.39	7.33	7.28
17	8.33	8.28	8.22	8.17	8.11	8.06	8.00	7.94	7.89	7.83
16	8.89	8.83	8.78	8.72	8.67	8.61	8.56	8.50	8.44	8.39
+15	-9.44	-9.39	-9.33	-9.28	-9.22	-9.17	-9.11	-9.06	-9.00	-8.94
14	10.00	9.94	9.89	9.83	9.78	9.72	9.67	9.61	9.56	9.50
13	10.56	10.50	10.44	10.39	10.33	10.28	10.22	10.17	10.11	10.06
12	11.11	11.06	11.00	10.94	10.89	10.83	10.78	10.72	10.67	10.61
11	11.67	11.61	11.56	11.50	11.44	11.39	11.33	11.28	11.22	11.17
+10	-12.22	-12.17	-12.11	-12.06	-12.00	-11.94	-11.89	-11.83	-11.78	-11.72
9	12.78	12.72	12.67	12.61	12.56	12.50	12.44	12.39	12.33	12.28
8	13.33	13.28	13.22	13.17	13.11	13.06	13.00	12.94	12.89	12.83
7	13.89	13.83	13.78	13.72	13.67	13.61	13.56	13.50	13.44	13.39
6	14.44	14.39	14.33	14.28	14.22	14.17	14.11	14.06	14.00	13.94
+5	-15.00	-14.94	-14.89	-14.83	-14.78	-14.72	-14.67	-14.61	-14.56	-14.50
4	15.56	15.50	15.44	15.39	15.33	15.28	15.22	15.17	15.11	15.06
3	16.11	16.06	16.00	15.94	15.89	15.83	15.78	15.72	15.67	15.61
2	16.67	16.61	16.56	16.50	16.44	16.39	16.33	16.28	16.22	16.17
1	17.22	17.17	17.11	17.06	17.00	16.94	16.89	16.83	16.78	16.72
+0	17.78	17.72	17.67	17.61	17.56	17.50	17.44	17.39	17.33	17.28
-0	-17.78	-17.83	-17.89	-17.94	-18.00	-18.06	-18.11	-18.17	-18.22	-18.28
1	18.33	18.39	18.44	18.50	18.56	18.61	18.67	18.72	18.78	18.83
2	18.89	18.94	19.00	19.06	19.11	19.17	19.22	19.28	19.33	19.39
3	19.44	19.50	19.56	19.61	19.67	19.72	19.78	19.83	19.89	19.94
4	20.00	20.06	20.11	20.17	20.22	20.28	20.33	20.39	20.44	20.50
-5	-20.56	-20.61	-20.67	-20.72	-20.78	-20.83	-20.89	-20.94	-21.00	-21.06
6	21.11	21.17	21.22	21.28	21.33	21.39	21.44	21.50	21.56	21.61
7	21.67	21.72	21.78	21.83	21.89	21.94	22.00	22.06	22.11	22.17
8	22.22	22.28	22.33	22.39	22.44	22.50	22.56	22.61	22.67	22.72
9	22.78	22.83	22.89	22.94	23.00	23.06	23.11	23.17	23.22	23.28
-10	-23.33	-23.39	-23.44	-23.50	-23.56	-23.61	-23.67	-23.72	-23.78	-23.83
11	23.89	23.94	24.00	24.06	24.11	24.17	24.22	24.28	24.33	24.39
12	24.44	24.50	24.56	24.61	24.67	24.72	24.78	24.83	24.89	24.94
13	25.00	25.06	25.11	25.17	25.22	25.28	25.33	25.39	25.44	25.50
14	25.56	25.61	25.67	25.72	25.78	25.83	25.89	25.94	26.00	26.06
-15	-26.11	-26.17	-26.22	-26.28	-26.33	-26.39	-26.44	-26.50	-26.56	-26.61
16	26.67	26.72	26.78	26.83	26.89	26.94	27.00	27.06	27.11	27.17
17	27.22	27.28	27.33	27.39	27.44	27.50	27.56	27.61	27.67	27.72
18	27.78	27.83	27.89	27.94	28.00	28.06	28.11	28.17	28.22	28.28
19	28.33	28.39	28.44	28.50	28.56	28.61	28.67	28.72	28.78	28.83
-20	-28.89	-28.94	-29.00	-29.06	-29.11	-29.17	-29.22	-29.28	-29.33	-29.39
21	29.44	29.50	29.56	29.61	29.67	29.72	29.78	29.83	29.89	29.94
22	30.00	30.06	30.11	30.17	30.22	30.28	30.33	30.39	30.44	30.50
23	30.56	30.61	30.57	30.72	30.78	30.83	30.89	30.94	31.00	31.06
24	31.11	31.17	31.22	31.28	31.33	31.39	31.44	31.50	31.56	31.61
-25	-31.67	-31.72	-31.78	-31.83	-31.89	-31.94	-32.00	-32.06	-32.11	-32.17
26	32.22	32.28	32.33	32.39	32.44	32.50	32.56	32.61	32.67	32.72
27	32.78	32.83	32.89	32.94	33.00	33.06	33.11	33.17	33.22	33.28
28	33.33	33.39	33.44	33.50	33.56	33.61	33.67	33.72	33.78	33.83
29	33.89	33.94	34.00	34.06	34.11	34.17	34.22	34.28	34.33	34.39
-30	-34.44	-34.50	-34.56	-34.61	-34.67	-34.72	-34.78	-34.83	-34.89	-34.94
31	35.00	35.06	35.11	35.17	35.22	35.28	35.33	35.39	35.44	35.50
32	35.56	35.61	35.67	35.72	35.78	35.83	35.89	35.94	36.00	36.06
33	36.11	36.17	36.22	36.28	36.33	36.39	36.44	36.50	36.56	36.61
34	36.67	36.72	36.78	36.83	36.89	36.94	37.00	37.06	37.11	37.17

AGM10144

HEIGHT CONVERSION
FEET TO METERS

1 foot = 0.3048006 meter
1 meter = 3.2808 feet

Feet	0	1	2	3	4	5	6	7	8	9
	m.	m.	m.	m.	m.	m.	m.	m.	m.	m.
0-	0.000	0.305	0.610	0.914	1.219	1.524	1.829	2.134	2.438	2.743
10-	3.048	3.353	3.658	3.962	4.267	4.572	4.877	5.182	5.486	5.791
20-	6.096	6.401	6.706	7.010	7.315	7.620	7.925	8.230	8.534	8.839
30-	9.144	9.449	9.754	10.058	10.363	10.668	10.973	11.278	11.582	11.887
40-	12.192	12.497	12.802	13.106	13.411	13.716	14.021	14.326	14.630	14.935
50-	15.240	15.545	15.850	16.154	16.459	16.764	17.069	17.374	17.678	17.983
60-	18.288	18.593	18.898	19.202	19.507	19.812	20.117	20.422	20.726	21.031
70-	21.336	21.641	21.946	22.250	22.555	22.860	23.165	23.470	23.774	24.079
80-	24.384	24.689	24.994	25.298	25.603	25.908	26.213	26.518	26.822	27.127
90-	27.432	27.737	28.042	28.346	28.651	28.956	29.261	29.566	29.870	30.175

Feet	0	10	20	30	40	50	60	70	80	90
100-	30.48	33.53	36.58	39.62	42.67	45.72	48.77	51.82	54.86	57.91
200-	60.96	64.01	67.06	70.10	73.15	76.20	79.25	82.30	85.34	88.39
300-	91.44	94.49	97.54	100.58	103.63	106.68	109.73	112.78	115.82	118.87
400-	121.92	124.97	128.02	131.06	134.11	137.16	140.21	143.26	146.30	149.35
500-	152.40	155.45	158.50	161.54	164.59	167.64	170.69	173.74	176.78	179.83
600-	182.88	185.93	188.98	192.02	195.07	198.12	201.17	204.22	207.26	210.31
700-	213.36	216.41	219.46	222.50	225.55	228.60	231.65	234.70	237.74	240.79
800-	243.84	246.89	249.94	252.98	256.03	259.08	262.13	265.18	268.22	271.27
900-	274.32	277.37	280.42	283.46	286.51	289.56	292.61	295.66	298.70	301.75
1000-	304.80	307.85	310.90	313.94	316.99	320.04	323.09	326.14	329.18	332.23
1100-	335.28	338.33	341.38	344.42	347.47	350.52	353.57	356.62	359.67	362.71
1200-	365.76	368.81	371.86	374.90	377.95	381.00	384.05	387.10	390.14	393.19
1300-	396.24	399.29	402.34	405.38	408.43	411.48	414.53	417.58	420.62	423.67
1400-	426.72	429.77	432.82	435.86	438.91	441.96	445.01	448.06	451.10	454.15
1500-	457.20	460.25	463.30	466.34	469.39	472.44	475.49	478.54	481.58	484.63
1600-	487.68	490.73	493.78	496.82	499.87	502.92	505.97	509.02	512.07	515.11
1700-	518.16	521.21	524.26	527.31	530.35	533.40	536.45	539.50	542.55	545.59
1800-	548.64	551.69	554.74	557.79	560.83	563.88	566.93	569.98	573.03	576.07
1900-	579.12	582.17	585.22	588.27	591.31	594.36	597.41	600.46	603.51	606.55
2000-	609.60	612.65	615.70	618.75	621.79	624.84	627.89	630.94	633.99	637.03
2100-	640.08	643.13	646.18	649.23	652.27	655.32	658.37	661.42	664.47	667.51
2200-	670.56	673.61	676.66	679.71	682.75	685.80	688.85	691.90	694.95	697.99
2300-	701.04	704.09	707.14	710.19	713.23	716.28	719.33	722.38	725.43	728.47
2400-	731.52	734.57	737.62	740.67	743.71	746.76	749.81	752.86	755.91	758.95
2500-	762.00	765.05	768.10	771.15	774.19	777.24	780.29	783.34	786.39	789.43
2600-	792.48	795.53	798.58	801.63	804.67	807.72	810.77	813.82	816.87	819.91
2700-	822.96	826.01	829.06	832.11	835.15	838.20	841.25	844.30	847.35	850.39
2800-	853.44	856.49	859.54	862.59	865.63	868.68	871.73	874.78	877.83	880.87
2900-	883.92	886.97	890.02	893.07	896.11	899.16	902.21	905.26	908.31	911.35
3000-	914.40	917.45	920.50	923.55	926.59	929.64	932.69	935.74	938.79	941.83
3100-	944.88	947.93	950.98	954.03	957.07	960.12	963.17	966.22	969.27	972.31
3200-	975.36	978.41	981.46	984.51	987.55	990.60	993.65	996.70	999.75	1002.79
3300-	1005.84	1008.89	1011.94	1014.99	1018.03	1021.08	1024.13	1027.18	1030.23	1033.27
3400-	1036.32	1039.37	1042.42	1045.47	1048.51	1051.56	1054.61	1057.66	1060.71	1063.75
3500-	1066.80	1069.85	1072.90	1075.95	1078.99	1082.04	1085.09	1088.14	1091.19	1094.23
3600-	1097.28	1100.33	1103.38	1106.43	1109.47	1112.52	1115.57	1118.62	1121.67	1124.71
3700-	1127.76	1130.81	1133.86	1136.91	1139.95	1143.00	1146.05	1149.10	1152.15	1155.19
3800-	1158.24	1161.29	1161.34	1167.39	1170.43	1173.48	1176.53	1179.58	1182.63	1185.67
3900-	1188.72	1191.77	1194.82	1197.87	1200.91	1203.96	1207.01	1210.06	1213.11	1216.15

AGM10145

HEIGHT CONVERSION
FEET TO METERS
CONTINUED

Feet	0	10	20	30	40	50	60	70	80	90
4000-	1219.2	1222.3	1225.3	1228.3	1231.4	1234.4	1237.5	1240.5	1243.6	1246.6
4100-	1249.7	1252.7	1255.8	1258.8	1261.9	1264.9	1268.0	1271.0	1274.1	1277.1
4200-	1280.2	1283.2	1286.3	1289.3	1292.4	1295.4	1298.5	1301.5	1304.5	1307.6
4300-	1310.6	1313.7	1316.7	1319.8	1322.8	1325.9	1328.9	1332.0	1335.0	1338.1
4400-	1341.1	1344.2	1347.2	1350.3	1353.3	1356.4	1359.4	1362.5	1365.5	1368.6
4500-	1371.6	1374.7	1377.7	1380.7	1383.8	1386.8	1389.9	1392.9	1396.0	1399.0
4600-	1402.1	1405.1	1408.2	1411.2	1414.3	1417.3	1420.4	1423.4	1426.5	1429.5
4700-	1432.6	1435.6	1438.7	1441.7	1444.8	1447.8	1450.9	1453.9	1456.9	1460.0
4800-	1463.0	1466.1	1469.1	1472.2	1475.2	1478.3	1481.3	1484.4	1487.4	1490.5
4900-	1493.5	1496.6	1499.6	1502.7	1505.7	1508.8	1511.8	1514.9	1517.9	1521.0
5000-	1524.0	1527.1	1530.1	1533.1	1536.2	1539.2	1542.3	1545.3	1548.4	1551.4
5100-	1554.5	1557.5	1560.6	1563.6	1566.7	1569.7	1572.8	1575.8	1578.9	1581.9
5200-	1585.0	1588.0	1591.1	1594.1	1597.2	1600.2	1603.3	1606.3	1609.3	1612.4
5300-	1615.4	1618.5	1621.5	1624.6	1627.6	1630.7	1633.7	1636.8	1639.8	1642.9
5400-	1645.9	1649.0	1652.0	1655.1	1658.1	1661.2	1664.2	1667.3	1670.3	1673.4
5500-	1676.4	1679.5	1682.5	1685.5	1688.6	1691.6	1694.7	1697.7	1700.8	1703.8
5600-	1706.9	1709.9	1713.0	1716.0	1719.1	1722.1	1725.2	1728.2	1731.3	1734.3
5700-	1737.4	1740.4	1743.5	1746.5	1749.6	1752.6	1755.7	1758.7	1761.7	1764.8
5800-	1767.8	1770.9	1773.9	1777.0	1780.0	1783.1	1786.1	1789.2	1792.2	1795.3
5900-	1798.3	1801.4	1804.4	1807.5	1810.5	1813.6	1816.6	1819.7	1822.7	1825.8
6000-	1828.8	1831.9	1834.9	1837.9	1841.0	1844.0	1847.1	1850.1	1853.2	1856.2
6100-	1859.3	1862.3	1865.4	1868.4	1871.5	1874.5	1877.6	1880.6	1883.7	1886.7
6200-	1889.8	1892.8	1895.9	1898.9	1902.0	1905.0	1908.1	1911.1	1914.1	1917.2
6300-	1920.2	1923.3	1926.3	1929.4	1932.4	1935.5	1938.5	1941.6	1944.6	1947.7
6400-	1950.7	1953.8	1956.8	1959.9	1962.9	1966.0	1969.0	1972.1	1975.1	1978.2
6500-	1981.2	1984.3	1987.3	1990.3	1993.4	1996.4	1999.5	2002.5	2005.6	2008.6
6600-	2011.7	2014.7	2017.8	2020.8	2023.9	2026.9	2030.0	2033.0	2036.1	2039.1
6700-	2042.2	2045.2	2048.3	2051.3	2054.4	2057.4	2060.5	2063.5	2066.5	2069.6
6800-	2072.6	2075.7	2078.7	2081.8	2084.8	2087.9	2090.9	2094.0	2097.0	2100.1
6900-	2103.1	2106.2	2109.2	2112.3	2115.3	2118.4	2121.4	2124.5	2127.5	2130.6
7000-	2133.6	2136.7	2139.7	2142.7	2145.8	2148.8	2151.9	2154.9	2158.0	2161.0
7100-	2164.1	2167.1	2170.2	2173.2	2176.3	2179.3	2182.4	2185.4	2188.5	2191.5
7200-	2194.6	2197.6	2200.7	2203.7	2206.8	2209.8	2212.9	2215.9	2218.9	2222.0
7300-	2225.0	2228.1	2231.1	2234.2	2237.2	2240.3	2243.3	2246.4	2249.4	2252.5
7400-	2255.5	2258.6	2261.6	2264.7	2267.7	2270.8	2273.8	2276.9	2279.9	2283.0
7500-	2286.0	2289.1	2292.1	2295.1	2298.2	2301.2	2304.3	2307.3	2310.4	2313.4
7600-	2316.5	2319.5	2322.6	2325.6	2328.7	2331.7	2334.8	2337.8	2340.9	2343.9
7700-	2347.0	2350.0	2353.1	2356.1	2359.2	2362.2	2365.3	2368.3	2371.3	2374.4
7800-	2377.4	2380.5	2383.5	2386.6	2389.6	2392.7	2395.7	2398.8	2401.8	2404.9
7900-	2407.9	2411.0	2414.0	2417.1	2420.1	2423.2	2426.2	2429.3	2432.3	2435.4
8000-	2438.4	2441.5	2444.5	2447.5	2450.6	2453.6	2456.7	2459.7	2462.8	2465.8
8100-	2468.9	2471.9	2475.0	2478.0	2481.1	2484.1	2487.2	2490.2	2493.3	2496.3
8200-	2499.4	2502.4	2505.5	2508.5	2511.6	2514.6	2517.7	2520.7	2523.7	2526.8
8300-	2529.8	2532.9	2535.9	2539.0	2542.0	2545.1	2548.1	2551.2	2554.2	2557.3
8400-	2560.3	2563.4	2566.4	2569.5	2572.5	2575.6	2578.6	2581.7	2584.7	2587.8
8500-	2590.8	2593.9	2596.9	2599.9	2603.0	2606.0	2609.1	2612.1	2615.2	2618.2
8600-	2621.3	2624.3	2627.4	2630.4	2633.5	2636.5	2639.6	2642.6	2645.7	2648.7
8700-	2651.8	2654.8	2657.9	2660.9	2664.0	2667.0	2670.1	2673.1	2676.1	2679.2
8800-	2682.2	2685.3	2688.3	2691.4	2694.4	2697.5	2700.5	2703.6	2706.6	2709.7
8900-	2712.7	2715.8	2718.8	2721.9	2724.9	2728.0	2731.0	2734.1	2737.1	2740.2

	0	100	200	300	400	500	600	700	800	900
9000-	2743	2774	2804	2835	2865	2896	2926	2957	2987	3018
10000-	3048	3078	3109	3139	3170	3200	3231	3261	3292	3322
11000-	3353	3383	3414	3444	3475	3505	3536	3566	3597	3627
12000-	3658	3688	3719	3749	3780	3810	3840	3871	3901	3932
13000-	3962	3993	4023	4054	4084	4115	4145	4176	4206	4237
14000-	4267	4298	4328	4359	4389	4420	4450	4481	4511	4542

AGM10146

NAUTICAL MILES TO KILOMETERS CONVERSION
FOR WIND SPEED AND VISIBILITY

1 nautical mile	=	6076.11549 feet
	=	1.852 kilometers
	=	1.15078 statute miles
1 kilometer	=	3281 feet
	=	0.6215 statute miles
	=	0.53946 nautical miles
1 statute mile	=	5280 feet
	=	0.8689763 nautical miles
	=	1.609344 kilometers

Nautical Miles	Kilometers	Nautical Miles	Kilometers	Nautical Miles	Kilometers
0	0	26	48.2	52	96.2
1	1.8	27	50.0	53	98.1
2	3.7	28	51.8	54	99.9
3	5.6	29	53.7	55	101.8
4	7.4	30	55.5	56	103.6
5	9.3	31	57.4	57	105.5
6	11.1	32	59.2	58	107.3
7	13.0	33	61.1	59	109.2
8	14.8	34	62.9	60	111.0
9	16.7	35	64.8	61	112.9
10	18.5	36	66.6	62	114.7
11	20.4	37	68.5	63	116.6
12	22.2	38	70.3	64	118.4
13	24.1	39	72.2	65	120.3
14	25.9	40	74.0	66	122.1
15	27.8	41	75.9	67	124.0
16	29.6	42	77.7	68	125.8
17	31.5	43	79.6	69	127.7
18	33.3	44	81.4	70	129.5
19	35.2	45	83.3	71	131.4
20	37.0	46	85.1	72	133.2
21	38.9	47	87.0	73	135.1
22	40.7	48	88.8	74	136.9
23	42.6	49	90.7	75	138.8
24	44.4	50	92.5		
25	46.3	51	94.4		

AGM10147

INCHES TO HECTOPASCALS (507)

Inches	0.00	0.01	0.02	0.03	0.04	0.05	0.06	0.07	0.08	0.09
	HECTOPASCALS									
28.0	948.2	948.5	948.9	949.2	949.5	949.9	950.2	950.6	950.9	951.2
28.1	951.6	951.9	952.3	952.6	952.9	953.3	953.6	963.9	954.3	954.6
28.2	955.0	955.3	955.6	956.0	956.3	956.7	957.0	957.3	957.7	958.0
28.3	958.3	958.7	959.0	959.4	959.7	960.0	960.4	960.7	961.1	961.4
28.4	961.7	962.1	962.4	967.7	963.1	963.4	963.8	964.1	964.4	964.8
28.5	965.1	965.6	965.8	966.1	966.5	966.8	967.2	967.5	967.8	968.2
28.6	968.5	968.8	969.2	969.5	969.9	970.2	970.5	970.9	971.2	971.6
28.7	971.9	972.2	972.6	972.9	973.2	973.6	973.9	974.3	974.6	974.9
28.8	975.3	975.6	976.0	976.3	976.6	977.0	977.3	977.6	978.0	978.3
28.9	978.7	979.0	979.3	979.7	980.0	980.4	980.7	981.0	981.4	981.7
29.0	982.1	982.4	982.7	983.1	983.4	983.7	984.1	984.4	984.8	985.1
29.1	985.4	985.8	986.1	986.5	986.8	987.1	987.5	987.8	988.1	988.5
29.2	988.8	989.2	989.5	989.8	990.2	990.5	990.9	991.2	991.5	991.9
29.3	992.2	992.5	992.9	993.2	993.6	993.9	994.2	994.6	994.9	995.3
29.4	995.6	995.9	996.3	996.6	997.0	997.3	997.6	998.0	998.3	998.6
29.5	999.0	999.3	999.7	1000.0	1000.3	1000.7	1001.0	1001.4	1001.7	1002.0
29.6	1002.4	1002.7	1003.0	1003.4	1003.7	1004.1	1004.4	1004.7	1005.1	1005.4
29.7	1005.8	1006.1	1006.4	1006.8	1007.1	1007.4	1007.8	1008.1	1008.5	1008.8
29.8	1009.1	1009.5	1009.8	1010.2	1010.5	1010.8	1011.2	1011.5	1011.9	1012.2
29.9	1012.5	1012.9	1013.2	1013.5	1013.9	1014.2	1014.6	1014.9	1015.2	1015.6
30.0	1015.9	1016.3	1016.6	1016.9	1017.3	1017.4	1017.9	1018.3	1018.6	1019.0
30.1	1019.3	1019.6	1020.0	1020.3	1020.7	1021.0	1021.3	1021.7	1022.0	1022.3
30.2	1022.7	1023.0	1023.4	1023.7	1024.0	1024.4	1024.7	1025.1	1025.4	1025.7
30.3	1026.1	1026.4	1026.8	1027.1	1027.4	1027.8	1028.1	1028.4	1028.8	1029.1
30.4	1029.5	1029.8	1030.1	1030.5	1030.8	1031.2	1031.5	1031.8	1032.2	1032.5
30.5	1032.8	1033.2	1033.5	1033.9	1034.2	1034.5	1034.9	1035.2	1035.6	1035.9
30.6	1036.2	1036.6	1036.9	1037.2	1037.6	1037.9	1038.3	1038.6	1038.9	1039.3
30.7	1039.6	1040.0	1040.3	1040.6	1041.0	1041.3	1041.7	1042.0	1042.3	1042.7
30.8	1043.0	1043.3	1043.7	1044.0	1044.4	1044.7	1045.0	1045.4	1045.7	1046.1
30.9	1046.4	1046.7	1047.1	1047.4	1047.7	1048.1	1048.4	1048.8	1049.1	1049.4

HECTOPASCALS TO INCHES (507)

HECTOPASCALS	0	1	2	3	4	5	6	7	8	9
	Inches									
940	27.76	27.79	27.82	27.84	27.88	27.91	27.94	27.96	27.99	28.02
950	28.05	28.08	28.11	28.14	28.17	28.20	28.23	28.26	28.29	28.32
960	28.35	28.38	28.41	28.44	28.47	28.50	28.53	28.56	28.59	28.61
970	28.64	28.67	28.70	28.73	28.76	28.79	28.82	28.85	28.88	28.91
980	28.94	28.97	29.00	29.03	29.06	29.09	29.12	29.15	29.18	29.21
990	29.23	29.26	29.29	29.32	29.35	29.28	29.41	29.44	29.47	29.50
1000	29.53	29.56	29.59	29.62	29.65	29.68	29.71	29.74	29.77	29.80
1010	29.83	29.85	29.88	29.91	29.94	29.97	30.00	30.03	30.06	30.09
1020	30.12	30.15	30.18	30.21	30.24	30.27	30.30	30.33	30.36	30.39
1030	30.42	30.45	30.47	30.50	30.53	30.56	30.59	30.62	30.65	30.68
1040	30.71	30.74	30.77	30.80	30.83	30.86	30.89	30.92	30.95	30.98
1050	31.01	31.04	31.07	31.10	31.12	31.15	31.18	31.21	31.24	31.27

Proportional Parts	Inches	.001	.002	.003	.004	.005	.006	.007	.008	.009
	HECTOPASCALS 00	.1	.1	.1	.2	.2	.2	.3	.3	

AGM10148

APPENDIX III

TIME ZONES

For reckoning time, the surface of the globe has been divided into 24 zones; each is bounded by meridians of 15° of arc, and each is 1 hour in longitude. The initial time zone lies between 7 1/2°E and 7 1/2°W of the Prime Meridian; it is called ZONE ZERO. Each zone, in turn, is designated by the number that represents the difference between local zone time and Coordinated Universal Time (UTC). See figure AIII-1. Zones have been modified near land to accord with the boundaries of the countries or the regions using corresponding time.

The zones lying east of zone zero are numbered 1 through 12 and are designated minus. And for each of the minus zones, the zone number is subtracted from the local time to obtain UTC. The zones west of zone zero are also numbered from 1 through 12 but are designated plus, since the zone number must be added to the local time to get UTC. The twelfth zone is divided by the 180th meridian, the minus half lying in east longitude and the plus half in west longitude. The zone number preceded by a plus or a minus sign constitutes the zone description. In addition to the time zone number, each zone is also designated by a letter *A* through *M* (*J* omitted) for the minus zones, and *N* through *Y* for the plus zones. (See top of figure AIII-1.)

Date/time groups (DTGs) are frequently used to express times of specific events, such as the time a message was written. A date/time group always contains 6 digits; The first 2 digits are the day of the month, and the last 4 digits are the time, using the 24-hour clock. The appropriate time zone letter designator follows the 6-digit date/time.

Local times may be used in the text of a message or letter. When local time is used, it must be accompanied by the zone letter-such as 08124. If local time is referred to frequently in the text, the suffix may be omitted provided a covering expression, such as "ALL TIMES QUEBEC", is used.

When it is necessary to indicate a date alone in a message, it is expressed by the day of the month, the three-letter abbreviation of the month, and (if necessary) the last two figures of the year: 3 FEB or 3 FEB 97. A night is expressed by the word *night* and the two dates over which it extends: NIGHT 3/4 FEB 97.

TIME CONVERSION TABLE

The time conversion table (table AIII-1) is useful for converting the time in one zone to the time in any other zone. Vertical columns indicate time zones. Zone *Z* is UTC. Time in each successive zone to the right of zone *Z* is 1 hour later, and to the left of zone *Z* is 1 hour earlier.

To calculate time in zone I when it is 1200 hours in zone U, for example, proceed as follows: Find 1200 in column U and locate the corresponding time in the line in column I (0500). Since both of the times (1200 in zone U and 0500 in zone I) are not in the "same day" area of table AIII-1, the time 0500 in zone I is tomorrow to the time 1200 in zone U. In other words, when it is 1200 in zone U, it is 0500 tomorrow in zone I; or when it is 0500 in zone I, it is 1200 yesterday in zone U.

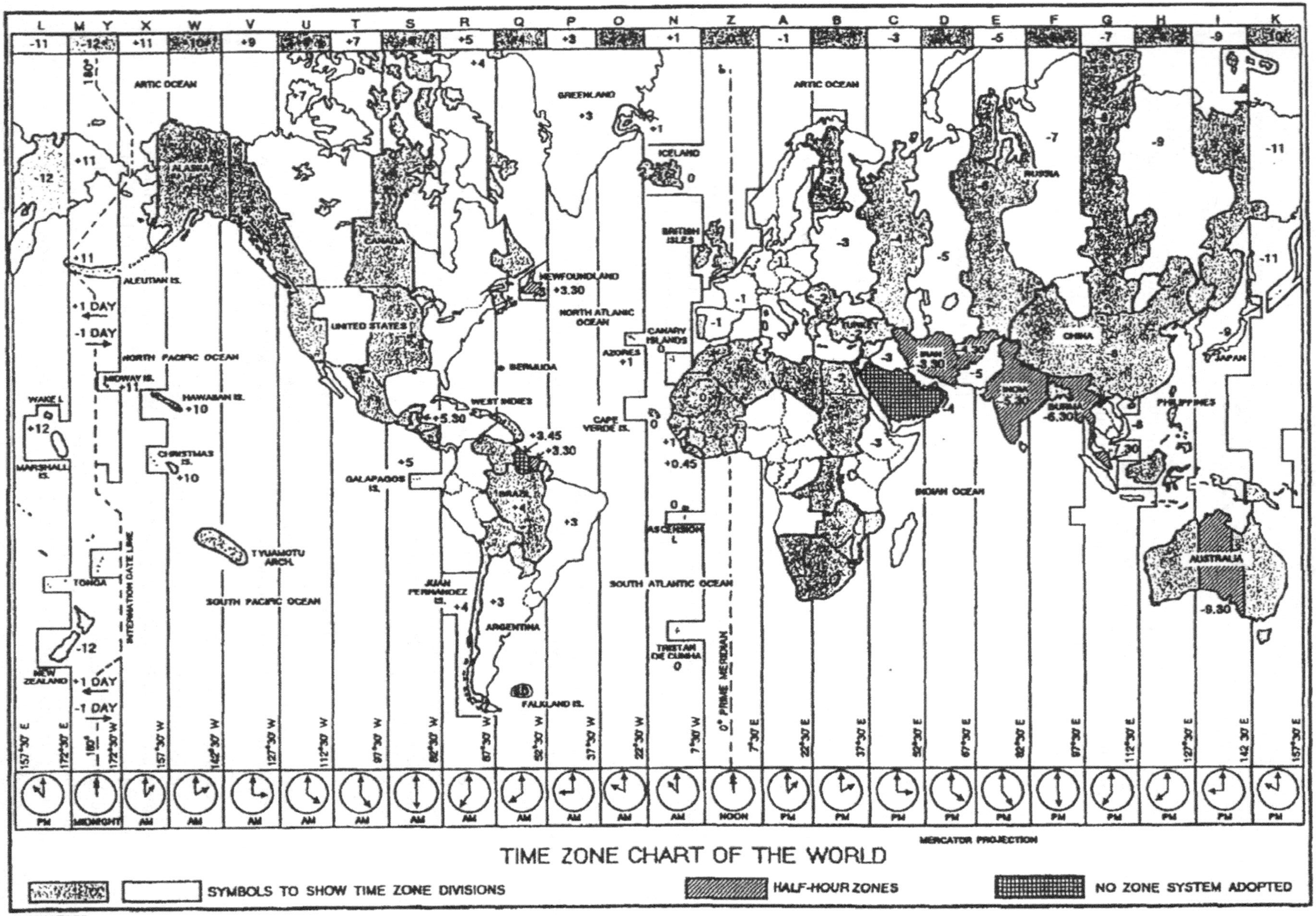

Figure AIII-1.—Time zones of the world.

Table AIII-1.—Time Conversion Table

PREVIOUS DAY	1800	1900	2000	2100	2200	2300	2400	0100	0200	0300	0400	0500	0600	0700	0800	0900	1000	1100	1200	1300	1400	1500	1600	1700	1800	SAME DAY
	1900	2000	2100	2200	2300	2400	0100	0200	0300	0400	0500	0600	0700	0800	0900	1000	1100	1200	1300	1400	1500	1600	1700	1800	1900	
	2000	2100	2200	2300	2400	0100	0200	0300	0400	0500	0600	0700	0800	0900	1000	1100	1200	1300	1400	1500	1600	1700	1800	1900	2000	
	2100	2200	2300	2400	0100	0200	0300	0400	0500	0600	0700	0800	0900	1000	1100	1200	1300	1400	1500	1600	1700	1800	1900	2000	2100	
	2200	2300	2400	0100	0200	0300	0400	0500	0600	0700	0800	0900	1000	1100	1200	1300	1400	1500	1600	1700	1800	1900	2000	2100	2200	
	2300	2400	0100	0200	0300	0400	0500	0600	0700	0800	0900	1000	1100	1200	1300	1400	1500	1600	1700	1800	1900	2000	2100	2200	2300	
	2400	0100	0200	0300	0400	0500	0600	0700	0800	0900	1000	1100	1200	1300	1400	1500	1600	1700	1800	1900	2000	2100	2200	2300	2400	
SAME DAY	0100	0200	0300	0400	0500	0600	0700	0800	0900	1000	1100	1200	1300	1400	1500	1600	1700	1800	1900	2000	2100	2200	2300	2400	0100	NEXT DAY
	0200	0300	0400	0500	0600	0700	0800	0900	1000	1100	1200	1300	1400	1500	1600	1700	1800	1900	2000	2100	2200	2300	2400	0100	0200	
	0300	0400	0500	0600	0700	0800	0900	1000	1100	1200	1300	1400	1500	1600	1700	1800	1900	2000	2100	2200	2300	2400	0100	0200	0300	
	0400	0500	0600	0700	0800	0900	1000	1100	1200	1300	1400	1500	1600	1700	1800	1900	2000	2100	2200	2300	2400	0100	0200	0300	0400	
	0500	0600	0700	0800	0900	1000	1100	1200	1300	1400	1500	1600	1700	1800	1900	2000	2100	2200	2300	2400	0100	0200	0300	0400	0500	
	0600	0700	0800	0900	1000	1100	1200	1300	1400	1500	1600	1700	1800	1900	2000	2100	2200	2300	2400	0100	0200	0300	0400	0500	0600	
	0700	0800	0900	1000	1100	1200	1300	1400	1500	1600	1700	1800	1900	2000	2100	2200	2300	2400	0100	0200	0300	0400	0500	0600	0700	
	0800	0900	1000	1100	1200	1300	1400	1500	1600	1700	1800	1900	2000	2100	2200	2300	2400	0100	0200	0300	0400	0500	0600	0700	0800	
	0900	1000	1100	1200	1300	1400	1500	1600	1700	1800	1900	2000	2100	2200	2300	2400	0100	0200	0300	0400	0500	0600	0700	0800	0900	
	1000	1100	1200	1300	1400	1500	1600	1700	1800	1900	2000	2100	2200	2300	2400	0100	0200	0300	0400	0500	0600	0700	0800	0900	1000	
	1100	1200	1300	1400	1500	1600	1700	1800	1900	2000	2100	2200	2300	2400	0100	0200	0300	0400	0500	0600	0700	0800	0900	1000	1100	
	1200	1300	1400	1500	1600	1700	1800	1900	2000	2100	2200	2300	2400	0100	0200	0300	0400	0500	0600	0700	0800	0900	1000	1100	1200	
	1300	1400	1500	1600	1700	1800	1900	2000	2100	2200	2300	2400	0100	0200	0300	0400	0500	0600	0700	0800	0900	1000	1100	1200	1300	
	1400	1500	1600	1700	1800	1900	2000	2100	2200	2300	2400	0100	0200	0300	0400	0500	0600	0700	0800	0900	1000	1100	1200	1300	1400	
	1500	1600	1700	1800	1900	2000	2100	2200	2300	2400	0100	0200	0300	0400	0500	0600	0700	0800	0900	1000	1100	1200	1300	1400	1500	
	1600	1700	1800	1900	2000	2100	2200	2300	2400	0100	0200	0300	0400	0500	0600	0700	0800	0900	1000	1100	1200	1300	1400	1500	1600	
	1700	1800	1900	2000	2100	2200	2300	2400	0100	0200	0300	0400	0500	0600	0700	0800	0900	1000	1100	1200	1300	1400	1500	1600	1700	
	Y	X	W	V	U	T	S	R	Q	P	O	N	Z	A	B	C	D	E	F	G	H	I	K	L	M	
	+12	+11	+10	+9	+8	+7	+6	+5	+4	+3	+2	+1	0	-1	-2	-3	-4	-5	-6	-7	-8	-9	-10	-11	-12	

AGM1F252

APPENDIX IV

WMO CODE TABLES

This appendix contains graphic or tabular representations of many of the code tables referenced in this training manual. The wording of the code tables has been simplified in some cases to clarify the meaning.

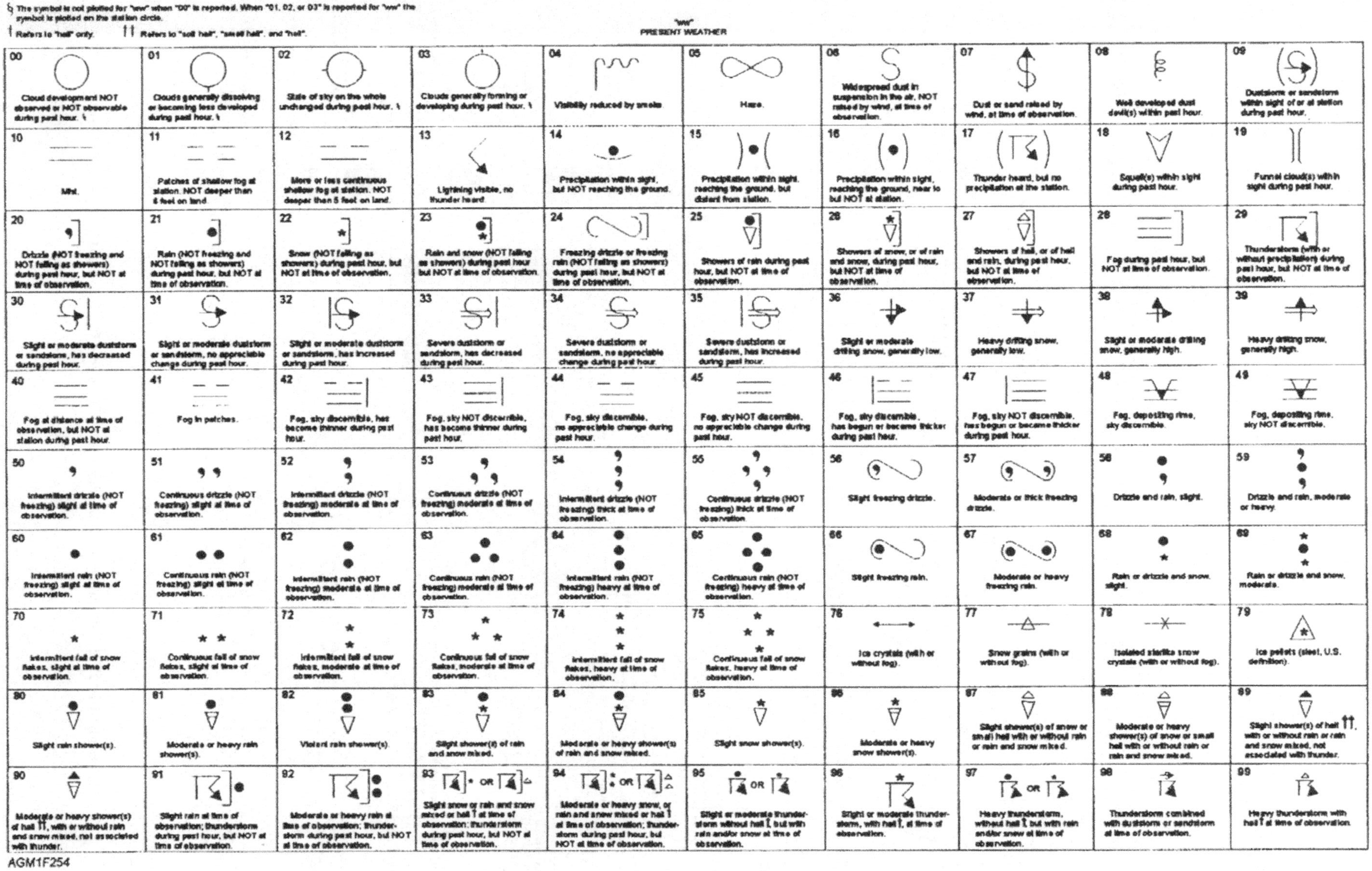

Figure AIV-1.—Present weather plotting symbols for code figure ww based on WMO Code table 4677.

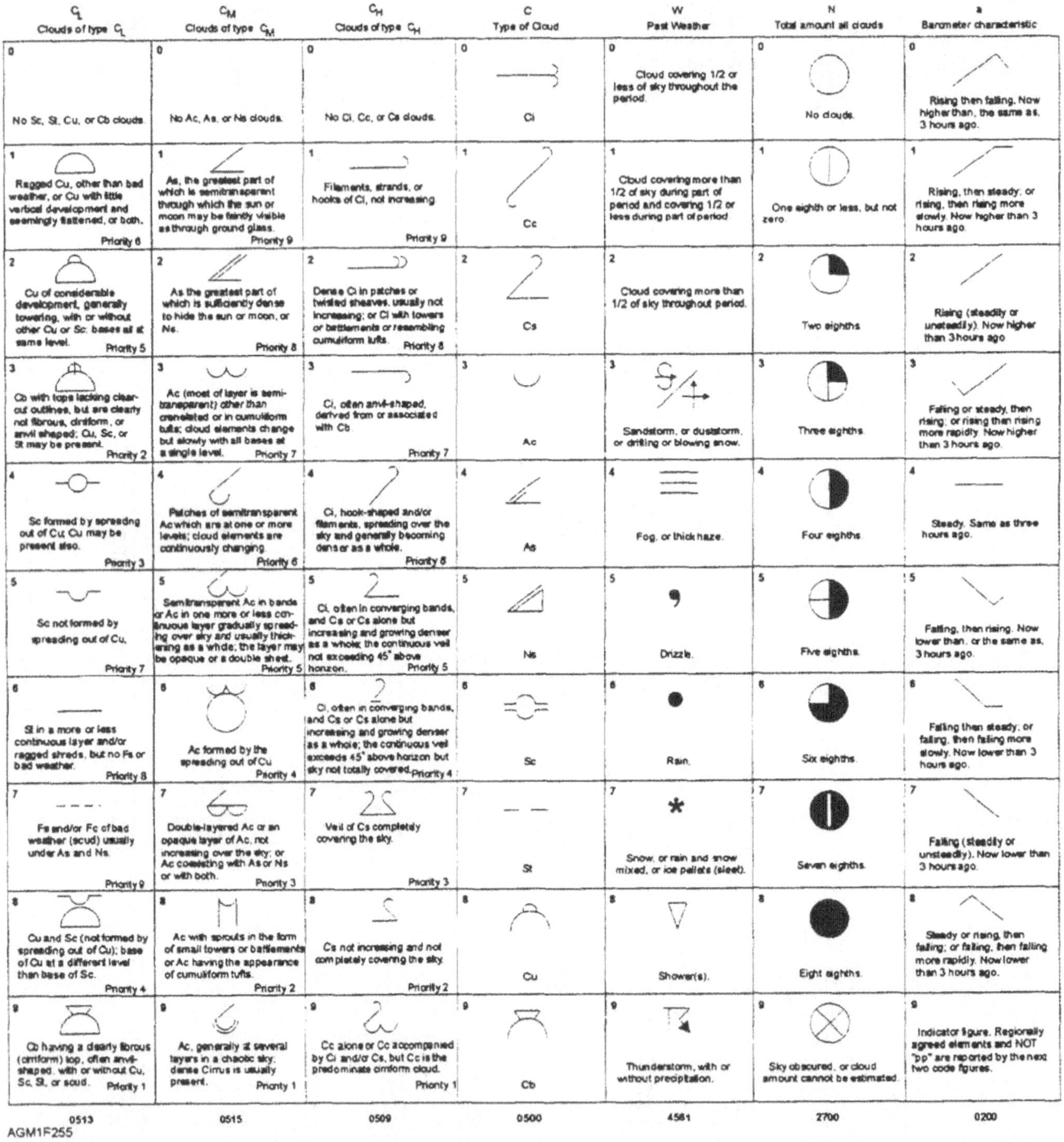

Figure AIV-2.—Plotting symbols used for code figures C_L, C_M, C_H, C, W, N, a.

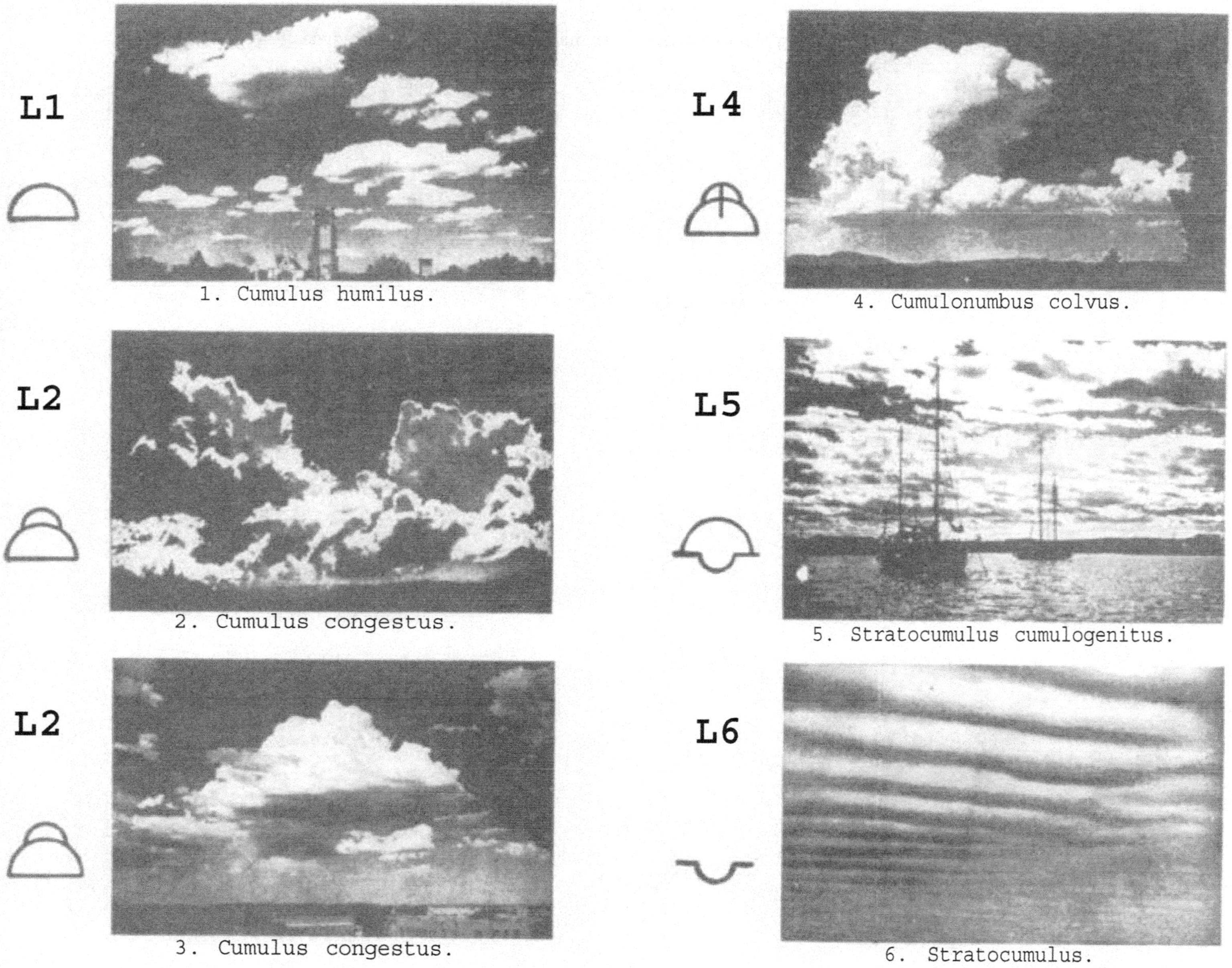

1. Cumulus humilus.

2. Cumulus congestus.

3. Cumulus congestus.

4. Cumulonumbus colvus.

5. Stratocumulus cumulogenitus.

6. Stratocumulus.

209.451

Figure AIV-3.—Examples of the 9 low-level sky states reported for code C_L, based on WMO Code table 0513.

L6

7. Stratus.

L7

8. Cumulus fractus of bad weather.

L8

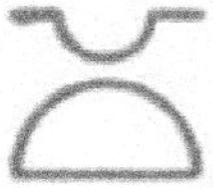

9. Cumulus humilis and stratocumulus.

L8

10. Cumulus congestus and stratocumulus.

L9

11. Cumulonimbus capillatus.

L9

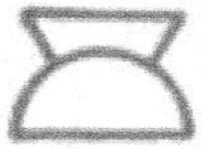

12. Cumulonimbus capillatus.

209.451

Figure AIV-3.—Examples of the 9 low-level sky states reported for code C_L, based on WMO Code table 0513—Continucd.

M1

13. Altostratus translucidus.

M2

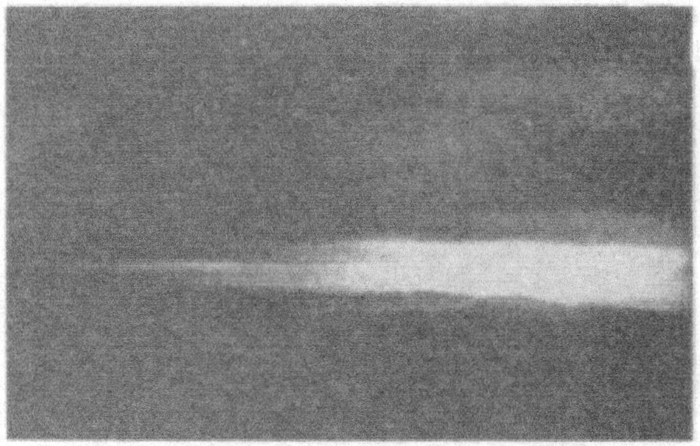

14. Altostratus opacus.

M3

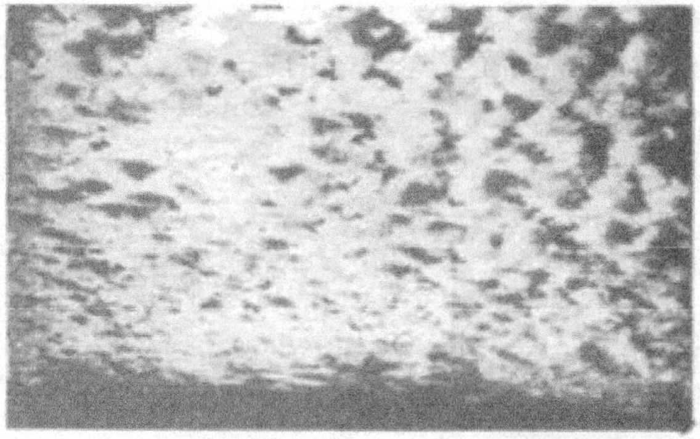

15. Altocumulus translucidus.

M4

16. Altocumulus lenticularis.

M5

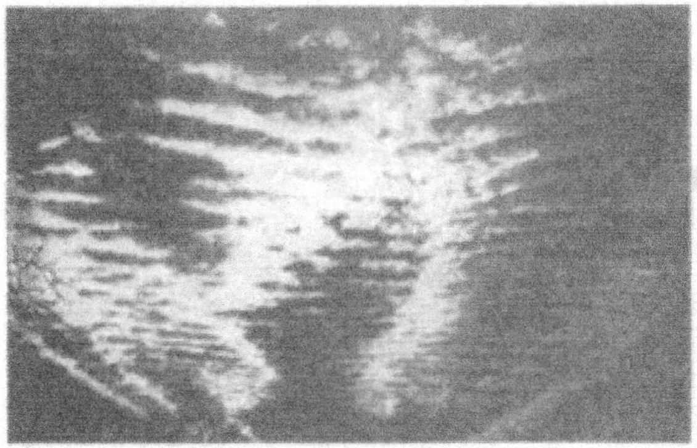

17. Altocumulus translucidus undulatus.

M6

18. Altocumulus cumulogenitus

Figure AIV-4.—Examples of the 9 mid-level sky states reported for code C_M, based on WMO Code table 0515.

209.452

M6

19. Altocumulus cumulonimbogenitus.

M7

20. Altocumulus duplicatus.

M7

21. Altocumulus opacus.

M8

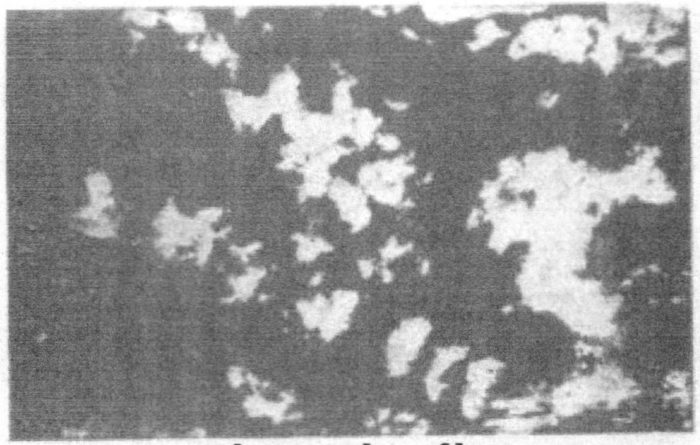

22. Altocumulus floccus.

M8

23. Altocumulus castellanus.

M9

24. Altocumulus of a chaotic sky.

Figure AIV-4.—Examples of the 9 mid-level sky states reported for code C_M, based on WMO Code table 0515—Continued.

209.42

H1

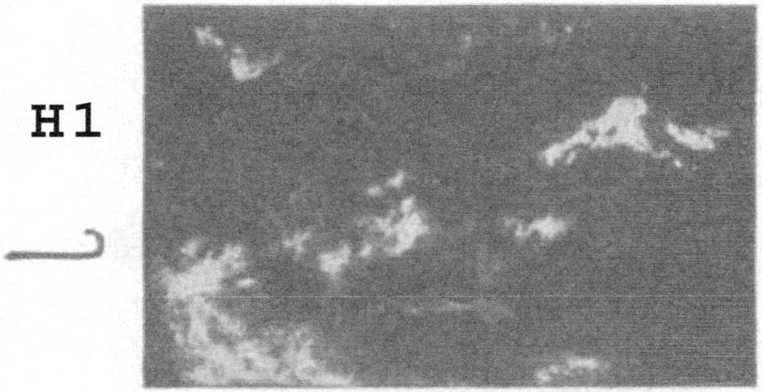

25. Cirrus fibratus.

H3

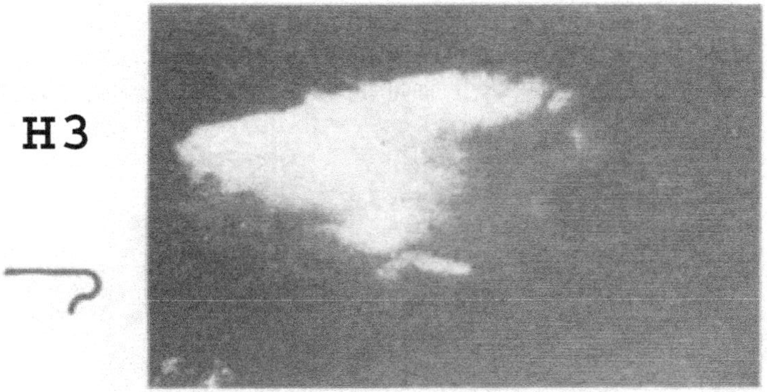

28. Cirrus spissatus cumulonimbogenitus.

H1

26. Cirrus fibratus.

H3

29. Cirrus spissatus cumulonimbogenitus.

H2

27. cirrus spissatus.

H4

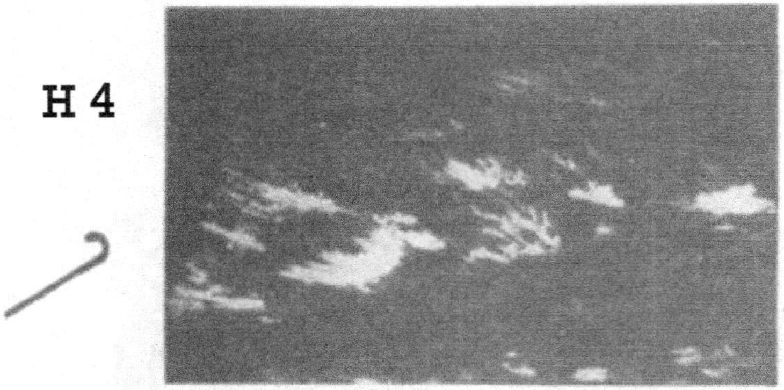

30. Cirrus uncinus.

209.453

Figure AIV-5.—Examples of the 9 high-level sky states reported for code C_H, based on WMO Code table 0509.

H5

31. Cirrus below 45°.

H6

32. Cirrus above 45°.

H7

33. Cirrostratus covering the whole sky.

H8

34. Cirrostratus not covering the while sky.

H8

35. Cirrostratus not covering the whole sky.

H9

36. Cirrocumulus.

209.453

Figure AIV-5.—Examples of the 9 high-level sky states reported for code C_H, based on WMO Code table 0509—Continued.

Table AIV-1.—WMO CODE 0700

WMO CODE 0700 Direction or brearing in one figure.	
Reported as towards: D_a, D_e, D_l, D_S	Reported as from: D, D_H, D_M, D_L, D_K, D_p,

Code 7 Northwest	Code 8 North	Code 1 Northeast
Code 6 West	Code 0 Stationary or overhead	Code 2 East
Code 5 Southwest	Code 4 south	Code 3 Southeast

Code 9 - All directions, confused, variable, or unknown.
Code / - Not reported.

Table AIV-2.—WMO CODE 0975

WMO CODE 0975
E' - State of the ground with snow or measurable ice cover.

Code	PREDOMINENT GROUND COVER
0	Ground predominatly covered by ice.
1	Compact or wet snow - 0.1 to 0.4 ground covered.
2	Compact or wet snow - 0.5 to 0.9 ground covered.
3	Compact or wet snow - 1.0 coverage - EVEN layer.
4	Compact or wet snow - 1.0 coverage - UNEVEN layer.
5	Loose dry snow - 0.1 to 0.4 ground covered.
6	Loose dry snow - 0.5 to 0.9 ground covered.
7	Loose dry snow - complete coverage - EVEN layer.
8	Loose dry snow - complete coverage - UNEVEN layer.
9	Snow - complete coverage - DEEP DRIFTS.

Table AIV-3.—WMO CODE 4451

WMO CODE 4451
v_S, - Ship's average speed made good during the past 3 hours.

Code	Speed
0	0 kt
1	1 - 5 kt
2	6 - 10 kt
3	11 - 15 kt
4	16 - 20 kt
5	21 - 25 kt
6	26 - 30 kt
7	31 - 35 kt
8	36 - 40 kt
9	over 40 kts
/	Not reported

Table AIV-4.—WMO CODE 0901

WMO CODE 0901 E - State of the mound without snow or measurable ice cover.	
Code	STATE OF THE GROUND
0	Dry (without cracks, dust, or lose sand).
1	Moist.
2	Wet (standing puddles).
3	Flooded.
4	Frozen.
5	Glaze ice.
6	Patchy loose dust or sand.
7	Thin layer of loose dust or sand covers ground completely.
8	Moderate or thick layer of loose dust or sand covers ground completely.
9	Extremely dry with cracks.

Table AIV-5.—WMO CODE 1600

WMO CODE 1600	Code	Meters	*Feet
h - Height above surface of the base of the lowest clouds seen. * U.S. observers use the modified scale in feet.	0	0 to 49	0 to 99
	1	50 to 99	100 to 299
	2	100 to 199	300 to 699
	3	200 to 299	700 to 999
	4	300 to 599	1000 to 1999
	5	600 to 999	2000 to 3299
	6	1000 to 1499	3300 to 4899
	7	1500 to 1999	4900 to 6499
	8	2000 to 2499	6500 to 7999
	9	No clouds or ≥2500m	No clouds or ≥ 8000ft
	/	Unknown	Unknown

Table AIV-6.—WMO CODE 1677

WMO CODE 1677

h_sh_s - Height of base of cloud layer or mass indicated by genus C.

h_th_t - Height of tops of low cloud or fog layer.

* U.S. observers use the modified scale in feet.

code	meters	feet	code	meters	feet	code	meters	feet	code	meters	feet
00	<30	<100	25	750	2500	50	1500	5000	75	7500	25000
01	30	100	26	780	2600	51	-		76	7800	26000
02	60	200	27	810	2700	52	not		77	8100	27000
03	90	300	28	840	2800	53	-		78	8400	28000
04	120	400	29	870	2900	54	used		79	8700	29000
05	150	500	30	900	3000	55	-		80	9000	30000
06	180	600	31	930	3100	56	1800	6000	81	10500	35000
07	210	700	32	960	3200	57	2100	7000	82	12000	40000
08	240	800	33	990	3300	58	2400	8000	83	13500	45000
09	270	900	34	1020	3400	59	2700	9000	84	15000	50000
10	300	1000	35	1050	3500	60	3000	10000	85	16500	55000
11	330	1100	36	1080	3600	61	3300	11000	86	18000	60000
12	360	1200	37	1110	3700	62	3600	12000	87	19500	65000
13	390	1300	38	1140	3800	63	3900	13000	88	2 1000	70000
14	420	1400	39	1170	3900	64	4200	14000	89	>2 1000	>70000
15	450	1500	40	1200	4000	65	4500	15000			
16	480	1600	41	1230	4100	66	4800	16000			
17	510	1700	42	1260	4200	67	5100	17000			
18	540	1800	43	1290	4300	68	5400	18000			
19	570	1900	44	1320	4400	69	5700	19000			
20	600	2000	45	1350	4500	70	6000	20000			
21	630	2100	46	1380	4600	71	6300	21000			
22	660	2200	47	1410	4700	72	6600	22000			
23	690	2300	48	1440	4800	73	6900	23000			
24	720	2400	49	1470	4900	74	7200	24000			

Table AIV-7.—WMO CODE 1690

WMO CODE 1690

$h_Bh_Bh_B$ - Height of lowest level of turbulence.
$h_ih_ih_i$ - Height of lowest level of icing.
$h_fh_fh_f$ - Altitude of 0°C isotherm.
$h_th_th_t$ - Altitude of cloud layer or mass.
$h_xh_xh_x$ - Altitude to which temperature and wind refer.
$h_sh_sh_s$ - Height of base of cloud layer or vertical visibility.

Code figures 000 to 099 are in single-unit increments, code figures 100 to 990 are in ten-unit increments.

Code figures are the number of 30-meter units of height or altitude. This is nearly identical with the U.S. convention of reporting heights or altitudes in hundreds of feet. A few examples follow:

code 000 = <30 meters = <100 feet
code 001 = 1 30-meter increment (30 meters), used as 100 feet
code 002 = 2 30-meter increments (60 meters), used as 200 feet
code 099 = 99 30-meter increments (2970 meters), used as 9,900 feet
code 990 = 990 30-meter increments (29,700 meters), used as 99,000 feet

Table AIV-8.—WMO CODE 2700

WMO CODE 2700

N - Total cloud cover.
N_h - Amount of cloud in a layer.
N_s - Amount of individual cloud layer with genus indicated by C.
N' - Amount of cloud with base below station level.

Code	Eights	Tenths
0	0	0
1	≤1	≤1
2	2	2-3
3	3	4
4	4	5
5	5	6
6	6	7-8
7	7	9
8	8	10
9	Sky obscured	
/	Sky condition not observed	

Table AIV-9.—WMO CODE 4019

WMO CODE 4019	
t_R - Duration of reference period for the amount of precipitation ending at the time of the report.	
Code	Period preceeding observation
1	6 hours
2	12 hours
3	18 hours
4	24 hours
5	1 hour
6	2 hours
7	3 hours
8	9 hours
9	15 hours

Table AIV-10.—WMO CODE 1819

WMO CODE 1819	
i_R - Indicator for inclusion or ommission of precipitation data.	
Code	Data group $6RRRt_R$ is:
0	Included in section 1 & 3.
1	Included in section 1.
2	Included in section 3.
3	Omitted - Precipitation amount = 0).
4	Omitted - Measurement not available.

Table AIV-11.—WMO CODE 1855

WMO CODE 1855		
i_w - Indicator for source and units of wind speed.		
Code	Source	Units
0	Estimated.	Meters per second.
1	From anemometer.	
3	Estimated.	Knots.
4	From anemometer.	

Table AIV-12.—WMO CODE 1860

WMO CODE 1860		
i_x - Indicator for type of station operation and inclusion of present and past weather data.		
Code	Station Type	Group $7wwW_1W_2$ is:
1	Manned.	Included.
2	Manned.	Omitted - No significant weather.
3	Manned.	Omitted - Not observed.
4	Automatic.	Included (Codes 4677 & 4561).
5	Automatic.	Omitted - No significant weather.
6	Automatic.	Omitted - Not observed.
7	Automatic.	Included (Codes 4680 & 4531).

Table AIV-13—WMO CODE 0822

WMO CODE 0822	
D_T - Amount of temperature change	
Code	Change
0	10°C
1	11°C
2	12°C
3	13°C
4	≥ 14°C
5	5°C
6	6°C
7	7°C
8	8°C
9	9°C

TABLE AIV-14.—WMO CODE 4377

WMO CODE 4377										
VV- Horizontal visibility at the surface.										
KM	MI	CODE	KM	MI	CODE	KM	MI	CODE	KM	CODE
<0.1	<1/16	00	2.6	1 5/8	26	N U		51	26	76
0.1	1/16	01	2.7		27	O S		52	27	77
0.2	1/8	02	2.8	1 3/4	28	T E		53	28	78
0.3	3/16	03	2.9		29	D		54	29	79
0.4	1/4	04	3.0	1 7/8	30			55	30	80
0.5	5/16	05	3.1		31	---		---	---	---
0.6	3/8	06	3.2	2	32	6	4	56	35	81
0.7		07	3.3		33	7		57	40	82
0.8	1/2	08	3.4		34	8	5	58	45	83
0.9		09	3.5		35	9		59	50	84
1.0	5/8	10	3.6	2 1/4	36	10	6	60	55	85
1.1		11	3.7		37	11	7	61	60	86
1.2	3/4	12	3.8		38	12		62	65	87
1.3		13	3.9		39	13	8	63	70	88
1.4	7/8	14	4.0	2 1/2	40	14	9	64	>70	89
1.5		15	4.1		41	15		65	---	---
1.6	1	16	4.2		42	16	10	66	<0.05	90
1.7		17	4.3		43	17		67	0.05	91
1.8	1 1/8	18	4.4	2 3/4	44	18	11	68	0.2	92
1.9		19	4.5		45	19	12	69	0.5	93
2.0	1 1/4	20	4.6		46	20		70	1	94
2.1		21	4.7		47	21	13	71	2	95
2.2	1 3/8	22	4.8	3	48	22		72	4	96
2.3		23	4.9		49	23	14	73	10	97
2.4	1 1/2	24	5.0		50	24	15	74	20	98
2.5		25				25		75	>50	99

Table AIV-15.—WMO CODE 1751

WMO CODE 1751	
I_s - Ice accretion on ships.	
Code	Source of Ice
1	Ice from ocean spray.
2	Ice from fog.
3	Ice from spray and fog.
4	Ice from rain.
5	Ice from spray and rain.

Table AIV-16.—WMO CODE 3551

WMO CODE 3551	
R_s - Rate of ice accretion on ships.	
Code	Ice increase or decrease
0	Ice not building up.
1	Building up slowly.
2	Building up rapidly.
3	Melting or breaking up slowly.
4	Melting or breaking up rapidly.

Table AIV-17.—WMO CODE 0639

WMO CODE 0639			
c_i - Concentration or arrangement of sea ice.			
Code	Description		
0	No ice in sight.		
1	Ship in open lead more than 1 nm wide, or ship in fast ice with ice edge out of sight.		
2	VERY OPEN PACK ICE - less than .3 sea covered.	Ice concentration uniform in observatin area.	Ship in ice or within .5 nm of ice edge.
3	OPEN PACK ICE - .3 to .6 sea covered.		
4	CLOSE PACK ICE - .7 to .8 sea covered.		
5	VERY CLOSE PACK ICE - .9 sea covered.		
6	Strips or patches of pack ice with open water between.	Ice concentration not uniform.	
7	Strips or patches of close or very-close pack ices with lesser concentration between.		
8	Fast ice with open water or open pack ice seaward.		
9	Fast ice with close or very close pack ice seaward.		
/	Unable to report due to darkness, low visibility, or because ship is more than .5 nm from ice edge.		

Table AIV-18.—WMO CODE 5239

<table>
<tr><th colspan="4">WMO CODE 5239
z_i - Present ice situation and trend of conditions over preceeding 3 hours.</th></tr>
<tr><th>Code</th><th colspan="3">Description</th></tr>
<tr><td>0
/</td><td colspan="3">Ship in open water with floating ice in sight.
Unable to report; low visibility or darkness.</td></tr>
<tr><td>1
2
3
4
5</td><td colspan="2">Ice easily penetrable - condition improving.
Ice easily penetrable - condition unchanging.
Ice easily penetrable - condition worsening.
Ice difficult to penetrate - condition improving.
Ice difficult to penetrate - condition unchanged.</td><td rowspan="2">Ship in ice.</td></tr>
<tr><td>6

7
8

9</td><td>Ice forming and floes freezing together.
Ice under slight pressure.
Ice under moderate or severe pressure.
Ship beset.</td><td>Ice difficult to penetrate, conditions worsening.</td></tr>
</table>

Table AIV-19.—WMO CODE 0439

WMO CODE 0439

b_i - Ice of land origin.

code	Description
0	No ice of land origin.
1	1 - 5 bergs, no growlers or bergy bits.
2	6 - 10 bergs, no growlers or bergy bits.
3	11 - 20 bergs, no growlers or bergy bits.
4	1 - 10 growlers or bergy bits, no bergs.
5	>10 growlers or bergy bits, no bergs.
6	1 - 5 bergs with growlers or bergy bits.
7	6 - 10 bergs with growlers or bergy bits.
8	11 - 20 bergs with growlers or bergy bits.
9	>20 bergs with growlers or bergy bits - major navigation hazard.
/	Unable to report because of darkness or visibility.

Table AIV-20.—WMO CODE 0739

WMO CODE 0739		
D_i - True bearing of principal ice edge.		
Code 0 - Ship inshore or in flaw lead.		
Code 7 Toward NW	Code 8 Toward N	Code 1 Toward NE
Code 6 - Toward W		Toward E - Code 2
Code 5 Toward SW	Code 4 Toward S	Code 3 Toward SE
Code 9 - Not determined (ship in ice). Code / - Unable to report or only ice of land origin present.		

Table AIV-21.—WMO CODE 3739

WMO CODE 3739

S_i - Stage of (sea ice) developement.

Code	Sea ice observed
0	New ice only (frazil, grease, slush, shuga).
1	Nilas or ice rind, less than 10 cm thick.
2	Young ice (grey, grey-white), 10 to 30 cm thick.
3	Mostly new and/or young ice with some first-year ice.
4	Mostly thin first-year ice with some new or young ice.
5	All thin first-year ice, 30 to 70 cm thick.
6	Mostly medium first-year ice (70 to 120 cm thick) and thick first-year ice (120 cm thick), with some thinner first-year ice.
7	All medium and thick first-year ice.
8	Mostly medium and thick first-year ice, with some old ice (usually more than 2 meters thick.
9	Mostly old ice.
/	Unable to report: Low visibility, darkness, only ice of land origin present, or ship is more than 0.5 nm from ice edge.

Table AIV-22.—WMO CODE 0264

WMO CODE 0264

a_3 - Standard isobaric surface for which the geopotential height is reported

Code	Level
1	1000 hPa
2	925 hPa
5	500 hPa
7	700 hPa
8	850 hPa

Table AIV-23.—WMO CODE 0265

WMO CODE 0265 a_4 - Type of equipment used.	
Code	Equipment
0	Pressure instrument in sonde
1	Optical theodolite
2	Radio theodolite
3	Radar
4	Pressure instrument in sonde, but sensor failed
5	VLF-Omega
6	Loran-C
7	Wind profiler
8	Satellite navigation
9	Reserved

Table AIV-24.—WMO CODE 1734

WMO CODE 1734 I_d - Indicator used to specify the hundreds of hectopascals figure for the level of the last wind report.		
Code	Part A	Part C
1	100 or 150 hPa	10 hPa
2	200 or 250 hPa	20 hPa
3	300 hPa	30 hPa
4	400 hPa	
5	500 hPa	50 hPa
7	700 hPa	70 hPa
8	850 hPa	
9	925 hPa	
0	1000 hPa	
/	No winds at any standard level	

Table AIV-25.—WMO CODE 3333

WMO CODE 3333 Q_c - Quadrant of the globe.		
Code	Latitude	Longitude
1	North	East
3	South	East
5	South	West
7	North	West

Table AIV-26.—WMO CODE 3850

WMO CODE 3850	
s_s - Indicator for the sign and type of measurement of sea-surface temperature.	
Code	Sign and method
0	positive or zero intake measurement
1	negative intake measurement
2	positive or zero bucket measurement
3	negative bucket measurement
4	positive or zero hull contact sensor
5	negative hull contact sensor
6	positive or zero neither intake, bucket, or hull
7	negative neither intake, bucket, or hull

Table AIV-27.—WMO CODE 3850

WMO CODE 3850	
s_w - Indicator for sign and type of wet-bulb temperature.	
Code	Sign and method
0	Positive or zero measured wet-bulb temp
1	Negative measured wet-bulb temp
2	Iced bulb measured wet-bulb temp
5	Positive or zero computed wet-bulb temp
6	Negative computed wet-bulb temp
7	Iced bulb computed wet-bulb temp

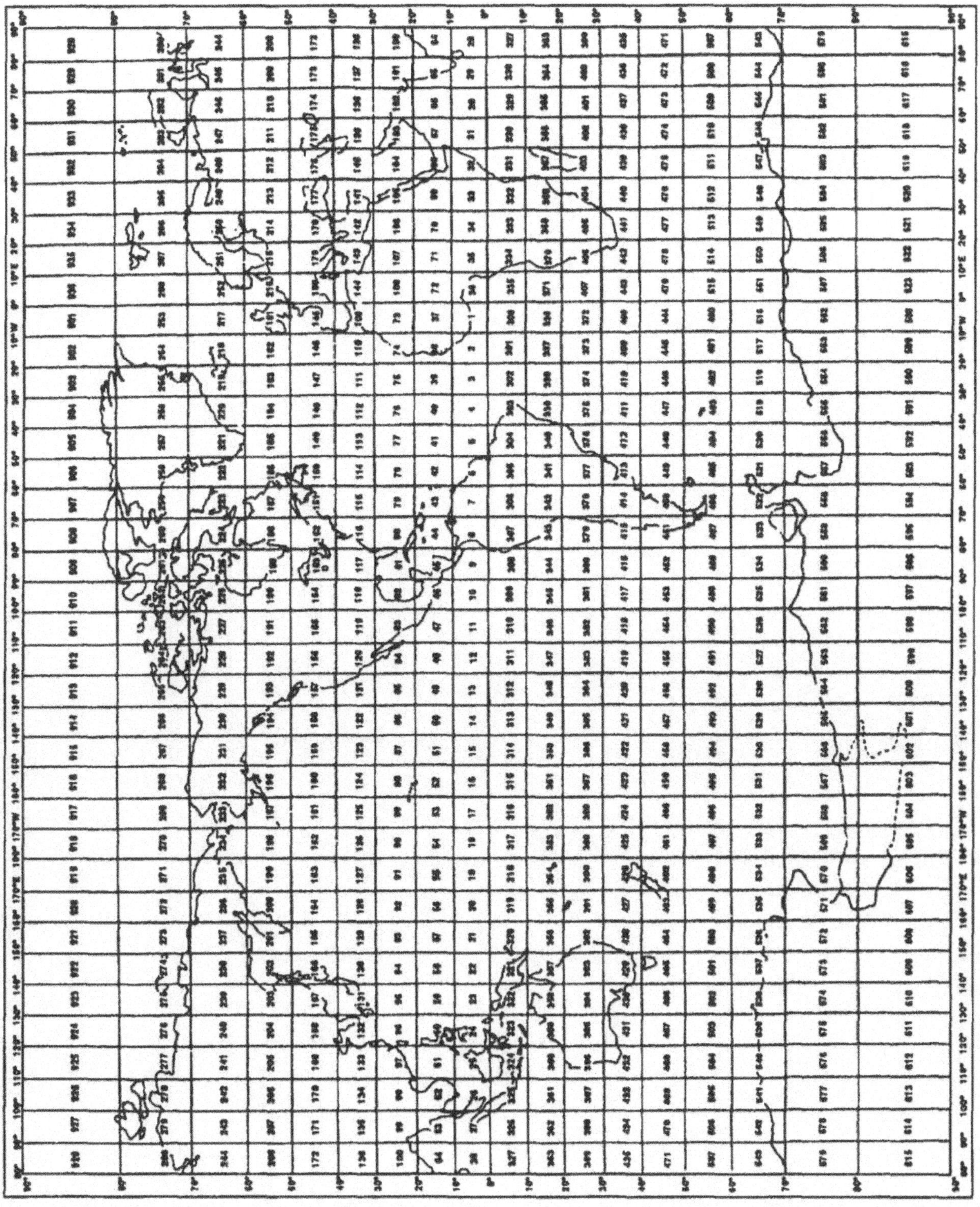

Figure AIV-6.—Marsden square number.

APPENDIX V

DESCRIPTIVE WINDS AND SEAS

WIND SPEED		SEAMAN'S TERM	WORLD METEOROLOGICAL ORGANIZATION	ESTIMATING WIND SPEED		STATE OF THE SEA		
KNOTS	KM PER HOUR			EFFECTS OBSERVED AT SEA	EFFECTS OBSERVED ON LAND	TERM	CODE	HEIGHT OF WAVES IN FEET
UNDER 1	UNDER 1	CALM	CALM	SEA LIKE MIRROR.	CALM; SMOKE RISES VERTICALLY.	CALM, GLASSY	0	0
1-3	1-5	LIGHT AIR	LIGHT AIR	RIPPLES WITH APPEARANCE OF SCALES; NO FOAM CRESTS.	SMOKE DRIFT INDICATES WIND DIRECTION; VANES DO NOT MOVE.			
4-6	6-11	LIGHT BREEZE	LIGHT BREEZE	SMALL WAVELETS; CRESTS OF GLASSY APPEARANCE, NOT BREAKING.	WIND FELT ON FACE; LEAVES RUSTLE; VANES BEGIN TO MOVE.	CALM, RIPPLED	1	0 - 1/3
7-10	12-19	GENTLE BREEZE	GENTLE BREEZE	LARGE WAVELETS; CRESTS BEGIN TO BREAK; SCATTERED WHITECAPS.	LEAVES, SMALL TWIGS IN CONSTANT MOTION; LIGHT FLAGS EXTENDED.	SMOOTH, WAVELETS	2	1/3 - 1 2/3
11-16	20-28	MODERATE BREEZE	MODERATE BREEZE	SMALL WAVES, BECOMING LONGER; NUMEROUS WHITECAPS.	DUST LEAVES, AND LOOSE PAPER RAISED UP; SMALL BRANCHES MOVE.	SLIGHT	3	2-4
17-21	29-38	FRESH BREEZE	FRESH BREEZE	MODERATE WAVES, TAKING LONGER FORM; MANY WHITECAPS; SOME SPRAY.	SMALL TREES IN LEAF BEGIN TO SWAY.	MODERATE	4	4-8
22-27	39-49	STRONG BREEZE	STRONG BREEZE	LARGER WAVES FORMING; WHITECAPS EVERYWHERE; MORE SPRAY.	LARGER BRANCHES OF TREES IN MOTION; WHISLING HEARD IN WIRES.	ROUGH	5	8-13
28-33	50-61	MODERATE GALE	NEAR GALE	SEA HEAPS UP; WHITE FOAM FROM BREAKING WAVES BEGINS TO BE BLOWN IN STREAKS.	WHOLE TREES IN MOTION; RESISTANCE FELT IN WALKING AGAINST WIND.	VERY ROUGH	6	13-20
34-40	62-74	FRESH GALE	GALE	MODERATELY HIGH WAVES OF GREATER LENGTH; EDGES OF CRESTS BEGIN TO BREAK INTO SPINDRIFT; FOAM IS BLOWN IN WELL MARKED STREAKS.	TWIGS AND SMALL BRANCHES BROKEN OFF TREES; PROGRESS GENERALLY IMPEDED.			
41-47	75-88	STRONG GALE	STRONG GALE	HIGH WAVES; SEA BEGINS TO ROLL; DENSE STREAKS OF FOAM; SPRAY MAY REDUCE VISIBILITY.	SLIGHT STRUCTURAL DAMAGE OCCURS; SLATE BLOWN FROM ROOFS.			
48-55	89-102	WHOLE GALE	STORM	VERY HIGH WAVES WITH OVERHANGING CRESTS; SEA TAKES WHITE APPEARANCE AS FOAM IS BLOWN IN VERY DENSE STREAKS; ROLLING IS HEAVY AND VISITIBILITY REDUCED.	SELDOM EXPERIENCED ON LAND; TREES BROKEN OR UPROOTED; CONSIDERABLE STRUCTURAL DAMAGE OCCURS.	HIGH	7	20-30
56-63	103-117	STORM	VIOLENT STORM	EXCEPTIONAL HIGH WAVES; SEA COVERED WITH WHITE FOAM PATCHES; VISIBILITY STILL MORE REDUCED.	VERY RARELY EXPERIENCED ON LAND; USUALLY ACCOMPANIED BY WIDESPREAD DAMAGE.	VERY HIGH	8	30-45
64+	118+	HURRICANE	HURRICANE	AIR FILLED WITH FOAM; SEA COMPLETELY WHITE DRIVING SPRAY; VISIBILITY GREATLY REDUCED.		PHENOMENAL	9	OVER 45

AGM1F263

WIND AND SEA SCALE FOR FULLY ARISEN SEA

WIND AND SEA SCALE FOR FULLY ARISEN SEA														
SEA - GENERAL		WIND				SEA								
						WAVE HEIGHT FEET								
SEA STATE	DESCRIPTION	BEAUFORT WIND FORCE	DESCRIPTION	RANGE (KNOTS)	WIND VELOCITY (KNOTS)	AVERAGE	SIGNIFICANT	AVERAGE 1/10 HIGHEST	SIGNIFICANT RANGE OF PERIODS (SECONDS)	(PERIOD OF MAXIMUM ENERGY OF SPECTRUM)	AVERAGE PERIOD	AVERAGE WAVE LENGTH	MINIMUM FETCH (NAUTICAL MILES)	MINIMUM DURATION (HOURS)
	SEA LIKE A MIRROR	0	CALM	<1	0	0	0	0	-	-	-	-	-	-
0	RIPPLES WITH THE APPERANCE OF SCALES ARE FORMED. BUT WITHOUT FOAM CRESTS.	1	LIGHT AIR	1 - 3	2	0.05	0.08	0.10	UP TO 1.2 SEC	0.7	0.5	10 IN	5	18 MIN
1	SMALL WAVELETS, STILL SHORT BUT MORE PRONOUNCED; CRESTS HAVE A GLASSY APPEARANCE, BUT DO NOT BREAK.	2	LIGHT BREEZE	4 - 6	5	0.18	0.29	0.37	0.4 - 2.8	2.0	1.4	6.7 FT	8	39 MIN
	LARGE WAVELETS, CREST BEGIN TO BREAK. FOAM OF A GLASSY APPEARANCE. PERHAPS SCATTERED WHITE HORSES.	3	GENTLE BREEZE	7 - 10	8.5	0.6	1.0	1.2	0.8-5.0	3.4	2.4	20	9.8	1.7 HRS
					10	0.88	1.4	1.8	1.0-6.0	4	2.9	27	10	2.4
2 / 3	SMALL WAVES BECOMING LARGER; FAIRLY FREQUENT WHITE HORSES.	4	MODERATE BREEZE	11 - 16	12	1.4	2.2	2.8	1.0-7.0	4.8	3.4	40	18	3.8
					13.5	1.8	2.9	3.7	1.4-7.6	5.4	3.9	52	24	4.8
					14	2.0	3.3	4.2	1.5-7.8	5.8	4.0	59	28	5.2
					16	2.9	4.6	5.8	2.0-8.8	6.5	4.6	71	40	6.6
4	MODERATE WAVES, TAKING A MORE PRONOUNCED LONG FORM; MANY WHITE HORSES ARE FORMED. (CHANCE OF SOME SPRAY).	5	FRESH BREEZE	17 - 21	18	3.8	6.1	7.8	2.5-10.0	7.2	5.1	90	55	8.3
					19	4.3	6.9	8.7	2.8-10.6	7.7	5.4	99	65	9.2
					20	5.0	8.0	10	3.0-11.1	8.1	5.7	111	75	10
5	LARGE WAVES BEGIN TO FORM; THE WHITE FOAM CRESTS ARE MORE EXTENSIVE EVERYWHERE. (PROBABLY SOME SPRAY).	6	STRONG BREEZE	22 - 27	22	6.4	10	13	3.4-12.2	8.9	6.3	134	100	12
					24	7.9	12	16	3.7-13.5	9.7	6.8	160	130	14
					24.5	8.2	13	17	3.8-13.6	9.9	7.0	164	140	15
6					26	9.6	15	20	4.0-14.5	10.5	7.4	188	180	17
	SEA HEAPS UP AND WHITE FOAM FROM BREAKING WAVES BEGINS TO BE BLOWN IN STREAKS ALONG THE DIRECTION OF THE WIND. (SPINDRIFT BEGINS TO BE SEEN).	7	NEAR GALE	28 - 33	28	11	18	23	4.5-15.5	11.3	7.9	212	230	20
					30	14	22	28	4.7-16.7	12.1	8.6	250	280	23
					30.5	14	23	29	4.8-17.0	12.4	8.7	258	290	24
					32	16	26	33	5.0-17.5	12.9	9.1	285	340	27
7	MODERATELY HIGH WAVES OF GREATER LENGTH; EDGES OF CRESTS BREAK INTO SPINDRIFT. THE FOAM IS BLOWN IN WELL MARKED STREAKS ALONG THE DIRECTION OF THE WIND. SPRAY AFFECTS VISIBILITY.	8	FRESH GALE	34 - 40	34	19	30	38	5.5-18.5	13.6	9.7	322	420	30
					36	21	35	44	5.8-19.7	14.5	10.3	363	500	34
					37	23	37	46.7	6.0-20.5	14.9	10.5	376	530	37
					38	25	40	50	6.2-20.8	15.4	10.7	392	600	38
					40	28	45	58	6.5-21.7	16.1	11.4	444	710	42
8	HIGH WAVES. DENSE STREAKS OF FOAM ALONG THE DIRECTION OF THE WIND. SEA BEGINS TO ROLL. VISIBILITY AFFECTED.	9	STRONG GALE	41 - 47	42	31	50	64	7.0-23.0	17.0	12.0	492	830	47
					44	36	58	73	7.0-24.2	17.7	12.5	534	960	52
					46	40	64	81	7.0-25.0	18.6	13.1	590	1110	57
9	VERY HIGH WAVES WITH LONG OVERHANGING CRESTS. THE RESULTING FOAM IS IN GREAT PATCHES AND IS BLOWN IN DENSE WHITE STREAKS ALONG THE DIRECTION OF THE WIND. ON THE WHOLE THE SURFACE OF THE SEA TAKES A WHITE APPEARANCE. THE ROLLING OF THE SEA BECOMES HEAVY AND SHOCK-LIKE. VISIBILITY IS AFFECTED.	10	STORM	48 - 55	48	44	71	90	7.5-26.0	19.4	13.8	650	1250	63
					50	49	78	99	7.5-27.0	20.2	14.3	700	1420	69
					51.5	52	83	106	8.0-28.2	20.8	14.7	738	1560	73
					52	54	87	110	8.0-28.5	21.0	14.8	750	1610	75
					54	59	95	121	8.0-29.5	21.8	15.4	810	1800	81
	EXCEPTIONALLY HIGH WAVES (SMALL AND MEDIUM-SIZED SHIPS MIGHT FOR A LONG TIME BE LOST TO VIEW BEHIND THE WAVES). THE SEA IS COMPLETELY COVERED WITH LONG WHITE PATCHES OF FOAM LYING ALONG THE DIRECTION OF THE WIND. EVERYWHERE THE EDGES OF THE WAVE CRESTS ARE BLOWN INTO FROTH. VISIBILITY AFFECTED.	11	VIOLENT STORM	56 - 63	56	64	103	130	8.5-31.0	22.6	16.3	910	2100	88
					59.5	73	116	148	10.0-32.0	24.0	17.0	985	2500	101
	AIR FILLED WITH FOAM AND SPRAY. SEA COMPLETELY WHITE WITH DRIVING SPRAY; VISIBILITY VERY SERIOUSLY AFFECTED.	12	HURRICANE	64+	>64	>80	>128	>164	101(35)	(26)	(18)	~	~	~

AGM1F264

APPENDIX VI

REFERENCES USED TO DEVELOP THE TRAMAN

NOTE: Although the following references were current when this TRAMAN was published, their continued currency cannot be assured. W-hen consulting these references, keep in mind that they may have been revised to reflect new technology or revised methods, practices, or procedures. You therefore need to ensure that you are studying the latest references.

Chapter 1

AWS/FM-300/005, Single Station Analysis and Forecasting Conversion Tables, Miscellaneous Charts, HQ, Air Weather Service, Scott AFB, Ill., 1983.

Cloud Types For *Observers,* Her Majesty's Meteorological Office, Publication 7 16 (Met. O. 716), Her Majesty's Stationary Office, London, 1982.

Federal Meteorological Handbook Number 1, FCM-Hl, *Surface Weather Observations and Reports,* U.S. Department of Commerce/National Oceanic and Atmospheric Administration, 1995 Revised Edition.

Huschke, Ralph E., *Glossary of Meteorology,* (NAVAIR 50-1B-504), American Meteorological Society, Boston, MA, 1959.

List, R.J., *Smithsonian Meteorological Tables,* 6th ed., NA-50-1B-521, The Smithsonian Institution, Washington, D.C., 1951.

Maneuvering Board Manual, 4th ed., DMA Pub. 217, Defense Mapping Agency Hydrographic/Topographic Center, Washington, D.C., 1984.

Manual of Barometry, NAVWEPS 50-1D-510, Bureau of Naval Weapons, Washington, D.C., 1963.

NATOPS General Flight and Operating Instructions, OPNAVINST 3710.7(Q), Office of the Chief of Naval Operations, Washington, D.C., 1995.

Nautical Almanac - 1996, U.S. Naval Observatory, Washington, D.C., 1996.

Naval Warfare Publication, *Underway Replenishment,* NWP-4-01.4 (Rev. E), Office of the Chief of Naval Operations, Washington, D.C., 1996.

Navy Occupational Safety and Health (NAVOSH) Program Manual for Forces Afloat, Vol. 1, OPNAVINST 5100.l9C, Office of the Chief of Naval Operations, Washington, D.C., 1994.

Operator's Manual, *Tactical Environmental Support System (TESS (3.0)) and Shipboard Meteorological and Oceanographic Observing System (SMOOS), Vol. III,* NAVELEXCEN VJO 14203-0302428A, NISE WEST, Vallejo, CA., 1993.

Physical Readiness Program, OPNAVINST 6110.1D, Office of the Chief of Naval Operations, Washington, D.C., 1990.

Pierson, W.J., et. al., *Practical Methods for Observing and Forecasting Ocean Waves,* H.O. Pub. 603, NAVWEPS 50-1P-540, Hydrographic Office, Washington, D.C., 1955.

Surface METAR Observations User's Manual, NAVMETOCCOMINST 3141.2, Commander, Naval Meteorology and Oceanography Command, Stennis Space Center, MS, 1996.

United States Navy Manual for Ship's Surface Weather Observations, NAVMETOCCOMINST 3144.lD, Commander, Naval Meteorology and Oceanography Command, Stennis Space Center, MS, 1996.

United States Navy Meteorological and Oceanographic Support System Manual, NAVMETOCCOMINST 3140.1 (K), Commander, Naval Meteorology and Oceanography Command, Stennis Space Center, MS, 1996.

WMO Publication No. 306, Manual on Codes, Vol. I, *International Codes* 1995 Edition, Secretariat of the World Meteorological Organization, Geneva, Switzerland, 1995.

WMO Publication No. 407, Manual on the Observation of Clouds and Other Meteors, Vol. I, *International Cloud Atlas* Revised Edition 1975, Secretariat of the World Meteorological Organization, Geneva, Switzerland, 1975.

Chapter 2

AAI Corporation, Manufacturer's Information and Equipment Specifications for ASOS, Baltimore, MD, n.p., n.d. 1991.

Automated Surface Observing System (ASOS), Ready Reference Guide, RSM1005-00038 (Rev. D), AAI Corporation, 1992.

Commander, Naval Meteorology and Oceanography Command letter 13950, Ser 5/294, of 26 July 1990, subject: Proper Interpretation of the GIFFT Ceiling Height Recorder.

Commander, Space and Naval Warfare Command letter 3140 of 6 October 1989, subject: Digital Altimeter Setting Indicator (DASI), Washington, D.C.

Federal Meteorological Handbook Number 1, FCM-H1, *Surface Weather Observations and Reports,* U.S. Department of Commerce/National Oceanic and Atmospheric Administration, 1995 Revised Edition.

GIFFT Corporation, Inc., *Technical Publication, RBC Ceilometer Recorder,* Anaheim, CA, 1984.

Handbook of Operation and Maintenance Instructions for Wind Measuring Sets AN/UMQ-5C and AN/UMQ-5D, NAVAER 50-30FR-525, Chief of the Bureau of Aeronautics, Washington, D.C., 1959.

Handbook of Operation and Service Instructions for the Wind Direction and Speed Recorder RD 108/UMQ-5, NAVAER 50-30FR-505; Chief of the Bureau of Aeronautics, Washington, D.C., 1953.

Handbook of Operation, Service, and Overhaul Instructions for the Marine Barograph, NAVAER 50-30BIC-1, Chief of the Bureau of Aeronautics, Washington, D.C., 1960.

Handbook of Operation, Service and Overhaul Instructions with Illustrated Parts Breakdown for the Automatic Weather Station (AN/GMQ-29A), NAVAIR 50-30 GMQ-29-2, Commander, Naval Air Systems Command, Washington, D.C., 1978.

Handbook, Operation and Care of Aerological Calculators, Computers and Evaluators, NAVAER 50-30FR-523, Chief of the Bureau of Aeronautics, Washington, D.C., 1959.

Handbook, Operation and Maintenance Instructions, Ceiling Light Projector ML-121 and Clinometer ML-119, NAVAER 50-30FR-521, Chief of the Bureau of Aeronautics, Washington, D.C., 1956.

Handbook, Operation and Service Instructions, with Illustrated Parts Breakdown, Cloud Height Set AN/GMQ-13(C), NAVAER 50-30GMQ13-1, Chief of the Bureau of Aeronautics, Washington, D.C., 1959.

Maintenance and Material Management (3-M) Systems Policies and Procedures, NAVMETOCCOMINST 4790.2A, Commander, Naval Meteorology and Oceanography Command, Stennis Space Center, MS, 1996.

Manual of Barometry, NAVWEPS 50-1D-510, Bureau of Naval Weapons, Washington, D.C., 1963.

Manual of Operation and Service Instructions for the Wind Measuring Set AN/PMQ-3C, NAVAIR 50-30PMQ3C-1, Commander, Naval Air Systems Command, Washington, D.C., 1968.

Mercury, Mercury Compounds, and Components Containing Mercury or Mercury Compounds; Control of, NAVSEAMST 5100.3B, Naval Sea Systems Command, Washington, D.C., 1980.

Meteorological Equipment Management and Planning Policy, NAVMETOCCOMINST 13950.1H, Commander, Naval Meteorology and Oceanography Command, Stennis Space Center, MS, 1994.

Naval Oceanography Command Barometer Calibration Program, NAVMETOCCOMINST 13950.3C, Commander, Naval Meteorology and Oceanography Command, Stennis Space Center, MS, 1986.

Operation and Maintenance Instructions, Digital Altimeter Setting Indicator (DASI), NAVELEX EM-450-AA-OMI-010-DASI, Naval Electronics Command, Charleston, SC, 1989.

Operation and Maintenance Instructions, Wind Speed and Direction Indicators ID-2446/U, ID-2447/U, and ID-2447(A)/U, NAVELEX EE250-CD-OMI-010/AIR 41734A-WSDI, Naval Electronics Command, Charleston, SC, 1989.

Operation and Maintenance, Standard Air, Maximum and Minimum Thermometers, Townsend Support, Sling and Rotor Psychrometers, and Instrument Shelters, NAVAIR 50-30FR-518, Chief of the Bureau of Aeronautics, Washington, D.C., 1956.

Operation Manual, Transmissometer AN/GMQ-32, Air Force Technical Order T.O. 31M1-2GMQ32-1 (NA-50-30GMQ1050I), Air Force Communications Command, Scott Air Force Base, IL, 1979.

Operator and Maintenance Instructions with Illustrated Parts Breakdown, OA-7900A/GMQ-10, NAVAIR 50-30GMQ10-7 preliminary ed., Naval Air Systems Command, Washington, D.C., n.d.

Operator's Manual, *Tactical Environmental Support System (TESS (3.0)) and Shipboard Meteorological and Oceanographic Observing System (SMOOS), Vol. III,* NAVELEXCEN VJO 14203-0302428A, NISE WEST, Vallejo, CA., 1993.

Ships' Maintenance and Material Management (3-M) Manual, OPNAVINST 4790.4C, Chief of Naval Operations, Washington, D.C., 1994.

Surface METAR Observations User's Manual, NAVMETOCCOMINST 3141.2, Commander, Naval Meteorology and Oceanography Command, Stennis Space Center, MS, 1996.

Technical Manual, Operation and Maintenance Instructions with Illustrated Parts Breakdown AN/GMQ-10B and AN/GMQ-10C, NAVWEPS 50-30GMQ-10-2, Chief of the Bureau of Naval Weapons, Washington, D.C., 1960.

Technical Manual, Operation, Maintenance, and Overhaul Instructions with Illustrated Parts Breakdown, (Shipboard) Elevator Clinometer ML-591/U, NAVWEPS 50-30-FR-13, Chief of the Bureau of Naval Weapons, Washington, D.C., 1965.

United States Navy Manual for Ship's Surface Weather Observations, NAVMETOCCOMINST 3144.1D, Commander, Naval Meteorology and Oceanography Command, Stennis Space Center, MS, 1996.

United States Navy Meteorological and Oceanographic Support System Manual, NAVMETOCCOMINST 3140.1 (K), Commander, Naval Meteorology and Oceanography Command, Stennis Space Center, MS, 1996.

Chapter 3

Commander, Naval Meterology and Oceanography Command letter 3140, Ser 3/040, of 7 February 1991, subject: "World Meteorological Organization (WMO) Code Changes," Stennis Space Center, MS.

Commander, Naval Meteorology and Oceanography Command letter 3140, Ser 3/544, of 10 September 1991, subject: "World Meteorological Organization (WMO) Code Changes," Stennis Space Center, MS.

Federal Meteorological Handbook Number 1, FCM-Hl, *Surface Weather Observations and Reports,* U.S. Department of Commerce/National Oceanic and Atmospheric Administration, Washington, D.C., 1995 Revised Edition.

Federal Meteorological Handbook Number 2, FCM-H2, *Surface Synoptic Codes,* U.S. Department of Commerce/National Oceanic and Atmospheric Administration, Washington, D.C., 1988.

Surface METAR Observations User's Manual, NAVMETOCCOMINST 3141.2, Commander, Naval Meteorology and Oceanography Command, Stennis Space Center, MS, 1996.

United States Navy Manual For Ship's Surface Weather Observations, NAVMETOCCOMINST 3144.1D, Commander, Naval Meteorology and Oceanography Command, Stennis Space Center, MS, 1996.

WMO Publication No. 306, Manual on Codes, Vol. I, *International Codes* 1995 Edition, Secretariat of the World Meteorological Organization, Geneva, Switzerland, 1995.

WMO Publication No. 306, Manual on Codes, Vol. II, *Regional Codes and National Coding Practices,* Secretariat of the World Meteorological Organization, Geneva, Switzerland, 1982.

Chapter 4

Bowditch, N., et. al., *The American Practical Navigator,* Vol. 1, DMA Pub. No. 9, (NVPUB9V1) Defense Mapping Agency Hydrographic/Topographic Center, Bethesda, MD, 1995.

DOD Flight Information Publication (En route), *IFR - Supplement United States,* ENRXXUSIFRSP, Defense Mapping Agency Aerospace Center, St. Louis, MO, 1996.

Facsimile Products, National Weather Service Forecasters Handbook No. 1; U.S. Department of Commerce/NOAA/NWS, Washington, D.C., 1993.

Flight Services, FM Order 7110.10L, U.S. Department of Transportation/Federal Aviation Administration, Strategic Operations/Procedures (ATO-100), Washington, D.C., 1996.

Location Identifiers, FAA Order 7350.6S, U.S. Department of Transportation/Federal Aviation Administration Office of Air Traffic System Management (ATM), Air Traffic Rules and Procedures Service (ATP), Washington, D.C., 1995.

Mapping, Charting, and Geodesy Handbook, AVDEP-HDBK-12, Commander, Naval Air Systems Command (Naval Aviation Systems Team), Washington, DC., 1995.

Naval Warfare Publication, *Operational Reports,* NWP-1-03.1, Office of the Chief of Naval Operations, Washington, D.C., 1987.

Navy Oceanographic Data Distribution System, FLENUMMETOCCENINST 3147.1, Fleet Numerical Meteorology and Oceanography Center, Monterey, CA, 1993.

Operator's Manual, *Tactical Environmental Support System (TESS (3.0)) and Shipboard Meteorological and Oceanographic Observing System (SMOOS), Vol. III,* NAVELEXCEN VJO 14203-0302428A, NISE WEST, Vallejo, CA., 1993.

Target Intelligence Specialist, Vol 2, Chap 3, AF CDC 20151, Air University Extension Course Institute, Gunter Air Force Station, AL, 1984.

The Use of the Skew T Log P Diagram in Analysis and Forecasting, NA 50-1P-5 (AWS/TR-79/006), Air Weather Service, Scott AFB, IL, 1979.

U.S. Department of Transportation, *Aeronautical Information Manual,* Federal Aviation Administration, Washington, D.C., 1995.

WMO Publication *No. 306,* Manual on Codes, Vol. I, *International Codes* 1995 Edition, Secretariat of the World Meteorological Organization, Geneva, Switzerland, 1995.

INDEX

D

E

I

K

L

Q

R

S

U

Assignment Questions

Information: The text pages that you are to study are provided at the beginning of the assignment questions.

ASSIGNMENT 1

Textbook Assignment: "Surface Observation Elements", chapter 1, pages 1-1 through 1-34.

1-1. What publication is the most comprehensive manual for U.S. Navy shipboard surface aviation weather observations?

1. NAVMETOCCOMINST 3144.1
2. NAVMETOCCOMINST 3141.2
3. FMH-1
4. FMH-2

1-2. What is the freezing point of water on the Kelvin temperature scale?

1. 273.16°K
2. 100.16°K
3. 32.00°K
4. 0.00°K

1-3. What is the standard time zone designation for the east coast of the United States?

1. -4
2. -5
3. +4
4. +5

1-4. If the Coordinated Universal Time is 1200Z, what is the local standard time in zone "W"?

1. 0200 next day
2. 0200 same day
3. 2200 next day
4. 2200 same day

1-5. What two days mark the beginning and end, respectively, of daylight savings time in the United States?

1. Last Sunday in October, and first Sunday in April
2. First Sunday in October, and last Sunday in April
3. First Sunday in April, and last Sunday in October
4. Last Sunday in April, and first Sunday in October

1-6. To prevent confusion on meteorological records, which of the following time formats should be used?

1. 1:10 PM
2. 1310 EST
3. 1310 L
4. 1310 UTC

1-7. Which of the following statements best defines the term state-of-the-sky?

1. Code numbers that equate only to types of clouds recognized by the United States
2. Code numbers that equate to intemationaly recognized sky states
3. Code numbers for the lowest broken or overcast layer of clouds
4. Code numbers for the description of the lowest cloud layer

1-8. Cumuliform clouds are generally associated with what type of air?

1. Dry and stable
2. Moist and stable
3. Dry and unstable
4. Moist and unstable

A. OROGRAPHIC B. FRONTAL C. CONVECTIVE D. TURBULENT

Figure 1-A

IN ANSWERING QUESTIONS 1-9 THROUGH 1-12, SELECT THE TYPE OF LIFT FROM FIGURE 1-A THAT MATCHES THE DESCRIPTION GIVEN. RESPONSES ARE USED ONLY ONCE.

1-9. Lift caused by air density differences between air masses along frontal zones.

1. A
2. B
3. C
4. D

1-10. Lift caused by friction between adjacent layers.

1. A
2. B
3. C
4. D

1-11. Lift caused by features on the earth's surface, such as mountains.

1. A
2. B
3. C
4. D

1-12. Lift caused by heating near the earth's surface.

1. A
2. B
3. C
4. D

1-13. In the middle latitudes, the top of the mid-etage is considered to extend to what altitude?

1. 6,500 ft
2. 10,000 ft
3. 18,500 ft
4. 23,000 ft

1-14. Of the following cloud genera or genus, which one is found in the low-etage?

1. Cirrus
2. Altocumulus
3. Cumulus
4. Cirrocumulus

1-15. What WMO classification identifies the size, shape, or form of the elements within a cloud layer?

1. Genera or genus
2. Species
3. Variety
4. Etage

A TARNSLUCIDUS B. DUPLICATUS C. UNDULATUS D. RADIATUS

Figure 1-B

IN ANSWERING QUESTIONS 1-16 THROUGH 1-19, SELECT THE CLOUD VARIETY IN FIGURE 1-B THAT MATCHES THE DESCRIPTION GIVEN. RESPONSES ARE USED ONLY ONCE.

1-16. Variety used to identify overlapping layers of cloud.

1. A
2. B
3. C
4. D

1-17. Variety used to identify a pattern known as "Abraham's Tree."

1. A
2. B
3. C
4. D

1-18. Variety that identifies a condition that will allow the outline of the sun or moon to be seen.

1. A
2. B
3. C
4. D

1-19. Variety that identifies a pattern called "wind-row" or "wave clouds."

1. A
2. B
3. C
4. D

1-20. Which of the following terms is correctly used to further identify a cirrus cloud when the cloud fibers are entangled or crossing one another?

1. Opaqus
2. Lacunosus
3. Intortus
4. Vertebratus

1-21. Of the following supplementary features, which one is another name for waterspouts and tornadoes?

1. Virga
2. Arcus
3. Incus
4. Tuba

1-22. Moderately developed cumulus clouds may produce precipitation.

1. True
2. False

1-23. Which of the following descriptions is associated with virga?

1. A dark, fuzzy area immediately below the base of a cumulus cloud
2. A dark, fuzzy area caused by precipitation that decreases in intensity below the cloud
3. A rain shower that evaporates before reaching the surface
4. Each of the above

1-24. What cloud in the genus cumulus is described as having several layers of cauliflower-life buildups?

1. Cumulus humilis
2. Cumulus mediocris
3. Cumulus congestus
4. Comulonimbus

1-25. The towering cumulus cloud is a special form of which of the following cloud species?

1. Cumulus congestus
2. Cumulus mediocris
3. Cumulonimbus capillatus
4. Comulonimbus calvus

1-26. In the mid-latitudes, cirrus blowoff from a cumulonimbus anvil top will occur most commonly at a maximum of how many feet?

1. 20,000
2. 25,000
3. 30,000
4. 45,000

1-27. Which of the following is an identifying feature of a cumulonimbus calvus?

1. A dark gray base
2. An anvil top
3. Thunder
4. Lightning

1-28. Which of the following statements is true regarding cumulonimbus mamma?

1. A strong indicator that funnel clouds are present
2. A strong indicator that conditions are favorable for severe weather
3. Consists of many rounded bulges from the top of dense cirrus blowoff
4. Consists of many rounded bulges from the base of thin cirrus blowoff

1-29. Of the following meteorological phenomena, which one is associated with an outflow boundary?

1. The funnel cloud
2. Virga
3. Mamma
4. Low-level wind shear

1-30. Roll clouds indicate that thunderstrom down-rush has occurred and that LLWS may be present.

1. True
2. False

1-31. A slowing rotating wall cloud indicates the possible development of what condition?

1. Funnel cloud
2. Cold air funnel
3. Dust cloud
4. Microburst

1-32. Ships should always avoid contact with waterspouts.

1. True
2. False

1-33. In relation to the cumulonimbus cloud, where can hail be found?

1. The top and sides of building and mature CB cells
2. Up to 25 miles from the CB cell
3. Under the cloud base
4. All of the above locations

1-34. What cloud genera is sometimes mistaken for altocumulus?

1. Cumulus
2. Stratocumulus
3. Cirrostratus
4. Stratus

1-35. What is the most common type of precipitation produced by stratocumulus clouds?

1. Snow flurries
2. Light rain showers or drizzle
3. Heavy rain showers
4. Sleet

1-36. Which of the following is a feature of a stratus cloud?

1. Regular dark patches on the cloud base
2. A very smooth and uniform base
3. Easily recognized cloud cells
4. Virga

1-37. What type of cloud layer may display the outline of the sun as blurred or fuzzy?

1. Stratus
2. Stratus opaqus
3. Altostratus
4. Altocumulus

1-38. Of the following conditions, which one may produce a corona?

1. Ice crystals in a low-etage cloud
2. Ice crystals in a mid-etage cloud
3. Liquid water droplets in a low-etage cloud
4. Liquid water droplets in a mid-etage cloud

1-39. If a halo is seen in a stratiform cloud, which of the following cloud genera is most likely present?

1. Stratus
2. Altostratus
3. Cirrostratus
4. Each of the above

1-40. Which of the following indicators may be used to reclassify altostratus clouds as nimbostratus?

1. The cloud base lowers to less than 6,500 ft
2. Stratus fractus clouds form under an altostratus base
3. Precipitation begins
4. All of the above

A. ALTOCUMULUS STRATIFORMUS
B. ALTOCUMULUS FLOCCUS
C. ALTOCUMULUS CASTELLANUS
D. ALTOCUMULUS LENTICULARIS

Figures 1-C

IN ANSWERING QUESTIONS 1-41 THROUGH 1-44, REFER TO FIGURE 1-C. SELECT THE ALTOCUMULUS SPECIES THAT MATCHES THE DESCRIPTION GIVEN. RESPONSES ARE USED ONLY ONCE.

1-41. The species that is an orographic cloud form.

1. A
2. B
3. C
4. D

1-42. The species that resembles small ragged, cumulus humilis clouds.

1. A
2. B
3. C
4. D

1-43. The species with towers or turrets extending upward from the cloud base.

1. A
2. B
3. C
4. D

1-44. The most comon species of altocumulus cloud.

1. A
2. B
3. C
4. D

1-45. The dense blowoff from the top of a cumulonimbus, often referred to as dense cirrus, is what cloud species?

1. Cirrus spissatus
2. Cirrus floccus
3. Cirrus uncinus
4. Cirrus castellanus

1-46. What cloud type appears as a thin veil over the sky without any distinguishable features?

1. Cirrostratus fibratus
2. Cirrostratus nebulosus
3. Altostratus
4. Altocumulus

1-47. Which of the following is an important indicator that a cumuliform cloud belongs to the genera cirrocumulus?

1. The cloud layerse are arranged in loosely packed rows
2. Each element is smaller than 1° of the sky
3. The small cells cover a large portion of the sky
4. The cloud elements are unlike high altocumulus clouds

1-48. Which of the following statements about orographic clouds is NOT true?

1. Orographic clouds are generally stationary
2. Orographic clouds are never associated with dangerous turbulence
3. Orographic clouds form when strong winds blow across mountain ranges
4. There are three significant orographic cloud forms

A. LENTICULARIS
B. FOEHNWALL
C. ROTOR
D. MOUNTAIN WAVE

Figure 1-D

IN ANSWERING QUESTIONS 1-49 THROUGH 1-52, REFER TO FIGURE 1-D. SELECT THE TERM THAT MATCHES THE DESCRIPTION GIVEN. RESPONSES ARE USED ONLY ONCE.

1-49. Not a cloud type, but a name for the condition that produces certain orographic clouds.

1. A
2. B
3. C
4. D

1-50. An orographic cloud that is cap-shaped and forms downwind from a mountain range.

1. A
2. B
3. C
4. D

1-51. An orographic cloud that is cap-shaped and forms on a mountain top.

1. A
2. B
3. C
4. D

1-52. An orographic cloud that is cat-eye shaped, with a windswept appearance.

1. A
2. B
3. C
4. D

1-53. Of the following descriptions, which one best describes cloud layer coverage?

1. An estimate of the total amount of clouds present in the sky
2. An estimate of the total amount of clouds at generally the same level in the sky
3. An estimate of the total amount of a particular cloud genus
4. An estimate of the total amount of a particular cloud species

1-54. When observing layer coverage, a partial obscuration hiding less than 1/8 of the sky may be ignored.

1. True
2. False

1-55. Of the following statements, which one is correct when you evaluate an area or patch of cloud for layer coverage?

1. Layer coverage includes opaque portions of the cloud, transparent portions of the cloud, and small areas of blue sky between individual cloud cells
2. Layer coverage includes opaque portions of the cloud and transparent portions of the cloud only
3. Layer coverage includes only opaque portions of a cloud

1-56. Which of the following evaluations is a concept used to determine cloud ceiling?

1. Total sky cover
2. Layer coverage
3. Summation sky coverage
4. Each of the above

1-57. When 4/8 of the sky is covered by clouds in a single layer, what is the layer classified?

1. Few
2. Scattered
3. Broken
4. Overcast

Three layers of clouds are present in the sky as follows:

Lower layer:	Total 2/8
Middle layer:	Total 3/8
Higher layer:	Total 3/8

Figure 1-E

IN ANSWERING QUESTIONS 1-58 AND 1-59, REFER TO FIGURE 1-E.

1-58. What is the correct summation coverage for the higher layer?

1. Few
2. Scattered
3. Broken
4. Overcast

1-59. Which of the following layers, if any, constitutes a ceiling?

1. The low layer
2. The middle layer
3. The higher layer
4. None of the above

1-60. If the base of a layer of clouds is measured at 9,750 feet above the ground, what is the correct reportable cloud height?

1. 10,000 ft
2. 9,950 ft
3. 9,900 ft
4. 9,500 ft

1-61. The greatest distance that objects may be seen and identified throughout half or more of the horizon circle is what type of visibility?

1. Sector
2. Variable
3. Differing level
4. Prevailing

1-62. Which of the following terms is most closely associated with how far a pilot can see at the moment the aircraft lands?

1. Sector visibility
2. Prevailing visibility
3. Tower visibility
4. Runway visual range

A. HAZE
B. SMOKE
C. DUST
D. SAND

Figure 1-F

IN ANSWERING QUESTIONS 1-63 THROUGH 1-66, REFER TO FIGURE 1-F. SELECT THE LITHOMETEOR THAT MATCHES THE DESCRIPTION GIVEN. RESPONSES ARE USED ONLY ONCE.

1-63. Gives distance objects or the sky a tan hue, even when the relative humidity is low.

1. A
2. B
3. C
4. D

1-64. Gives distant objects a blue hue when viewed against dark backgrounds, or gives the sky a yellow tinge.

1. A
2. B
3. C
4. D

1-65. Hazardous to aircraft and is normally only present in the atmosphere when winds are 21 knots or higher and conditions are dry.

1. A
2. B
3. C
4. D

1-66. When dispersed in the atmosphere, gives the air a red tinge, especially at sunrise and sunset.

1. A
2. B
3. C
4. D

A. FOG
B. MIST
C. GROUND FOG
D. SHALLOW FOG

Figure 1-G

IN ANSWERING QUESTIONS 1-67 THROUGH 1-70, SELECT THE HYDROMETEOR IN FIGURE 1-G THAT MATCHES THE DESCRIPTION GIVEN. RESPONSES ARE USED ONLY ONCE.

1-67. Used to identify suspended liquid water droplets that reduce prevailing visibility from 1,000 meters to 9,000 meters.

1. A
2. B
3. C
4. D

1-68. Reduces visibility only within 20 feet of the ground to less than 5/8 mile.

1. A
2. B
3. C
4. D

1-69. Deeper than 20 feet and reduces prevailing visibility to less than 5/8 mile.

1. A
2. B
3. C
4. D

1-70. By definition does NOT restrict prevailing visibility.

1. A
2. B
3. C
4. D

1-71. The fact that precipitation is either liquid, freezing, or frozen is related to what precipitation classification?

1. Character
2. Type
3. Form
4. Intensity

1-72. Which precipitation type is identified by white, opaque, rounded or conical kernels of frozen water that occur(s) as showers?

1. Snow
2. Snow pellets
3. Snow grains
4. Ice pellets

1-73. Which of the following precipitation types is/are produced exclusively in thunderstorms?

1. Hail
2. Ice crystals
3. Snow grains
4. Ice pellets

1-74. Which of the following indicators is/are used as a guide for precipitation intensity?

1. Rate of accumulation
2. Visibility
3. Spray appearance over hard surfaces
4. All of the above

1-75. What are rences in precipitation character between showery precipitation and intermittent precipitation?

1. Showery precipitation falls from cumuliform clouds and changes intensity more abruptly than intermittent precipitation
2. Showery precipitation falls from stratiform clouds and changes intensity more abruptly than intermittent precipitation
3. Shower precipitation falls from cumuliform clouds and changes intensity more gradually than intermittent precipitation
4. Showery precipitation falls from stratiform clouds and changes intensity more gradually than intermittent precipitation

ASSIGNMENT 2

Textbook Assignment: "Surface Observation Elements" (continued), chapter 1, pages 1-34 through 1-65. "Surface Observation Equipment", chapter 2, pages 2-1 through 2-9.

2-1. After ice crystals or rain droplets form, what process is responsible for their growth in size?

1. Accretion
2. Sublimation
3. Condensation
4. Evaporation

2-2. What is the standard abbreviation for lightning that discharges from cloud to ground?

1. LTGCA
2. LTGCC
3. LTGCG
4. LTGIC

2-3. Which, if any, of the following photometeors is routinely reported in surface weather observations?

1. Corona
2. Rainbow
3. Mirage
4. None of the above

A.	SEA-LEVEL PRESSURE
B.	BAROMETRIC PRESSURE
C.	STATION PRESSURE
D.	ALTIMETER SETTING

Figure 2-A

IN ANSWERING QUESTIONS 2-4 THROUGH 2-7, SELECT THE TERM IN FIGURE 2-A THAT MATCHES THE DESCRIPTION GIVEN. RESPONSES ARE USED ONLY ONCE.

2-4. Pressure read directly from a barometer.

1. A
2. B
3. C
4. D

2-5. Identified with the Q-signal QFE.

1. A
2. B
3. C
4. D

2-6. Indicates a pressure value calculated by applying a constant additive correction factor.

1. A
2. B
3. C
4. D

2-7. Used by aircraft to allow correct determinations of height above mean sea level.

1. A
2. B
3. C
4. D

2-8. Which of the following pressure values is routinely used by aircraft flying at altitudes above 18,000 feet?

1. Standard pressure
2. Altimeter setting
3. Sea-level pressure
4. Pressure tendency

2-9. Which of the following time periods is NOT routinely used for pressure tendencies?

1. 3 hr
2. 6 hr
3. 12 hr
4. 24 hr

A.	DRY-BULB
B.	WET-BULB
C.	DEW-POINT
D.	FROST-POINT

Figure 2-B

IN ANSWERING QUESTIONS 2-10 THROUGH 2-13, SELECT THE TYPE OF TEMPERATURE FROM FIGURE 2-B THAT MATCHES THE DESCRIPTION GIVEN. RESPONSES ARE USED ONLY ONCE.

2-10. Ambient air temperature air temperature as measured by a thermometer or an automatic sensor.

1. A
2. B
3. C
4. D

2-11. An approximation that represents the saturation point of air when the temperature is below freezing.

1. A
2. B
3. C
4. D

2-12. Must be calculated, and represents the temperature to which air must be cooled to reach saturation.

1. A
2. B
3. C
4. D

2-13 The lowest temperature that may be reached due to evaporative cooling.

1. A
2. B
3. C
4. D

2-14 Which of the following statements is true concerning the use of a ship's seawater injection temperature as a sea-surface temperature reading?

1. It is the least accurate sea surface temp reading
2. It is the most accurate sea-surface temperature reading
3. No corrections are needed to use injection temperature as sea-surface temperature
4. In the middle latitudes, sea-surface temperature and injection temperature are nearly identical

2-15. Wind direction is normally reported and forecast to what degree?

1. Nearest degree
2. Nearest 5 degrees
3. Nearest 10 degrees
4. Nearest 15 degrees

2-16. Of the following directions, which one should be reported when the winds are blowing at 10 knots directly toward the south?

1. 000°
2. 180°
3. 360°
4. 270°

2-17. On a ship, which of the following relative bearing directions represents dead astern?

1. 0°
2. 90°
3. 180°
4. 270°

2-18. To convert magnetic wind direction to true wind direction, the appropriate magnetic declination must be added or subtracted.

1. True
2. False

2-19. What is the standard unit of measure for wind speeds used by U.S. military forces?

1. Meters per second
2. Kilometers per hour
3. Statute miles per hour
4. Knots

2-20. One knot equals how many kilometers per hour?

1. 1.852
2. 1.1507
3. 0.5144
4. 0.5394

2-21. Which of the following phenomena is NOT considered wind character?

1. Mean wind speed
2. Variable wind direction
3. Wind speed gusts
4. Peak wind gust

A. VARIABLE WIND
B. GUST
C. SQUALL
D. PEAK WIND SPEED

Figure 2-C

IN ANSWERING QUESTIONS 2-22 THROUGH 2-25, SELECT THE TERM IN FIGURE 2-C THAT MATCHES THE DESCRIPTION GIVEN. RESPONSES ARE USED ONLY ONCE.

2-22. Rapid fluctuation in wind speed with a variation between peaks and lulls of 10 knots or more.

1. A
2. B
3. C
4. D

2-23. A sudden increase in wind speed by 16 knots or more, and remains 22 knots or greater for at least 1 minute.

1. A
2. B
3. C
4. D

2-24. The highest observed instantaneous wind speed.

1. A
2. B
3. C
4. D

2-25. Wind direction fluctuating by 60° or more.

1. A
2. B
3. C
4. D

2-26. What phenomenon is described as any change in wind direction by 45° or more during a 15-minute time period?

1. Gust
2. Squall
3. Wind shift
4. Variable wind

2-27. Which of the following wind directions is computed for Foxtrot Corpin?

1. Relative
2. True
3. Magnetic
4. Variable

2-28. Which of the following is NOT a factor affecting wave height?

1. Wave length
2. Fetch length
3. Wind speed
4. Wind duration

A. HEIGHT
B. DIRECTION
C. PERIOD
D. LENGTH

Figure 2-D

IN ANSWERING QUESTIONS 2-29 THROUGH 2-32, REFER TO FIGURE 2-D. SELECT THE WAVE PARAMETER THAT MATCHES THE DESCRIPTION GIVEN. RESPONSES ARE USED ONLY ONCE.

2-29. A measurement routinely underestimated from larger ships.

1. A
2. B
3. C
4. D

2-30. A measurement that is NOT made during a wave observation.

1. A
2. B
3. C
4. D

2-31. Measured time from wave crest to wave crest, or from trough to next trough.

1. A
2. B
3. C
4. D

2-32. Best measured by using the gyroscope repeater on a pelorus column.

1. A
2. B
3. C
4. D

2-33. When observing significant wave height, which of the following observations is most correct?

1. The average height of 100 waves
2. The average height of 50 waves
3. The average height of only the highest 50 waves
4. The average height of only the highest 1/3 of the waves

2-34. Which of the following statements is true about swell waves?

1. They take on a sine-wave pattern
2. They are waves that have moved out and away from the area in which they were formed
3. They have distinct wave periods
4. Each of the above

2-35. Which of the following parameters are evaluated for Romeo Corpin?

1. Seas and winds
2. Seas and visibility
3. Winds and visibility
4. Winds and precipitation

2-36. Which of the following statements is true regarding the danger of shipboard ice accretion?

1. It is a safety hazard to personnel
2. It may cause breakage of wires and antenna
3. It may cause the ship to roll excessively or capsize
4. Each of the above

2-37. What is the average freezing point of seawater?

1. +4.0°C
2. 0.0°C
3. -1.9°C
4. -4.0°C

2-38. What is the term used to describe a sea ice floe 5 km across?

1. Giant floe
2. Vast floe
3. Big floe
4. Medium floe

2-39. What is the term used to identify a piece of sea ice 1.5 m across?

1. Small floe
2. Ice cake
3. Small ice cake
4. Pancake ice

A. FRACTURE
B. LEAD
C. THAW HOLE
D. POLYNA

Figure 2-E

IN ANSWERING QUESTIONS 2-40 THROUGH 2-43, SELECT THE TERM IN FIGURE 2-E THAT MATCHES THE DESCRIPTION GIVEN. RESPONSES ARE USED ONLY ONCE.

2-40. Any sizable area of seawater enclosed by ice.

1. A
2. B
3. C
4. D

2-41. A break or crack in the ice sheet.

1. A
2. B
3. C
4. D

2-42. A long unfrozen or refrozen break or crack in an ice sheet that may be navigated by a ship.

1. A
2. B
3. C
4. D

2-43. A melt hole through the ice.

1. A
2. B
3. C
4. D

2-44. Of the following methods, which one correctly identifies how sea ice topographic features are formed?

1. The accumulation and erosion of snowfall
2. The pressure when ice sheets are compressed
3. The stress when ice sheets are separated
4. They are largely man-made

2-45. Which of the following statements is true about the drift of pack ice?

1. The seasons pay a role
2. It drifts to the right of true wind in the Northern Hemisphere
3. The actual drift is about 30° from wind direction
4. Each of the above

2-46. In the Northern Hemisphere, the speed of drifting pack ice ranges from 1.4 to 2.4 percent of wind speed.

1. True
2. False

2-47. Where do most icebergs originate in the northern hemisphere?

1. Greenland
2. Norway
3. Kamchatka
4. Canada

2-48. An iceberg about the size of a truck is called

1. an iceberg
2. a bergy bit
3. a growler
4. a medium floe

2-49. Icebergs move more by the influence of the subsurface water currents than by the influence of the winds.

1. True
2. False

2-50. Of the following statements, which one is true concerning relative humidity?

1. It is the ratio of water vapor in the air at different pressure levels
2. It is the ratio of water vapor the air holds to the amount it can possibly hold
3. It is a comparison between the actual air temperature and a standard temperature where most people feel comfortable
4. It is an indicator for impending precipitation

IN ANSWERING QUESTION 2-51, REFER TO FIGURE 1-38 IN THE TRAMAN.

2-51. What is the approximate apparent temperature if the air temperature is 90°F and the relative humidity is 80 percent?

1. 115°F
2. 105°F
3. 99°F
4. 89°F

2-52. As a General Heat Stress Index measurement, what does an apparent temperature of 101°F mean?

1. Fatigue is possible with prolonged physical activity
2. Heat exhaustion or heatstroke is possible with prolonged physical activity
3. Heat exhaustion or heatstroke is likely with prolonged physical activity
4. Heatstroke imminent with physical activity

2-53. Which of the following factors does the wet-bulb globe temperature (WBGT) index consider that is not considered by the general heat stress index (GHSI)?

1. Temperature
2. Humidity
3. Radiant energy
4. Vapor pressure

2-54. Computation of wind chill equivalent temperature includes an adjustment for the presence of sunshine.

1. True
2. False

IN ANSWERING QUESTION 2-55, REFER TO FIGURE 1-39 IN THE TRAMAN.

2-55. If the air temperature is 20°F and the wind speed is 15 knots, what is the approximate wind chill temperature?

1. 8°F
2. 2°F
3. -5°F
4. -50°F

IN ANSWERING QUESTION 2-56, REFER TO FIGURE 1-40 IN THE TRAMAN.

2-56. If the temperature is 15°C, what is the approximate safe time for a person in the water?

1. .9 hr
2. 1.7 hr
3. 2.0 hr
4. 3.0 hr

2-57. The water temperature is 40°F and the wind chill index on the surface is 25°F. Which of the following protective clothing should be worn by aircrew personnel in these conditions?

1. Thermal undergarments only
2. Anti-immersion suits only
3. Thermal undergarments and anti-immersion suits

2-58. The water temperature is 55°F in the target area. The decision whether or not the aircrew should wear anti-immersion suits is based upon what factor?

1. Number of mission personnel
2. Type of mission aircraft
3. Rescue response time
4. Time of day

2-59. If the pressure altitude at sea level is +1,500 feet, which of the following statements is correct?

1. The air is less dense than standard
2. The air is near standard density
3. The air is more dense than standard

2-60. If the pressure altitude indicates the air density is more dense than standard, which of the following statements is correct?

1. An aircraft may takeoff with a greater load than normal
2. An aircraft may only takeoff with a normal load
3. An aircraft may only takeoff with a lighter-than-normal load

IN ANSWERING QUESTION 2-61, REFER TO TABLE 1-5 IN THE TRAMAN.

2-61. When the altimeter setting is 30.02 inches and the station elevation is 500 feet, what is the pressure altitude?

1. -591 ft
2. -91 ft
3. 409 ft
4. 500 ft

2-62. Density altitude is affected the most by changes in what factors?

1. Pressure and humidity
2. Humidity and altitude
3. Pressure and temperature
4. Humidity and temperature

2-63. Given an actual temperature of 25°C and a pressure altitude of zero, what is the virtual temperature (V_t)?

1. +10°C
2. -10°C
3. +34°F
4. -34°F

2-64. The mass of water vapor present in a unit mass of air is known as what kind of humidity?

1. Relative
2. Specific
3. General
4. Total

2-65. In the ASOS, how are individual sensors linked to the data collection package?

1. Fiber optics
2. Hard-wired
3. Telephone modem
4. Microwave

2-66. How is the ASOS ACU connected to maintenance personnel, the weather office, or other users?

1. Fiber optics
2. Hard-wired
3. Telephone modem
4. Microwave

2-67. What is the maximum cloud height measured by ASOS?

1. 5,000 ft
2. 8,500 ft
3. 10,000 ft
4. 12,000 ft

2-68. Which of the following data is NOT measured by SMOOS?

1. Cloud coverage
2. Cloud height
3. Visibility
4. Pressure

2-69. With visibility greater than 3 miles, the SMOOS cloud height detector can measure a maximum of how many cloud-base levels below 12,000 feet?

1. One
2. Two
3. Three
4. Four

2-70. What type of visibility is measured by the SMOOS?

1. Prevailing
2. Runway visual range
3. Tower
4. Equivalent

2-71. Of the following display methods, which one is used by the AN/GMQ-29 system to display most of the meteorological data?

1. Indicator dials
2. LED readouts
3. LCD readouts
4. Digital recorders

2-72. With each tip of the ML-588/GMQ-14 tipping bucket rain gauge, a signal is sent to the GMQ-29 display controller. What other instrument receives the signal?

1. ML-642 pressure sensor
2. RD-108/UMQ-5 recorder
3. ML-643 dew point sensor
4. GMQ-29 digital voltmeter

2-73. Which of the following equipment is NOT normally stored within a ML-41 instrument shelter?

1. An electric psychrometer
2. A rotor psychrometer
3. A maximum and minimum thermometer
4. A rain gauge

ASSIGNMENT 3

Textbook Assignment: "Surface Observation Equipment" (continued), chapter 2, pages 2-9 through 2-39. "Surface Observation Codes", chapter page 3-1 through 3-18.

3-1. Liquid-in-glass thermometers for meteorological applications should be read to what increment?

1. Whole degree
2. Half degree
3. Quarter degree
4. Tenth degree

3-2. You are using a wet-bulb thermometer in below freezing temperatures. After wetting the wet-bulb wick, how many minutes must you wait before reading the value?

1. 1
2. 5
3. 10
4. 15

3-3. What is the range of temperatures, from lowest to highest, that can be measured by an alcohol thermometer?

1. -115°C to +78.5°C
2. -115°F to +78.5°F
3. -20°C to +120.0°C
4. -20°F to +120.0°F

3-4. Of the following statements, which one correctly identifies why you should avoid using mercury thermometers?

1. Mercury thermometers are expensive
2. Alcohol thermometers are better suited for meteorological and oceanographic measurements
3. Mercury is prone to separation
4. Mercury is a hazardous material

3-5. When the temperature is above 85°F, how many minutes must an electric psychrometer be allowed to acclimatize prior to use?

1. 1 to 2 minutes
2. 5 to 10 minutes
3. 10 to 15 minutes
4. 15 to 20 minutes

3-6. You are using an electric psychrometer at night in an unlighted area. When, if ever, should you turn on the scale lights?

1. When turning on the fan motor
2. Only when actually reading the temperatures
3. Whenever light is needed in the observation area
4. Never; scale lights are provided for calibration only

3-7. What is the pressure element in an ML-448/UM precision aneroid barometer?

1. A Sylphon cell
2. A series of gears
3. An electronic circuit
4. A bellows

3-8. A precision aneroid barometer should be read to the nearest

1. 0.5 in.
2. 0.05 in.
3. 0.005 in.
4. 0.0005 in.

3-9. The primary use of the marine barograph is to determine

1. 3-hour pressure tendencies
2. sea-level pressure
3. altimeter setting
4. station pressure

3-10. How often should recording charts be replaced on the barograph?

1. Once a quarter
2. Once a week
3. Every 4 days
4. Every day

3-11. What type of purple ink should you use to reink a barograph pen?

1. Fountain-pen ink
2. Recording ink
3. Stamp-pad ink
4. Ball-point-pen ink

3-12. What is the maximum number of indicators that may be connected to each AN/UMQ-5 wind measuring system?

1. One
2. Two
3. Five
4. Six

3-13. Which of the following indicators/recorders is used primarily by air traffic control personnel?

1. ID-2447/U
2. RD-108/UMQ-5
3. ID-300/UMQ-5
4. ID-586/UMQ-5

3-14. While operating at the standard speed, the RD-108/UMQ-5 recorder can run for approximately how many days before the chart must be changed?

1. 30
2. 15
3. 10
4. 7

3-15. What kind of wind is measured by the Type B-3 wind indicator?

1. True
2. Total
3. Relative
4. Magnetic

3-16. If the apparent winds are read on the wind indicator as 120° at 27 knots, and the ship is moving toward True North, which of the following computed true winds is most likely correct?

1. 140° at 33 knots
2. 090° at 23 knots
3. 230° at 40 knots
4. 330° at 25 knots

3-17. What is the purpose for depressing the trigger on the AN/PMQ-3 hand-held anemometer?

1. To check calibration
2. To release the wind vane
3. To switch to the high scale
4. To switch to the low scale

3-18. How often should the wind-speed transmitter unit of the AN/PMQ-3 be oiled?

1. After each use
2. Once per month
3. Twice per quarter
4. Twice per year

3-19. The ML-217 plastic rain gauge is used only to measure liquid precipitation.

1. True
2. False

3-20. Runway visual range may be changed by turning up the intensity of the runway lighting.

1. True
2. False

A. R-437/GMQ-10
B. ML-461/GMQ-10
C. OA-7900/GMQ-10
D. ID-820/GMQ-10

Figure 3-A

IN ANSWERING QUESTIONS 3-21 THROUGH 3-23, REFER TO FIGURE 3-A. SELECT THE GMQ-10 COMPONENT THAT MATCHES THE DESCRIPTION GIVEN. NOT ALL RESPONSES ARE USED.

3-21. Provides a LED digital display of runway visual range in hundreds of feet.

1. A
2. B
3. C
4. D

3-22. The transmissometer indicator/recorder unit.

1. A
2. B
3. C
4. D

3-23. The transmissometer projector.

1. A
2. B
3. C
4. D

A. DAILY
B. WEEKLY
C. EVERY 2 WEEKS
D. MONTHLY

Figure 3-B

IN ANSWERING QUESTIONS 3-24 THROUGH 3-26, REFER TO FIGURE 3-B. NOT ALL RESPONSES ARE USED.

3-24. If the ID-353B/GMQ-10 recorder is operated continuously with the AN/GMQ-32 system, how often must the recorder chart be replaced?

1. A
2. B
3. C
4. D

3-25. How often must the ID-820/GMQ-10 recorder chart drive spring be rewound?

1. A
2. B
3. C
4. D

3-26. How often must the ID-820/GMQ-10 zero adjustment be readjusted?

1. A
2. B
3. C
4. D

3-27. Which of the three mode settings on the OA-7900A/GMQ-10 indicator provides runway visual range?

1. NORMAL only
2. NORMAL or DAY
3. NORMAL or NIGHT
4. DAY or NIGHT

3-28. When using the RO-546 recorder, you must apply the equipment correction factor to what reading(s)?

1. Recorded elevation angles for solid cloud bases under 45° elevation
2. Recorded elevation angles for solid cloud bases over 45° elevation
3. Recorded elevation angles for fog and stratus under 25° elevation
4. All recorded elevation angles regardless of cloud type or cloud height

3-29. In which of the following situations should a new correction factor be determined for the RO-546 recorder?

1. After replacing a projector bulb
2. After refocusing the projector lights
3. After cleaning the detector lens
4. After each of the above

3-30. What maintenance responsibilities, if any, do Aerographers have for the RO-546 recorder?

1. Replacing fuses and adjusting the power supply
2. Changing the chart paper and chart-marking stylus
3. Sending the unit out for calibration
4. None

3-31. What is the maximum range of the ML- 12 1 ceiling light projector?

1. 5,000 ft
2. 10,000 ft
3. 12,500 ft
4. 15,000 ft

3-32. What value does a clinometer actually measure?

1. Cloud elevation angle
2. Baseline distance
3. Cloud height
4. Height above waterline

3-33. Ceiling balloons provide reasonable measurements of cloud heights up to what maximum height?

1. 2,000 ft
2. 7,500 ft
3. 12,000 ft
4. 25,000 ft

3-34. After a 30-gram balloon is inflated, how much free lift, in grams, should the balloon have?

1. 75
2. 89
3. 126
4. 139

A. CP-718/UM DENSITY ALTITUDE COMPUTER
B. CP-165/UM PSYCHROMETRIC COMPUTER
C. CP-264/U TRUE WIND COMPUTER
D. CP-402/UM PRESSURE REDUCTION COMPUTER

Figure 3-C

IN ANSWERING QUESTIONS 3-35 THROUGH 3-38, REFER TO FIGURE 3-C. MATCH EACH HAND-HELD COMPUTER TO THE DESCRIPTION GIVEN. RESPONSES ARE USED ONLY ONCE.

3-35. Used to compute dew-point temperature or relative humidity.

1. A
2. B
3. C
4. D

3-36. Used to compute up to six different values including specific humidity.

1. A
2. B
3. C
4. D

3-37. Used to compute pressure altitude and altimeter setting.

1. A
2. B
3. C
4. D

3-38. Used to calculate a desired course and speed of a ship to recover or launch aircraft, but has other applications as well.

1. A
2. B
3. C
4. D

3-39. What should you do when you first become aware of an equipment outage?

1. Ignore it
2. Notify your supervisor
3. Call maintenance personnel
4. Unplug the power

3-40. While at sea, an equipment failure occurs that is beyond the capability of onboard technicians to repair. What maintenance resource should be used during the next in-port period?

1. 3-M
2. FTRs
3. METEM personnel
4. Naval supply system

3-41. Which of the following maintenance sources provides support for all shipboard equipment?

1. 3-M
2. FTRs
3. METEM personnel
4. Contractors

3-42. Asia is located in what WMO region?

1. I
2. II
3. III
4. IV

3-43. What type of code form is used only within a specific country?

1. National
2. Regional
3. International
4. Special

3-44. Which of the following forms is used to record shipboard weather observations?

1. FMH Form-10
2. CNMOC Form 3144.1
3. CNMOC Form 3140/12
4. CNMOC Form 3141/3

3-45. If you discover an error in an observation after it has just been transmitted, what, if anything, should you do?

1. Erase the incorrect data on the observation record and enter the correct data
2. Line out the incorrect data on the observation record and enter the correct data in red only
3. Line out the incorrect data, enter the correct data in red, transmit a correction, and enter the transmission time of the correction
4. Nothing

3-46. At the end of each month, observation forms must be forwarded to which of the following commands?

1. FNMOC, Monterey, CA
2. NLMOD, Pensacola, FL
3. FNMOD, Asheville, NC
4. CNMOC, Bay St. Louis, MS

3-47. If dangerous weather such as a tornado is seen, what type of observation should be transmitted?

1. METAR
2. SNYPOTIC
3. SPECI
4. RECORD SPECIAL

3-48. Which of the following observations should contain 3-hourly additive data?

1. 1300Z
2. 1500Z
3. 1800Z
4. 1750Z

(actual observation) METAR MXXX 221100Z 21017G34KT 2SM TSRA BKNO15CB OVC040 18/17 A2991 RMK FRQ LTGCC W TS W MOV NE SLP133; (symbolic format for this case) METAR CCC YYGGggZ dddffGf_mf_mKT VVVVVSM w'w' $N_sN_sN_sh_sh_sh_s$ $N_sN_sN_sh_sh_sh_s$ T'T'/T'$_d$T'$_d$ A$P_HP_HP_HP_H$ (supplemental);
(actual observation) METAR MYYY 221056Z 07005KT 1/2SM R03L/2400FT FG VV005 15/15 A2992 RMK SLP132; (symbolic format for this case) METAR CCCC YYGGggZ dddffKT VVVVVSM RD_RD_R/$V_RV_RV_RV_R$ft w'w' VV$h_sh_sh_s$ T'T'/T'$_d$T'$_d$ A$P_HP_HP_HP_H$ (supplemental);

Figure 3-D

IN ANSWERING QUESTIONS 3-49 THROUGH 3-54, REFER TO THE METAR CODED REPORTS FOR STATIONS MXXX AND MYYY IN FIGURE 3-D.

3-49. What is the wind direction at station MXXX?

1. 340°
2. 210°
3. 170°
4. 080°

3-50. What is the horizontal surface visibility reported at station MYYY?

1. 1/2 statute mile
2. 1/2 statute meter
3. 15 nautical miles
4. 2,400 feet

3-51. What is the runway visual range reported by station MYYY?

1. 300 feet
2. 2,400 meters
3. 2,400 feet
4. 0300 variable 2,400 feet

3-52. What weather is occurring at stations (a) MXXX and (b) MYYY?

1. (a) Thunderstorm with rain
   (b) smoke
2. (a) Thunderstorm with rain
   (b) fog
3. (a) Thunderstorm with hail
   (b) fog
4. (a) Thunderstorm with hail
   (b) smoke

3-53. What is the sea level pressure at MXXX?

1. 29.91 in.
2. 29.91 hPa
3. 1013.3 in.
4. 1013.3 hPa

3-54. How much sky cover is present over station MYYY?

1. 0/8
2. 1/8
3. 8/8
4. Unable to determine; no significant clouds cover was reported

3-55. In METAR observations, temperatures are reported to the nearest

1. tenth degree Fahrenheit
2. tenth degree Celsius
3. whole degree Fahrenheit
4. whole degree Celsius

3-56. When you are reporting altimeter settings, what is the meaning of "A," as in A2992?

1. The altimeter setting is an approximation
2. The altimeter setting is the maximum for the day
3. The altimeter setting is in inches
4. The altimeter setting is in hectopascals

3-57. Unlike shore stations, the METAR code aboard ship is used only to record data, not to encode data for transmission.

1. True
2. False

3-58. International code FM 12-XI SYNOP is used to report which of the following observations?

1. Land synoptic
2. Ship synoptic
3. Land aviation
4. Ship aviation

(actual observation)
AAXX 18004 72409 21660 72214 10147 20131 39986 49991 54000 70500 84121 91749 333 10147 20086 60001 70005;
(symbolic format for this case)
AAXX YYGGi_w IIiii $i_R i_X$hVV Nddff **1**S_nTTT **2**$S_n T_d T_d T_d$ **3**$P_O P_O P_O P_O$ **4**PPPP **5**appp **7**ww$W_1 W_2$ **8**$N_h C_L C_M C_H$ **9**ggGG **333** **1**$S_n T_X T_X T_X$ **2**$S_n T_n T_n T_n$ **6**RRRt_R **7**$R_{24} R_{24} R_{24} R_{24}$;

Figure 3-E

IN ANSWERING QUESTIONS 3-59 THROUGH 3-62, REFER TO FIGURE 3-E AT THE TOP OF THIS PAGE. REFER TO THE REGION IV LAND-STATION SYNOPTIC REPORT.

3-59. What is the station's block/station number?

1. 72409
2. 21660
3. 72213
4. 70005

3-60. What is the wind direction?

1. 072°
2. 130°
3. 210°
4. 220°

3-61. What is the dew-point temperature?

1. 14.7°F
2. 14.7°C
3. 13.1°F
4. 13.1°C

3-62. In WMO Region IV at 0000Z, what maximum and minimum temperatures are reported following the 333 indicator?

1. Previous 12-hour minimum and previous 18-hour maximum
2. Previous 24-hour maximum and previous 12-hour minimum
3. Previous 12-hour maximum and previous 18-hour minimum
4. Previous 24-hour maximum and minimum

3-63. If a station within the United States reports code groups after a "555" indicator, which of the following publications would provide the information necessary to decode these groups?

1. WMO Pub 306, Vol 1, International Codes
2. NAVMETOCCOMINST 3141.2
3. FMH-1 (Federal Meteorological Handbook Number 1)
4. FHM-2 (Federal Meteorological Handbook Number 2)

3-64. What publication is used to convert the IRCS to the ship's name and country of registration?

1. FMH-2
2. FMH-1
3. ACP-100
4. WMO Pub 306, Vol 1

The maritime data section of an actual ship synoptic report reads:
22212; 00123 20504 30207 40806 51004
The symbolic format of this section is:
222$D_S V_X$ 0$S_S T_W T_W T_W$ 2$P_W P_W H_W H_W$ 3$d_{W1} d_{W1} d_{W2} d_{W2}$ 4$P_{W1} P_{W1} H_{W1} H_{W1}$ 5$P_{W2} P_{W2} H_{W2} H_{W2}$;

Figure 3-F

IN ANSWERING QUESTIONS 3-65 THROUGH 3-69, REFER TO FIGURE 3-F.

3-65. Which group reports sea-surface temperature?

1. 22212
2. 00123
3. 20504
4. 30207

3-66. What is the sea-wave height and sea-wave period?

1. 4 ft, 5 sec
2. 5 ft, 4 sec
3. 7 ft, 5 sec
4. 8 ft, 4 sec

3-67. What is the primary swell-wave direction?

1. 020°
2. 030°
3. 070°
4. 080°

3-68. What is the secondary swell-wave period?

1. 4 sec
2. 6 sec
3. 8 sec
4. 10 sec

3-69. What is the primary swell-wave height?

1. 4 ft
2. 7 ft
3. 10 ft
4. 16 ft

3-70. What code form do meteorological buoys moored near coastal harbors use to report position?

1. Land Synoptic (SYNOP)
2. Ship Synoptic (SHIP)
3. SPECI
4. METAR

ASSIGNMENT 4

Textbook Assignment: "Plotting Standards"; chapter 4, pages 4-1 through 4-38.

4-1. In the geographical coordinate system, which coordinate lines extend (a) north/south, and (b) east/west?

1. (a) Parallels of latitude
   (b) meridians of longitude
2. (a) Parallels of longitude
   (b) meridians of latitude
3. (a) Meridians of longitude
   (b) parallels of latitude
4. (a) Meridians of latitude
   (b) parallels of longitude

4-2. On any chart, 60 nautical miles equals

1. 1° of longitude
2. 1° of latitude
3. 1° of longitude arc
4. 1° of latitude arc

4-3. When meteorological positions are converted during encoding, what would be the equivalent value of 27°16'12"?

1. 27.2
2. 27.3
3. 27.4
4. 28.0

4-4. In the Universal Transverse Mercator Grid system, grid zone boundaries are the latitude and longitude lines.

1. True
2. False

4-5. Which of the following is a designation for a position on the 10,000-meter grid?

1. 57T
2. 36MAB
3. 27ABC 28
4. 45IOX 10

4-6. When you are reading a grid position such as 27GHM 12345678, where is the position within the "HM" 100,000-meter grid?

1. East 12340 meters, North 56780 meters
2. East 1234 meters, North 5678 meters
3. East 56780 meters, North 12340 meters
4. East 5678 meters, North 1234 meters

4-7. Where are the grid lines truly parallel to the longitude lines?

1. At the equator
2. At the center of each grid zone
3. At the left side of each grid zone
4. At the right side of each grid zone

4-8. In the Universal Polar Sterographic Grid, what are the two grid zones used in the north polar region?

1. A and B
2. Y and Z
3. W and X
4. W and Z

4-9. A rhumb line track is a straight line drawn on what type of chart?

1. Gnomonic
2. Polar sterographic
3. Mercator
4. Lambert conformal

4-10. Of the following charts, which one is the smallest scale chart?

1. Chart measures 4 ft by 4 ft with a scale of 1:5,000
2. Chart measures 4 ft by 4 ft with a scale of 1:20,000
3. Chart measures 2 ft by 2 ft with a scale of 1:6,000
4. Chart measures 2 ft by 2 ft with a scale of 1:25,000

4-11. Which of the following federal agencies produces most of the maps and charts used by the U.S. military?

1. NAVOCEANO
2. Federal Aviation Administration (FAA)
3. The National Imagery and Mapping Agency/Topographic Center (NIMA)
4. Department of Transportation

4-12. In the complete WMO international block/station number, how many numbers identify the (a) block and (b) individual station?

1. (a) Two (b) three
2. (a) Two (b) four
3. (a) Three (b) three
4. (a) Three (b) four

4-13. Within the United States, which of the following letter or letter/number groups follows the format of a national station identifier?

1. 724090
2. KBIX
3. BIX
4. BIX9A

4-14. Which of the following station identifiers is an ICAO station identifier?

1. 724095
2. KLAX
3. JFK
4. 63W

4-15. Which of the following sources contains the most comprehensive listing of station identifiers cross-referenced to latitude and longitude?

1. DMA catalog
2. OPARS Data Base
3. FAA Order 7250.4, Location Identifiers
4. DoD Flight Information Publication (En route), IFR Supplement -United States

4-16. Which of the following terms refers to a display of conditions anticipated to occur at some time in the future?

1. Map
2. Chart
3. Analysis
4. Prognosis

4-17. What term refers to a forecaster's interpretation of alphanumeric or graphic products?

1. Prognosis
2. Forecast
3. Analysis
4. Briefing Aid

4-18. When two sets of history are placed on a chart, which of the following colors are used for (a) the oldest history and (b) the most recent history?

1. (a) Yellow, (b) orange
2. (a) Black, (b) yellow
3. (a) Orange, (b) yellow
4. (a) Yellow, (b) black

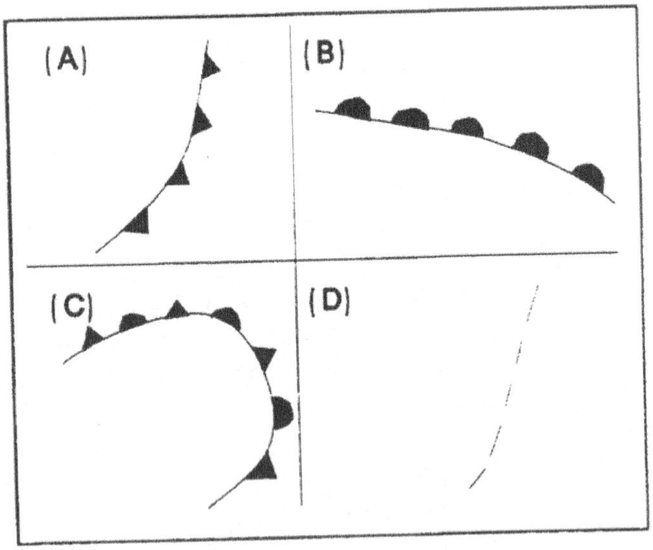

Figure 4-A

IN ANSWERING QUESTIONS 4-19 THROUGH 4-22, REFER TO FIGURE 4-A. RESPONSES ARE USED ONLY ONCE.

4-19. What symbol is used to depict the axis of a trough?

1. A
2. B
3. C
4. D

4-20. What symbol depicts a cold front at the surface?

1. A
2. B
3. C
4. D

4-21. What symbol depicts an occluded front at the surface?

1. A
2. B
3. C
4. D

4-22. What symbol depicts a warm front at the surface?

1. A
2. B
3. C
4. D

4-23. What color is used to indicate a convergent asymptote on a streamline chart?

1. Blue
2. Black
3. Orange
4. Red

4-24. What symbol is used to show moderate clear air turbulence?

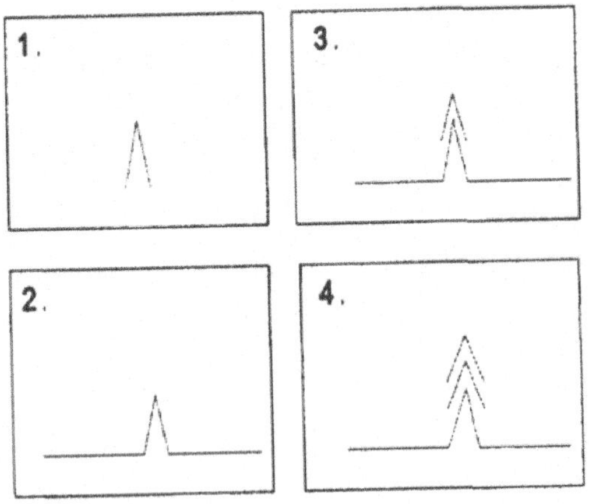

4-25. What symbol is used to show light clear icing?

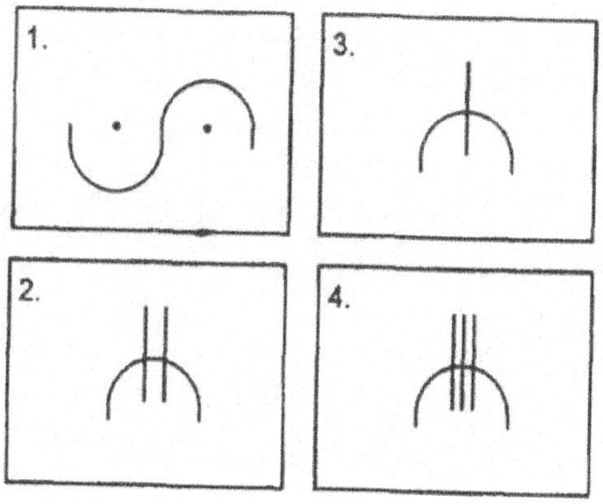

4-26. What is the isoline name for values of equal wind speed?

1. Isobar
2. Isotach
3. Isotherm
4. Isoheight

4-27. Isodrosothenns are drawn to connect lines of equal

1. thickness
2. dew-point depression
3. dew-point temperature
4. precipitation amounts

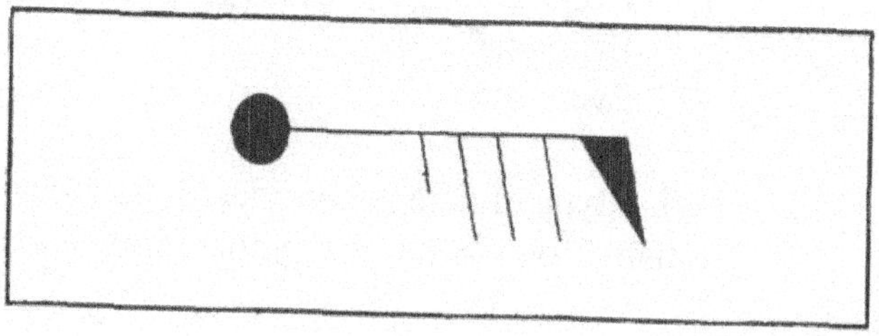

Figure 4-B

IN ANSWERING QUESTION 4-28, REFER TO FIGURE 4-B. ASSUME THAT TRUE NORTH IS AT THE TOP OF THIS PAGE.

4-28. What (a) wind direction and (b) wind speed is indicated by the plotted wind?

1. (a) 270° (b) 45 knots
2. (a) 270° (b) 85 knots
3. (a) 090° (b) 45 knots
4. (a) 090° (b) 85 knots

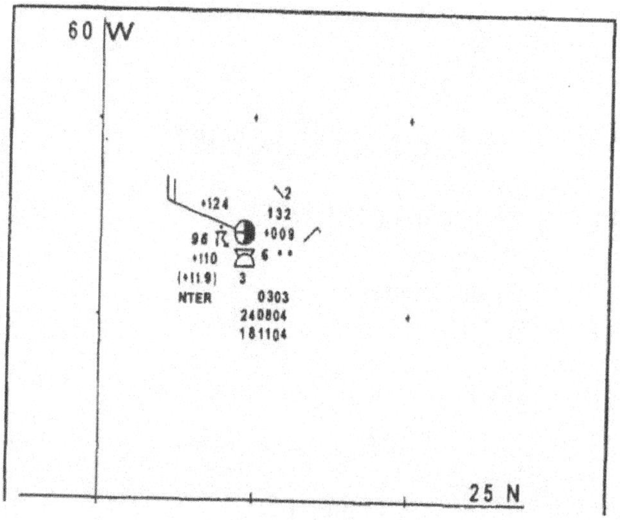

Figure 4-C

IN ANSWERING QUESTIONS 4-29 THROUGH 4-3 1, REFER TO FIGURE 4-C. THIS FIGURE IS A SURFACE SYNOPTIC SHIP OBSERVATION PLOT FOLLOWING THE WMO COMPLETE SYMBOLIC MODEL.

4-29. What value is plotted for sea level pressure?

1. 009
2. +124
3. 132
4. 303

4-30. What value is plotted for sea-surface temperature?

1. 96
2. +110
3. +11.9
4. +124

4-31. What is the reported code figure for low-cloud height?

1. 2
2. 3
3. 6
4. 9

925-hPa LEVEL REPORT	
AS CODED: 92606 06017 27513	
PPP	= 925 HpA
hhh	= 606 decameters
TTT	= 06.0°C
DD	= 1.7°
dd	= 275°

Figure 4-D

IN ANSWERING QUESTION 4-32, REFER TO FIGURE 4-D.

4-32. Which of the following 925-hPa level plots is correctly plotted for the information reported in figure 4-D?

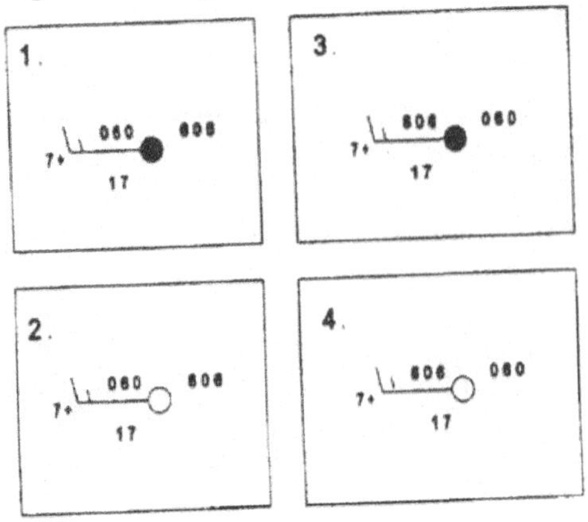

4-33. When are METAR observations usually plotted on a chart?

1. When received
2. When plotting an LAWC
3. When synoptic reports are considered inaccurate
4. When the forecaster is interested in an ocean area

4-34. When plotting the total cloud coverage (N) reported in a METAR observation report, how should "BKN" be plotted?

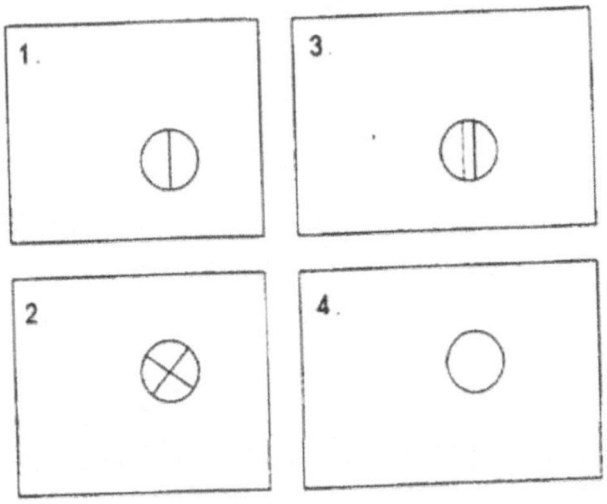

4-35. Isotherm contours on an SST analysis are usually drawn at what interval?

1. Every 2°C
2. Every 2°F
3. Every 4°C
4. Every 4°F

4-36. What type of data is considered the most reliable for sea-surface temperature charts?

1. All IR satellite data
2. Only calibrated IR satellite data from low-moisture, cloud-free areas
3. Bathythermograph observations
4. Ship reported SST

4-37. When arrows are drawn for wave directions, swell-wave directions are indicated by a "wavy" arrow from the station circle pointing toward the reported direction.

1. True
2. False

4-38. On the Skew T, Log P diagram, what is the orientation of isotherms?

1. Horizontal
2. Vertical
3. Right-diagonal
4. Left-diagonal

4-39. The saturation-adiabatic lapse rate is the rate at which

1. moist air cools as it rises
2. dry air cools as it rises
3. saturated air cools as it rises
4. actual air in the atmosphere cools as it rises

4-40. What do the dashed, right-diagonal green lines represent on a Skew T, Log P diagram?

1. Mixing ratio
2. Isobars
3. Saturation adiabats
4. U.S. standard atmosphere

4-41. When adding a pressure altitude scale to a Skew T, Log P diagram, altitude values are added to the isotherms so that each 10°C change in temperature equals how much of an altitude change?

1. 100 meters
2. 1,000 meters
3. 1,500 meters
4. 10,000 meters

4-42. In which of the following parts of the TEMP SHIP code would you find temperature and dew-point information for 756 hPa?

1. Part A
2. Part B
3. Part C
4. Part D

4-43. When you are plotting only the current upper-air observation on a Skew T, Log P diagram, what color(s) should you use?

1. Blue for all plots
2. Red for all plots
3. Red for temperature, blue for dew point, black for pressure altitude
4. Red for temperature, black for dew point, blue for pressure altitude

4-44. Of the three digits reported in the TEMP codes for the altitude of each pressure level, which of the following levels is prefixed with a "1" and suffixed with a "0" to obtain the altitude in meters?

1. 850-hPa
2. 700-hPa
3. 300-hpa
4. 250-hpa

4-45. Which of the following symbols is used to indicate a temperature plot on a Skew T, Log P diagram?

1. Triangle
2. Circle
3. Square
4. Diamond

4-46. What is the actual dew-point depression given a code figure of 57 in the Upper Air code?

1. 5.7°F
2. 7.0°F
3. 5.7°C
4. 7.0°C

4-47. When tropopause information is plotted, and group 88999 is reported, the block letters NO TROP are written just to the right of the colored portion of the diagram at what hPa level?

1. 100 hPa
2. 200 hPa
3. 300 hPa
4. 400 hPa

4-48. Should sea-surface temperature be plotted on the Skew T, Log P diagram? If so, where?

1. Yes; in the analysis block
2. Yes; in the top margin of the diagram
3. Yes; in the bottom margin of the diagram directly under the temperature trace
4. No; it is not plotted

4-49. As used in ship movement reports, what is the check-sum for the number group 079-28W?

1. 6
2. 7
3. 8
4. 9

4-50. The term "speed of advance" and "indicated speed" mean the same thing.

1. True
2. False

4-51. What type of background should you use to plot great circle routes?

1. Polar sterographic
2. Mercator projection
3. Lambert conformal
4. Rhumb line

Made in the
USA
Middletown, DE

76044226R00166